MOLECULAR
BIOLOGY
INTELLIGENCE
UNIT

SPHINGOLIPID-MEDIATED SIGNAL TRANSDUCTION

Yusuf A. Hannun, M.D.
Departments of Medicine and Cell Biology
Duke University Medical Center
Durham, North Carolina, U.S.A.

Springer
New York Berlin Heidelberg London Paris
Tokyo Hong Kong Barcelona Budapest

LANDES
BIOSCIENCE
AUSTIN, TEXAS
U.S.A.

Molecular Biology Intelligence Unit
SPHINGOLIPID-MEDIATED SIGNAL TRANSDUCTION

R.G. LANDES COMPANY
Austin, Texas, U.S.A.

Printed in the U.S.A.

Please address all inquiries to the Publishers:
R.G. Landes Company, 810 S. Church Street, Georgetown, Texas, U.S.A. 78626
Phone: 512/ 863 7762; FAX: 512/ 863 0081

International distributor (except North America):
Springer-Verlag GmbH & Co. KG
Tiergartenstrasse 17, D-69121 Heidelberg, Germany

Springer

International ISBN: 3-540-62714-6

Library of Congress Cataloging-in-Publication Data
CIP applied for but not received as of publication date.

PUBLISHER'S NOTE

R.G. Landes Bioscience Publishers produces books in six Intelligence Unit series: *Medical, Molecular Biology, Neuroscience, Tissue Engineering, Biotechnology* and *Environmental.* The authors of our books are acknowledged leaders in their fields. Topics are unique; almost without exception, no similar books exist on these topics.

Our goal is to publish books in important and rapidly changing areas of bioscience for sophisticated researchers and clinicians. To achieve this goal, we have accelerated our publishing program to conform to the fast pace at which information grows in bioscience. Most of our books are published within 90 to 120 days of receipt of the manuscript. We would like to thank our readers for their continuing interest and welcome any comments or suggestions they may have for future books.

Shyamali Ghosh
Publications Director
R.G. Landes Company

CONTENTS

ABBREVIATIONS

CDK	cyclin-dependent kinase
Cer	ceramide (N-fatty acyl sphingosine)
DAG	diacylglycerol
deNAcGM3	de-N-acetyl-GM3
DMS	N,N-dimethylsphingosine
EGF	epidermal growth factor
FGF	fibroblast growth factor
Gal	galactose
GlcCer	glucosylceramide (Glcβ1→1Cer)
GM3	NeuAcα2→3Galβ1→4Glcβ1→1Cer
GD3	NeuAca2→8NeuAcα2→3Galβ1→4Glcβ1→1Cer
Grp	glucose-regulated protein
GSL	glycosphingolipid
IP3	inositol triphosphate
LacCer	lactosylceramide (Galβ1→4Glcβ1→1Cer)
MAPK	mitogen-activated protein kinase
NGF	nerve growth factor
PA	phosphatidic acid
PC	phosphatidylcholine
PDGF	platelet-derived growth factor
PI	phosphatidylinositol
PKA	protein kinase A
PKC	protein kinase C
PLPS	plasmalopsychosine
RK	receptor kinase
Ser	serine
SL	sphingolipid
SM	sphingomyelin
SMase	sphingomyelinase

ABBREVIATIONS

SPC	sphingophosphorylcholine
SPG	sialosylparagloboside (NeuAcα2→3Galβ1→4GlcNAcβ1→3Galβ1→4Glcβ1→1Cer)
Sph	sphingosine
TNFα	tumor necrosis factor alpha
Tyr	tyrosine

EDITOR

Yusuf A. Hannun, M.D.
Departments of Medicine and Cell Biology
Duke University Medical Center
Durham, North Carolina, U.S.A.
Chapters 1, 2

CONTRIBUTORS

Abdelkarim Abousalham, Ph.D.
Department of Biochemistry
Signal Transduction Laboratories
and the Lipid and Lipoprotein Research Group
University of Alberta
Edmonton, Alberta, Canada
Chapter 8

Christopher Alexander, Ph.D.
Department of Chemistry
Emory University
Atlanta, Georgia, U.S.A.
Chapter 11

Leslie R. Ballou, Ph.D.
The Department of Veterans Affairs Medical Center
and the Departments of Medicine and Biochemistry
The University of Tennessee
Memphis, Tennessee, U.S.A.
Chapter 3

Robert M. Bell, Ph.D.
GlaxoWellcome, Inc.
Research Triangle Park, North Carolina, U.S.A.
Chapter 4

David N. Brindley, Ph.D., D.Sc.
Department of Biochemistry
Signal Transduction Laboratories
and the Lipid and Lipoprotein Research Group
University of Alberta
Edmonton, Alberta, Canada
Chapter 8

Olivier Cuvillier, Ph.D.
Department of Biochemistry and Molecular Biology
Georgetown University Medical Center
Washington, DC, U.S.A.
Chapter 9

Larry W. Daniel, Ph.D.
Department of Biochemistry
Bowman Gray School of Medicine
Wake Forest University
Winston-Salem, North Carolina, U.S.A.
Chapter 6

Ghassan Dbaibo, M.D.
Department of Pediatrics
American University of Beirut
Beirut, Lebanon
Chapter 2

Elena Fuior
Department of Biochemistry
and Molecular Biology
Georgetown University
Medical Center
Washington, DC, U.S.A.
Chapter 9

Jane Goh
Department of Chemistry
Emory University
Atlanta, Georgia, U.S.A.
Chapter 11

Sen-itiroh Hakomori, M.D.
Biomembrane Division
Pacific Northwest Research
Foundation
and Departments of Pathobiology
and Microbiology
University of Washington
Seattle, Washington, U.S.A.
Chapter 10

Mariana Nikolova-Karakashian, Ph.D.
Department of Biochemistry
Emory University
Atlanta, Georgia, U.S.A.
Chapter 11

Yutaka Kikuchi, Ph.D.
Department of Biochemistry
Signal Transduction Laboratories
and the Lipid and Lipoprotein
Research Group
University of Alberta
Edmonton, Alberta, Canada
Chapter 8

Joanna Y. Lee, Ph.D.
Departments of Medicine
and Cell Biology
Duke University Medical Center
and the VA GRECC
Durham, North Carolina, U.S.A.
Chapter 5

Dennis C. Liotta, Ph.D.
Department of Chemistry
Emory University
Atlanta, Georgia, U.S.A.
Chapter 11

Charles P. McKay, M.D.
Alfred I. duPont Institute
Wilmington, Delaware, U.S.A.
Chapter 12

David S. Menaldino, Ph.D.
Department of Biochemistry
Emory University
Atlanta, Georgia, U.S.A.
Chapter 11

Alfred H. Merrill, Jr., Ph.D.
Department of Biochemistry
Emory University
Atlanta, Georgia, U.S.A.
Chapter 11

Sheldon Milstien, Ph.D.
Laboratory of Cell Biology, NIMH
Bethesda, Maryland, U.S.A.
Chapter 9

Antonio Gómez-Muñoz, Ph.D.
Universidad del Pais Vasco
Faculdad de Ciencias
Bilbao, Spain
Chapter 8

Lina M. Obeid, M.D.
Departments of Medicine
and Cell Biology
Duke University Medical Center
and the VA GRECC
Durham, North Carolina, U.S.A.
Chapter 5

Norman S. Radin, Ph.D.
Nephrology Division
Department of Internal Medicine
University of Michigan
Ann Arbor, Michigan, U.S.A.
Chapter 7

James A. Shayman, M.D.
Nephrology Division
Department of Internal Medicine
University of Michigan
Ann Arbor, Michigan, U.S.A.
Chapter 7

Sarah Spiegel, Ph.D.
Department of Biochemistry
and Molecular Biology
Georgetown University
Medical Center
Washington, DC, U.S.A.
Chapter 9

Jay C. Strum, Ph.D.
GlaxoWellcome, Inc.
Research Triangle Park,
North Carolina, U.S.A.
Chapter 4

Katherine I. Swenson, Ph.D.
Department of Molecular
Cancer Biology
Duke University Medical Center
Durham, North Carolina, U.S.A.
Chapter 4

Teresa R. Vales
Department of Biochemistry
Emory University
Atlanta, Georgia, U.S.A.
Chapter 11

David W. Waggoner, Ph.D.
Department of Biochemistry
Signal Transduction Laboratories
and the Lipid and Lipoprotein
Research Group
University of Alberta
Edmonton, Alberta, Canada
Chapter 8

Elaine Wang
Department of Biochemistry
Emory University
Atlanta, Georgia, U.S.A.
Chapter 11

Susan P. Whitman
Department of Biochemistry
Bowman Gray School of Medicine
Wake Forest University
Winston-Salem,
North Carolina, U.S.A.
Chapter 6

PREFACE

"The fatty acids dissolve, while a body remains insoluble, which is of an alkaloidal nature, and to which, in commemoration of the many enigmas which it presented to the inquirer, I have given the name of Sphingosin."

– Thudichum, 1884.

Thus, the era of sphingolipid studies was ushered in with an enigmatic label aptly describing not only their initial discovery, but the various stages of scientific development of this area of study. Their complexity and structural variation ever-expanding, sphingolipids, as a class, may comprise at least 300-400 individual species, excluding "microheterogeneity" of structure within each species. What are the reasons underlying this complexity and diversity of structure? What are the functions of individual sphingolipids? What can we expect from the study of sphingolipids? These are some of the key questions that have propelled investigation of sphingolipids and their roles in biology. Intensive investigation over several decades resulted in the identification and determination of the structure and complexity of sphingolipids. Subsequent, and equally extensive, studies raised intriguing possibilities for roles of sphingolipids in immune modulation, cell-cell interaction, tumor modulation, embryogenesis and development, and as receptors and cofactors for cell surface receptors.

More recently, another intriguing aspect of sphingolipid biology unfolded with the increasing recognition that sphingolipids and their derived products play important roles in signal transduction and cell regulation. This area of research is predicated on the premise that sphingolipid metabolism is highly regulated in response to extracellular and intracellular signals and stimuli. The sphingolipid products of this regulated metabolism then serve as second messengers and bioeffector molecules that transmit signals and information and, therefore, effect or modulate a variety of cell responses.

The picture that has emerged over the last few years is one of novel functions for "old" and "new" sphingolipids and their derived products. These molecules appear to play critical roles in highly significant areas of research, such as the regulation of cell death (apoptosis), cell cycle arrest, tumor cell invasion, and cell senescence and aging. For example, regulated formation of ceramide may serve as an important determinant of whether cells undergo apoptosis, whereas sphingosine-1-phosphate may play an important role in regulating cell motility and tumor invasion. In essence, these studies have served to merge and tie two heretofore poorly studied areas of research: sphingolipid metabolism and the above mentioned areas of biology that have only received belated attention as the new frontiers of cell biology.

This book offers a current compilation and a critical discussion of these various areas of sphingolipid metabolism and biology by many of the pioneers and leaders in these fields of investigation. The study of regulated lipid metabolism and lipid-mediated signal transduction has always been complex and involved; the study of sphingolipids is no exception. The expert discussions in this book will hopefully aid the reader in negotiating the difficulties in the study of sphingolipid signal transduction and in discerning the emerging themes and exciting areas of investigation.

One may conclude that the study of sphingolipid signal transduction is only in its infancy, with much more to come. It is easy to anticipate the discovery of novel products of sphingolipid metabolism that may play key roles in various aspects of cell biology. Perhaps the blueprints outlined in these chapters will stimulate further research into this expanding and developing area.

Yusuf A. Hannun, M.D.

Acknowledgments

I would like to thank my past and present co-workers in my laboratory who have made our research possible and enjoyable. I would also like to thank the authors of the eleven chapters for their outstanding contributions.

CHAPTER 1

Introduction and Overview

Yusuf A. Hannun

INTRODUCTION: PARADIGMS OF SIGNAL TRANSDUCTION THROUGH LIPIDS

Lipids have unique biophysical properties that set them apart from all other small molecules of living cells. These properties (which include hydrophobicity, poor aqueous solubility, and self-assembly) are critical to the formation of membrane bilayers which, in turn, define the boundaries of the cells as well as intracellular compartments. The lipids of the plasma membrane (which comprise glycerolipids, sphingolipids, and cholesterol) provide for a fluid membrane that serves to delineate the cell and protect the intracellular milieu from the extracellular environment. The asymmetric charge distribution of membrane bilayers contributes to the membrane potential, whereas the hydrophobicity of the membrane provides an appropriate environment for hydrophobic proteins and hydrophobic segments of transmembrane and membrane-associated proteins. These physical and chemical properties of the membrane lipids may be considered as the key determinants of the structural functions of membrane lipids. For these structural functions, the simplest membrane necessitates the presence of at least two distinct lipids that maintain the bilayer and the charge asymmetry (for example, the cytoplasmic membranes of bacteria contain predominantly phosphatidylethanolamine, a zwitterionic lipid, and phosphatidylglycerol, an anionically-charged phospholipid).

The variety of individual lipid species of eukaryotic membranes and the diversity of their structural composition suggests a multiplicity of additional functions. It is a major thesis of contemporary studies on signal transduction and cell regulation that this diversity of membrane lipids reflects several important functions of these lipids in intercellular and intracellular communication and signal transduction. In its simplest formulation (Fig. 1.1), this paradigm, as applied to signal transduction, considers each individual membrane lipid as a source of unique information encoded by the structure of the lipid. Recognition of this lipid precursor as a substrate by a specific enzyme then allows for the controlled formation of products of its metabolism. These products are then recognized by specific targets (almost invariably proteins that are capable of recognizing the specific and often subtle chemical features of these lipid-derived products). Therefore, in its simplest formulation, this

Sphingolipid-Mediated Signal Transduction, edited by Yusuf A. Hannun.

paradigm requires the existence of a *regulated* enzyme of lipid metabolism that serves as the *input* point, a specific lipid substrate for this enzyme, specific products that are generated from the action of this enzyme, and specific protein targets that respond to the changing levels in these products and that constitute the *output* of this pathway. As such, these pathways allow for efficient signal transduction across the membrane bilayer, using the very key elements that form the structural basis of this bilayer. It is perhaps this strategic location in the plasma membrane that has rendered lipids attractive targets for information processing and signal transduction.

The historical development of the scientific underpinnings for this paradigm has evolved over several decades, with initial observations of regulated turnover of membrane glycerophospholipids in response to extracellular agents. In landmark studies in the 1950s, Hokin and Hokin observed that treatment of pancreatic tissue with acetylcholine resulted in turnover of inositol phospholipids.[1] Over the ensuing three decades, the key components and features of this "PI cycle" were determined, and insight into its possible functions in cell regulation and signal transduction became clarified. At this point in time, it is recognized that a number of extracellular agents, hormones, and growth factors act on their membrane receptors, and through different mechanisms cause the activation of PI-specific phospholipases. These phospholipases C then cleave inositol phospholipids and especially phosphatidylinositol bis-phosphate, causing the release of the head group inositol trisphosphate (IP_3) and the lipid backbone diacylglycerol (DAG). IP_3 interacts with an intracellular membrane receptor and causes the release of calcium into the cytosol whereas DAG activates protein kinase C (PKC) and possibly other targets, thus initiating a signal transduction pathway.[2-6]

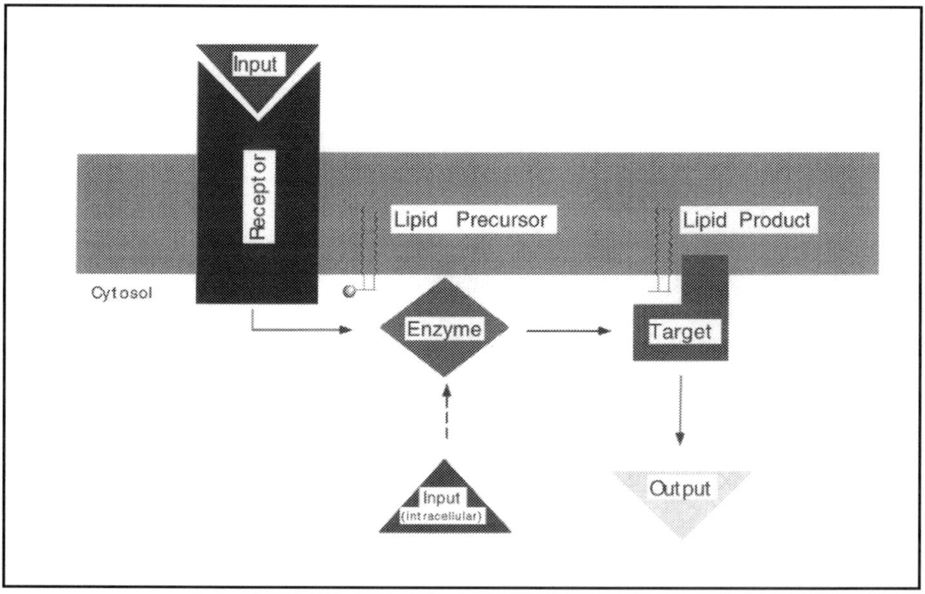

Fig. 1.1. Scheme of signaling through sphingolipids. The paradigm of signaling through sphingolipids requires the "deciphering" of an incoming signal (either from outside the cell or from within). This is accomplished through enzymes of sphingolipid metabolism that respond to these signals. The output of this signaling (or information processing) pathway requires the regulation of levels of the sphingolipid-derived product and the recognition of these changes by a target protein whose function is thus regulated by the changes in the levels of the candidate lipid mediator. This then serves as the proximal output of the system.

Although the PI cycle is the best-developed paradigm for lipid-mediated signal transduction, other examples have continued to accumulate as a result of important breakthroughs in deciphering the regulation of lipid metabolism and signal transduction.[7,8] For example, the activation of phospholipase D causes the formation of phosphatidic acid, which may serve as a second messenger on its own, or may serve as a precursor to formation of DAG.[9-11] Phospholipase A_2 causes the formation of lysophospholipids (such as lyphophosphatidic acid) and arachidonate, both of which may serve as second messengers or intracellular modulators.[7,12-14] Arachidonate is also further metabolized into a large array of eicosanoids, each of which appears to interact with a specific plasma membrane or internal receptor, thus initiating a specific arm of signal propagation.[15] PI-3 kinases, on the other hand, act on membrane inositol phospholipids to create specific species that are phosphorylated at the three position of the inositol ring. These enzymes (and by inference, their products) appear to play important roles in mitogenic regulation of cells.[8,16] Taken together, these emerging fields of investigation clearly point to a fundamental role of the glycerolipids and their derived products in signal transduction and cell regulation. Cholesterol may also be considered as a precursor to a large variety of intercellular hormones and messengers (such as glucocorticoids, estrogens, androgens, and mineralocorticoids). This leaves the third major class of membrane lipids, the sphingolipids, with a need for determining whether they also participate in signal transduction and cell regulation.

Membrane sphingolipids are more complex in their structures than the glycerolipids and may comprise at least 300 mammalian sphingolipid species.[17-19] The initial characterization and determination of sphingolipids came with the discovery of sphingosine by Thudicum, who dubbed the molecule in reference to the Greek Sphinx, as an indication of their enigmatic structure. Since then, a wealth of information has accumulated concerning the structure of the various sphingolipids, their expression in different tissues and during different phases of growth and development, and their effects on various cell types and especially in cell-cell recognition and interaction. The discovery of inhibition of PKC by sphingosine, the defining backbone of sphingolipid,[20] prompted a wider examination of the possible roles of sphingolipids and their derived products in signal transduction. This has resulted in intensified investigation on the role of sphingolipids in signal transduction, with emphasis on the biologic and biochemical activities of ceramide, sphingosine, sphingosine-1-phosphate, and other related molecules.[18,21-29] Importantly, such investigations have provided novel tools and insights into important areas of contemporary biologic research (such as programmed cell death, tumor invasiveness, inflammatory responses, and cell senescence) that otherwise have proven recalcitrant to biochemical and molecular approaches.

This chapter will provide the background on the structure and metabolism of sphingolipids and the emerging roles of sphingolipids and their derived products in signal transduction and cell regulation. This will be followed by a discussion of the tools and approaches currently employed in the study of sphingolipid signal transduction, and the chapter will conclude by a discussion of the significance of studying sphingolipid signal transduction and its future directions. Hopefully this will provide the essential background for the ensuing chapters that delve in detail into each of the areas presented here.

THE INFRASTRUCTURE OF SPHINGOLIPID SIGNAL TRANSDUCTION

For signal transduction to occur utilizing sphingolipids and their derived products (and, by extension, any other membrane lipid), there must be regulated metabolism of sphingolipid precursors, the generation of specific products, and the interaction of these products with specific targets (Fig. 1.1). One may consider signal transduction in a more global sense as *informa-*

tion processing. Processing of this information requires specificity of chemical structure at two levels: the level of the precursor sphingolipid and the level of the product. It also requires at least two regulated events: regulated enzymes that recognize the sphingolipid precursors, and targets that are regulated by the products of these enzymatic reactions. Therefore, the basic infrastructure of signaling through lipids requires familiarity with the structures of the involved molecules, the metabolism of these molecules, the regulation of this metabolism, and the mechanism of action of these products.

Structure of Sphingolipids

Structurally, sphingolipids are defined and distinguished by the presence of a sphingoid backbone (Fig. 1.2). In mammalian cells, this is usually sphingosine, (2S,3R,4E)-2-amino-1,3-dihydroxy-4-octadecene, whereas in yeast and plant cells this is usually phytosphingosine. Other variations include dihydrosphingosine, C_{20}-sphingosine, and possibly other multiply hydroxylated analogs of sphingosine. Collectively, sphingosine, dihydrosphingosine, phytosphingosine, and related long-chain amino bases are also termed sphingoid bases. The next building block in sphingolipid structure is ceramide, which is derived from sphingosine by acylation at the 2-amino position. This usually involves an N-stearoyl derivative for mammalian sphingolipids, with variations including various chain lengths as well as α- or ω-hydroxylated fatty acids. The acylated dihydrosphingosine and phytosphingosine are termed dihydroceramide and phytoceramide, respectively. Collectively, these three different species of ceramide may be termed as ceramides. In turn, these ceramides serve as the defining component for most, if not all, known sphingolipids. These sphingolipids are primarily derived from ceramide by incorporating different substituents at the 1-hydroxyl position. The simplest such sphingolipids include ceramide phosphate, which contains a phosphodiester group at the 1-hydroxyl position, and cerebrosides, which contain either a glucose or galactose substituent at the 1-hydroxyl position. Other phosphorylated sphingolipids include ceramide phosphoethanolamine, and sphingomyelin (ceramide phosphorylcholine). The majority of sphingolipids, however, are derived from the cerebrosides. Sulfatides are derived from galactosylceramide by the addition of a sulfate group in ester linkage with the sugar. The complex glycolipids derive from glucosylceramide by having two or more glycose units added at the 1-hydroxyl position. The gangliosides may be considered as a subclass of glycosphingolipids that contain neuraminic acid residues in their carbohydrate head group. Other minor variants of sphingolipids include 3-acylated derivatives and derivatives of ceramide and cerebroside that are acylated at the ω-hydroxyl group of the 2-fatty acyl group (which are found predominantly in skin).

A final group of sphingolipids consists of the N-deacylated derivatives. These include psychosine (or 1-galactocylsphingosine), glucosylsphingosine, sphingosine-1-phosphate, sphingosyl phosphocholine and, potentially, derivatives of any other sphingolipids. Depending on one's perspective, these N-deacylated derivatives may also be considered as derivatives of sphingosine with the same 1-hydroxyl derivatives as the parent sphingolipids. Thus, sphingosine 1-phosphate (which corresponds to the N-deacylated ceramide phosphate) and sphingocylphosphocholine (which corresponds to the N-deacylated derivative of sphingomyelin) may be considered as phosphorylated derivatives of sphingosine. These lysosphingolipids are found in very low concentrations in the cell, but they are increasingly recognized as important second messengers and regulatory molecules.

In mammalian cells, it has been estimated that the number of unique sphingolipids (defined primarily by the specific substituent at the 1-hydroxyl position) may exceed 300 individual species. This does not include the sphingoid bases, the lysosphingolipids, and other, yet to be discovered, molecules. Obviously, this tremendous structural variation provides for a unique source of information processing as each sphingolipid species may act as a specific reservoir of information recognized by specific enzymes or molecules that react with this particular sphingolipid.

In addition to these main defining features of individual sphingolipid species, there are multiple variations, including the hydrocarbon length of the sphingoid base, the length of the fatty acyl derivative, hydroxylation of the fatty acid, and possibly hydroxylations of the sphingoid backbone.

Yeast also contains sphingolipids which are primarily distinguished from the mammalian counterparts by having phytosphingosine as the major building block instead of sphingosine. Yeast also contains phytoceramide, but instead of the glycosphingolipids, they contain unique phosphorylated species which include ceramide phosphoinositol (IPC), mannosylinositol-phosphate ceramide (MIPC) and mannosyl diinositol-diphosphate-ceramide [$M(IP)_2C$].

B. General Metabolism of Sphingolipids

Both the biosynthetic and degradative pathways of sphingolipid metabolism proceed through stepwise metabolic reactions and intermediates. The biosynthetic pathway (Fig. 1.3) commences with the condensation of serine and palmitoyl Co-A to yield 3-ketodihydrosphingosine, which is then reduced to dihydrosphingosine. In turn, dihydrosphingosine serves as the precursor for ceramide synthase and is acylated to form dihydroceramide.[30,31]

At this point, the 4-5-trans double bond of sphingolipids may be introduced into dihydroceramide, yielding ceramide. Alternatively, the double bond may be introduced subsequent to incorporation of dihydroceramide into dihydrosphingomyelin or dihydrocerebroside. Ceramide may then serve as the precursor for the synthesis of sphingomyelin, galactosylceramide, glucosylceramide, ceramide phosphate, or ceramide phosphoethanolamine.[32,33] Galactosylceramide may serve as a precursor for sulfatides, whereas the glucosylceramide serves as the main building block for the glycosphingolipids and the gangliosides. Again, the latter proceeds through stepwise addition of individual glycose units or sialic acids (Fig. 1.3).

The breakdown of complex glycolipids and sphingolipids also proceeds stepwise through the action of several hydrolyses that remove individual units of the head group. For example, different glucosidases and galactosidases remove individual glycose units to reduce the complex glycolipids into cerebrosides. Cerebrosidases then break down cerebroside into ceramide. Sphingomyelinases break down sphingomyelin also to yield ceramide and choline phosphate. On the catabolic pathway, ceramide is then deacylated by one of several ceramidases to yield sphingosine. Sphingosine may then be phosphorylated to yield sphingosine-1-phosphate, which serves as a substrate for a specific lyase which breaks it down into ethanol amine phosphate and palmitaldehyde. Thus, the complete cycle of sphingolipid metabolism starts with a fatty acyl Co-A and serine and individual donors of head group substituents (such as UDP galactose, UDP glucose, and PAPS: 3'phosphoadenosine, 5'-phosphosulfate), and breakdown eventually results in the release of these individual head groups and the reformation of fatty acids and derivatives, as well as ethanolamine phosphate.

It is quite possible that additional pathways of sphingolipid metabolism exist that yield novel products and derivatives. For example, recently it has been shown that sphingosine may be acylated at the 2-amino position to result in the formation of N-acetylsphingosine (or C_2 ceramide).[34] In another study, it was also recently shown that C_2 ceramide is acylated at the 3-hydroxyl position, yielding a novel acylated ceramide.[35] Although the products of these pathways may exist in very low levels in the cell, it can not be discounted that they may play important roles in cell regulation.

Metabolism of sphingolipids in yeast is less complex because of the smaller number of yeast sphingolipids. These pathways seem to follow the same blueprint, in that the initial rate-limiting enzyme is serine-

A

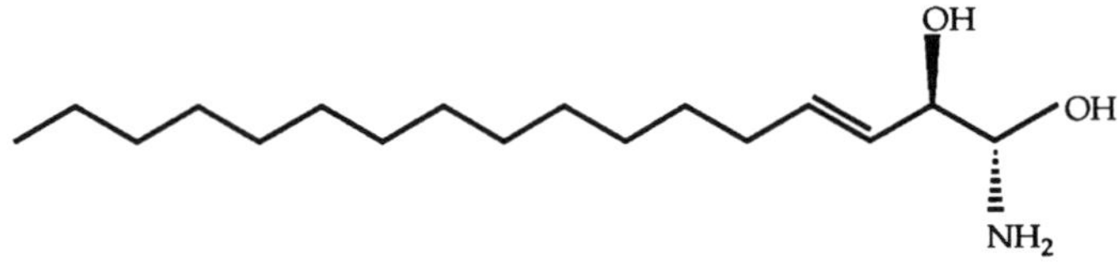

(2S,3R,4E)-2-Amino-1,3-dihydroxy-octadecene: *D-erythro-sphingosine*

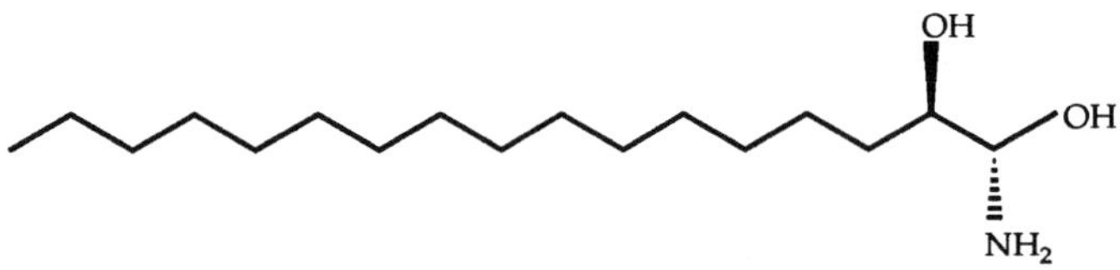

(2S,3R)-2-Amino-1,3-dihydroxy-octadecane: *D-erythro-dihydrophingosine*

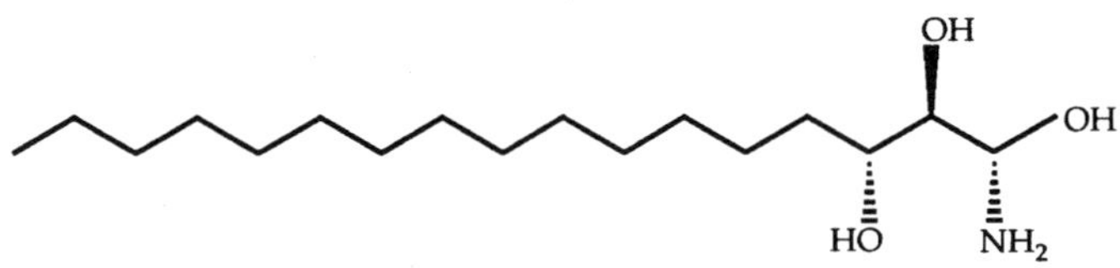

(2S,3S,4R)- D-ribo-2-Amino-1,3,4-trihydroxy-octadecane: *Phytosphingosine*

B

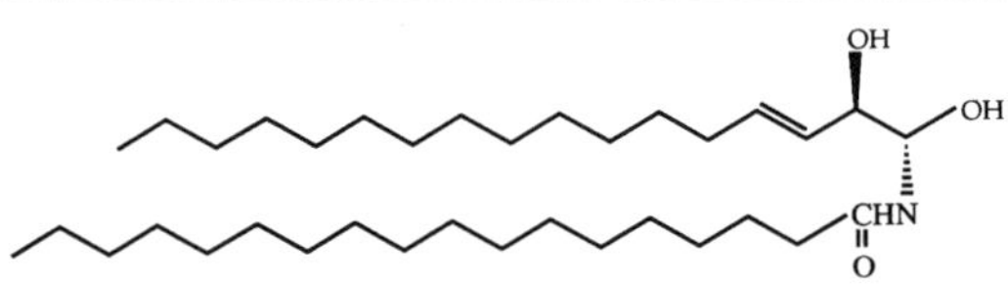

Ceramide

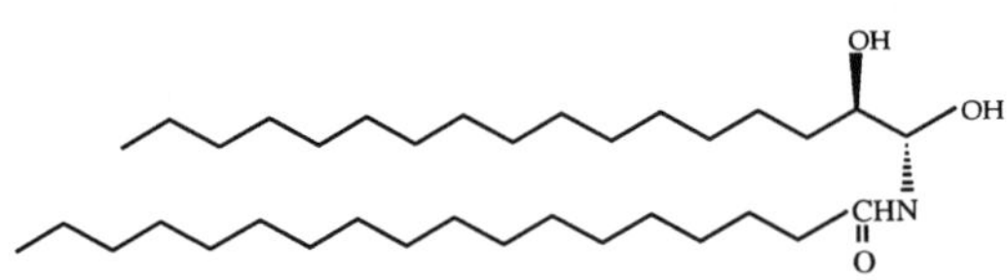

Dihydroceramide

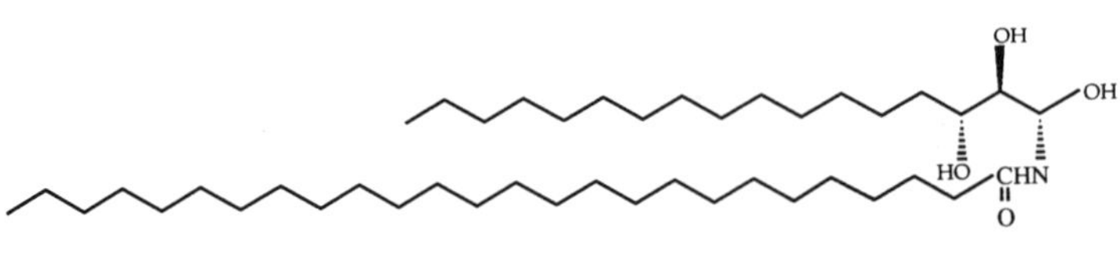

Phytoceramide

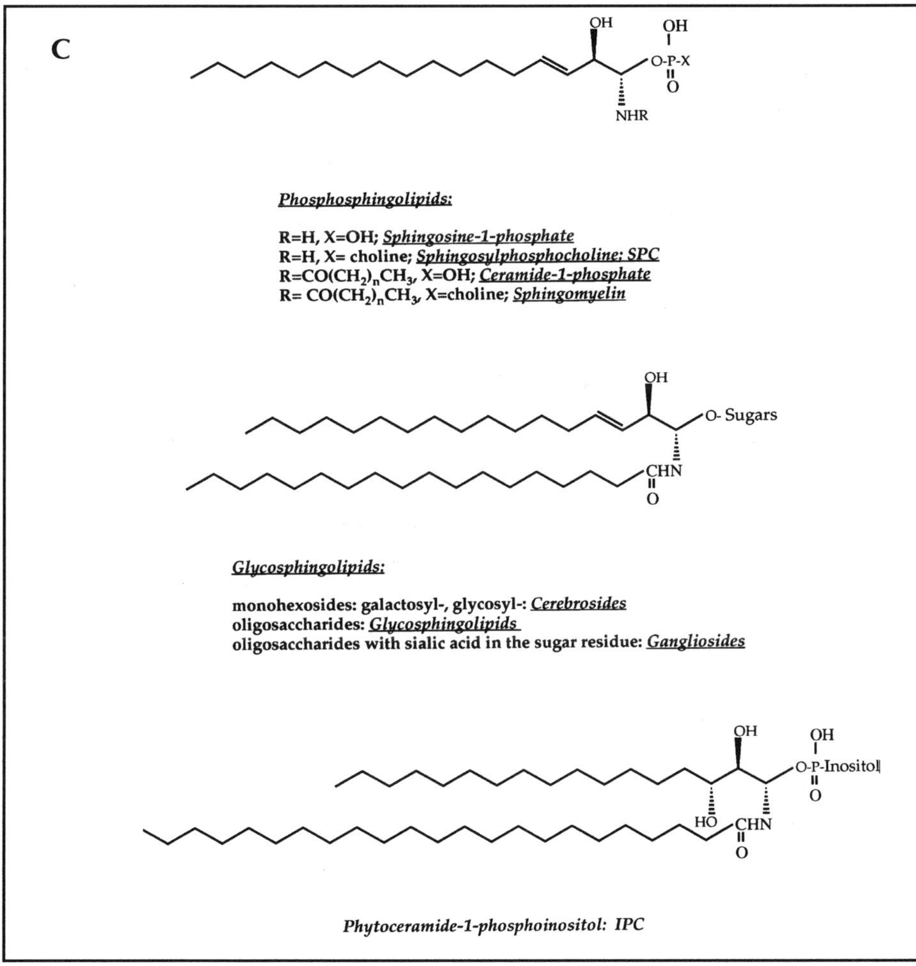

Fig. 1.2. Structures of key and representative sphingolipids.

palmitoyl transferase, which has been recently cloned from yeast and is known to exist as a product of two genes, each of which is necessary for the function of the enzyme. The remaining enzymes of yeast sphingolipid metabolism have not been cloned, although yeasts are known to contain ceramide synthase, sphingosine-1-phosphate lyase, and other enzymes necessary for the formation of IPC, MIPC, and $M(IP)_2C$.[36,37]

Very few of the enzymes of sphingolipid metabolism in mammalian cells have been purified, characterized, or cloned. Recently, acid sphingomyelinase and cerebroside synthase were cloned.[38,39] Alkaline and acid ceramidase have been purified and characterized.[40,41] Glyco-transferases and hydrolases involved in carbohydrate metabolism of the head group have also been cloned and characterized.

An interesting enzyme of sphingolipid metabolism is sphingomyelin synthase, which utilizes phosphatidylcholine and ceramide as precursors and transfers the cholinephosphate head group from phosphatidylcholine to ceramide. This results in the formation of sphingomyelin and DAG. This enzyme's action is thought to be reversible, and therefore it may regulate the levels of both ceramide and DAG.

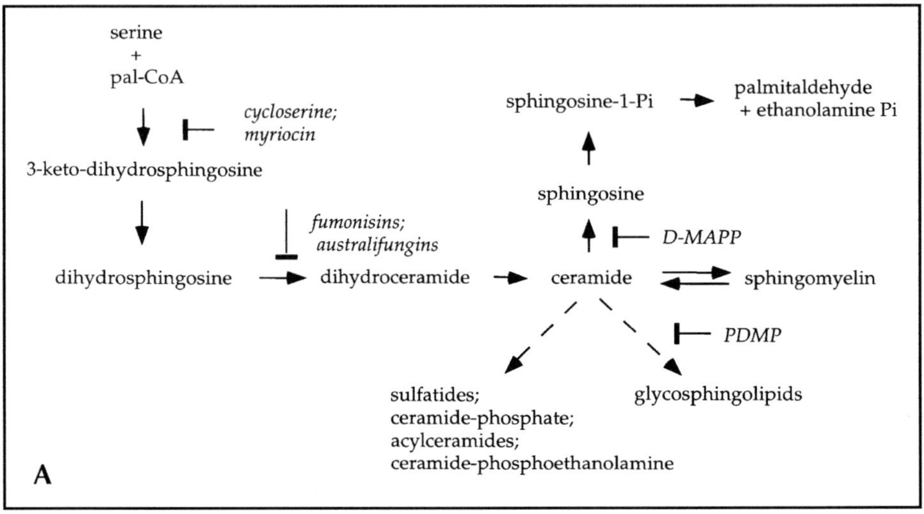

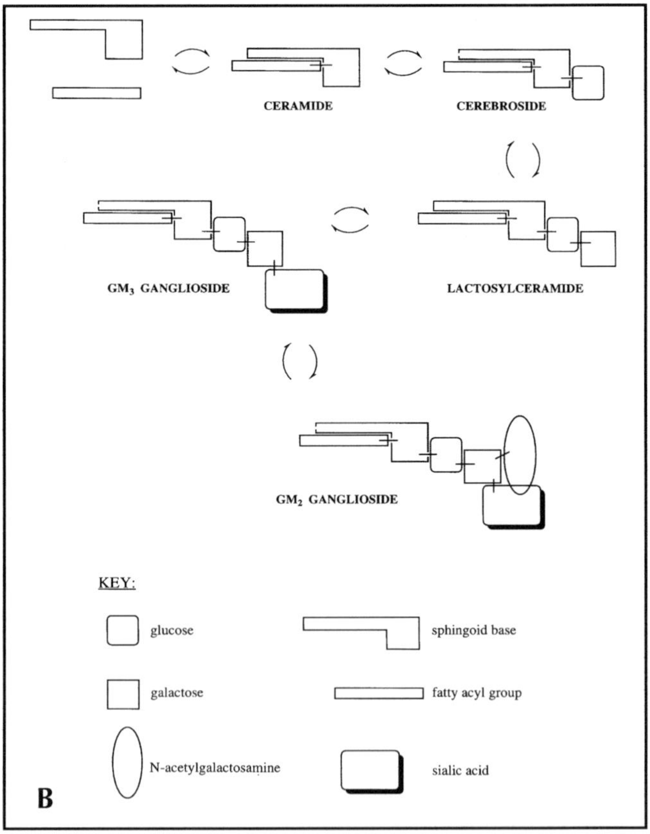

Fig. 1.3. Pathways of sphingolipid metabolism. (A) Pathways involved in the synthesis and degradation of sphingoid bases and ceramides. Most "higher" or complex sphingolipids derive from ceramide through addition of substituents at the 1-OH position of ceramide. (B) Schematic pathways for the synthesis of glycolipids and gangliosides illustrating an example of the complexity of head groups of sphingolipids in the case of the ganglioside GM_2. The addition and elimination of individual subunits in the head group proceeds through distinct and consecutive reactions.

Regulated Metabolism of Sphingolipids

The regulation of sphingolipid metabolism is probably one of the least studied areas of intermediary metabolism. However, it is the thesis of this chapter (and, indeed, of the whole book) that it is this regulated metabolism per se that underlies the significance of sphingolipids in signal transduction and cell regulation. According to the paradigm illustrated in Figure 1.1, in order to achieve signal transduction, an input signal must somehow result in modulation of the activity of a specific enzyme of sphingolipid metabolism. The activation of this enzyme (or, reciprocally, its inhibition) allows for regulation of the levels of precursor and product sphingolipids. This then serves as the driving force for information processing through the changes in levels of sphingolipid species and, consequently, through changes in the activity of enzymes or proteins that recognize these sphingolipids.

It is in this aspect that the regulated metabolism of sphingolipids assumes great significance. Indeed, the existing pathways of sphingolipid metabolism already offer multiple potential sites of regulation and numerous actual or potential bioactive products. What may be even more promising is the distinct possibility that many additional pathways of sphingolipid metabolism are yet to be discovered, whose products may play additional roles in cell biology.

This section will provide a general outline of known regulated enzymes of sphingolipid metabolism. The reader is referred to individual chapters in this book for detailed discussion of the regulation of these enzymes and the significance of such regulation to signaling through the involved sphingolipid.

Enzymes known to be regulated

Much attention has focused recently on sphingomyelinases, which appear to be critical enzymes regulating the levels of sphingomyelin and ceramide. Activation of these enzymes causes accumulation of ceramide either transiently or in a persistent fashion. Ceramide, in turn, is emerging as an important bioeffector molecule with roles in signal transduction, growth regulation, apoptosis, and inflammation. At least four different sphingomyelinases are known, and these include the neutral membrane sphingomyelinase, a cytosolic neutral sphingomyelinase, a lysosomal acid sphingomyelinase, and an alkaline sphingomyelinase. These enzymes, the pathways they regulate, and the role of the product ceramide in various biologic functions are discussed in detail in the chapters by Gómez-Munoz et al; Dbaibo and Hannun; Ballou; Lee and Obeid; Shayman and Radin; Strum, Swenson and Bell; and Whitman and Daniel.

Ceramidase is the enzyme involved in catabolizing ceramide. The immediate product of action of this enzyme is sphingosine. However, in cells, this enzyme may lead to the accumulation of sphingosine-1-phosphate or may simply lead to the clearance of ceramide. This obviously depends on the relative activity of ceramidase, sphingosine kinase, and sphingosine-1-phosphate lyase. The regulation and potential significance of this enzyme is discussed in the chapters by McKay and by Merrill et al.

Sphingosine kinase is the enzyme that converts sphingosine into the bioactive molecule sphingosine-1-phosphate. Therefore, activation of this enzyme may result in differential accumulation of sphingosine-1-phosphate, whereas its inhibition may cause accumulation of sphingosine and a drop in levels of sphingosine-1-phosphate. Because both sphingosine and sphingosine-1-phosphate are potentially important bioregulators, this enzyme assumes an important role in sphingolipid-mediated signal transduction. This enzyme is discussed in more detail in the chapters by Spiegel et al and by Hakomori.

Enzymes with potential regulatory mechanisms

At this point, it is quite conceivable that most enzymes involved in sphingolipid metabolism are regulated, or, at least, regulated isoforms of such enzymes may exist.

Regulation of these enzymes would then serve to determine the levels of the precursor and product of the respective enzymatic reactions. For example, regulation of the activity of sphingomyelin synthase would serve to regulate, in a reciprocal manner, the levels of ceramide and diacylglycerol. This may prove to be critical for regulation of growth and viability. This is discussed further in the chapter by Dbaibo and Hannun. Sphingosine-1-phosphate lyase, the enzyme that clears sphingosine-1-phosphate (and, consequently, clears the final step in sphingolipid catabolism) may play an important role in determining the levels of sphingosine-1-phosphate. Ceramide synthase may be activated in response to the chemotherapeutic agent, daunorubicin, and may provide a novel mechanism for modulating the levels of ceramide. This is discussed in the chapter by Merrill et al.

Cerebroside synthase is an enzyme that forms cerebroside from ceramide. As such, activation of this enzyme serves to lower ceramide levels and increase the levels of cerebroside whereas inhibition of this enzyme causes an elevation in ceramide levels and an attenuation of the levels of cerebroside and glycosphingolipids. The regulation and significance of this enzyme is discussed in the chapter by Shayman and Radin.

The interrelationships of pathways of sphingolipid and glycerolipid metabolism are discussed in the chapter by Gómez-Munoz et al.

Mechanism of Action of Individual Sphingolipids

For propagation of signal transduction through sphingolipids and their derived products, it is necessary to invoke the presence of specific targets that recognize the products of these regulated enzymatic pathways. As illustrated in Figure 1.1, these putative targets would serve as the initial output of these signaling pathways. In the field of signaling through lipids in general, identification of these targets and their specific relevance to signaling has been one of the most recalcitrant areas of study. Some of the best-determined targets are high-affinity receptors for eicosanoids and protein kinase C, which serves as a target for diacylglycerol. On the sphingolipid side, candidate targets include a receptor for sphingosine-1-phosphate, a ceramide-activated protein phosphatase, a ceramide-activated protein kinase, and a sphingosine-activated protein kinases (see chapters by Hakomori, Dbaibo and Hannun, Ballou, and Spiegel et al).

PROPOSED ROLES FOR SPHINGOLIPID SIGNAL TRANSDUCTION IN CELL REGULATION

The main emphasis of this book is to bring together the various aspects of currently investigated areas of sphingolipid-mediated cellular responses. It is, indeed, the significance of these areas of cell biology that has rendered the study of sphingolipids of prime importance. These various areas are summarized here:

Sphingolipids Are Essential for Both Mammalian and Yeast Cells

In mammalian systems, mutants have been obtained in the first rate-limiting enzyme of sphingolipid biosynthesis, serine-palmitoyl CoA transferase (SPT). These temperature-sensitive mutants are not viable at the restrictive temperature, and they require supplementation with sphingolipids to regain their viability function.[42,43] Mutants in the yeast *S. cerevisiae* have been obtained which are auxotrophs for sphingoid bases,[44] and these mutants were found to have a mutation in the same enzyme (yeast SPT).[36] Deletions of SPT caused loss of viability in yeast, which is, again, reconstituted by adding sphingoid bases. Taken together, these data point strongly to the essential role of sphingolipids in the viability of eukaryotic cells.

Role of Sphingolipids and Their Derived Products in Stress Responses

This is best illustrated with mammalian pathways regulating ceramide levels. These pathways appear to be activated by various stress responses, and ceramide has been proposed to be a main regulator of various growth responses and cell responses to stress. This aspect is discussed in more detail in the chapter by Dbaibo and Hannun, and by Shayman and Radin.

Role in Growth Suppression

A number of sphingolipids and their derived products, such as ceramide, sphingosine, and other lysosphingolipids, are potent suppressors of growth. In various cell systems, they induce apoptosis, cell cycle arrest, or cell senescence. This has led to the proposal that these sphingolipids may function as "tumor suppressor lipids."[45] These studies are discussed in detail in the chapters by Dbaibo and Hannun, Hakomori, Gómez-Munoz et al, and Whitman and Daniel.

Inflammation

The inflammatory response is an intricate and complex pathway of interrelated signal transduction events in a number of cell types that participate in inflammation. Sphingolipids, and in particular ceramides, have emerged as potentially key candidates in the mediation of inflammatory responses to some of the most important cytokines that regulate inflammation, such as interleukin-1 and TNFα. This is discussed in great detail in the chapter by Ballou, and by Merrill et al.

Tumor Cell Invasion

A key activity of sphingosine-1-phosphate appears to be the regulation of cell motility and invasion. These effects are discussed in detail in the chapters by Spiegel et al, and Hakomori.

Cell Senescence

A very exciting and emerging role for sphingolipids and ceramide is their involvement in the induction or propagation of the senescent phenotype. Cell senescence in tissue culture relates to the loss of ability of cells to undergo mitogenic cell division, and this parameter has been correlated closely to the aging of tissues. The chapter by Lee and Obeid discusses in detail the emerging understanding of the role of sphingolipids and ceramide in cell senescence.

Meiotic Maturation

Studies from several groups are beginning to implicate sphingomyelinase and ceramide in oocyte maturation and meiotic progression. This is discussed in detail in the chapter by Strum, Swenson, and Bell.

TOOLS, APPROACHES, AND HAZARDS IN THE STUDY OF SIGNALING THROUGH SPHINGOLIPIDS

Approaches and Tools

Building upon the paradigm of signaling through lipids as illustrated in Figure 1.1, it becomes clear that the minimum requirement for understanding the basic components of any signal transduction pathway through sphingolipids requires the following:

Determining the structure and levels of the precursor lipid and the structure of the product

In the context of signaling, it is the precise structural composition of the precursor that is the primary determinant of recognition by regulated enzymes. This serves as the input to this pathway, whereas the structure of the product determines its recognition by specific targets, which in turn constitutes the output of this signaling pathway. Fortunately, this area of research has developed to a high level of sophistication in the last several decades. Biochemists and chemists have put various techniques into good use in determining the structure of various sphingolipids and their products. The most widely used techniques relate to thin layer chromatography, NMR spectroscopy, GC mass spectroscopy, FAB mass spectroscopy, chemical synthesis and modification, recognition of specific sphingolipids and glycosphingolipids by specific antibodies, metabolic labeling studies, HPLC analysis and use of enzymatic methods. For example, ceramide and related species can be recognized by derivatization with *E. coli* diacylglycerol kinase, which yields the 1-phosphorylated ceramides (i.e., ceramide phosphate).[46,47] Derivatization of sphingoid bases with O-phthaldehyde, a method adapted from the characterization of amino acids, has proven to be a very powerful method for determining the levels of sphingoid bases,[48] which are usually exceedingly low in the cell. Sphingosine kinase has recently been utilized to quantitate the levels of sphingosine by converting it to sphingosine-1-phosphate.[49] Acylation of sphingoid bases and lysosphingolipids with radioactive acyl groups has also resulted in powerful methods to quantitate these sphingolipid species. TLC is best used for determining and analyzing the main species of sphingolipids, which usually serve as the substrates and precursors in these signal transduction pathways. For example, sphingomyelin levels may be determined by quantitating the phosphate mass in sphingomyelin isolated by TLC. Alternatively, metabolic labeling with choline (or phosphate or palmitate) followed by TLC results in a high throughput assay for quantitating sphingomyelin levels. A rapid assay for measuring sphingomyelin utilizes a combined approach of metabolic labeling and the use of bacterial sphingomyelinase.[50] Metabolic labeling with choline yields primarily phosphatidylcholine and sphingomyelin in the lipid extract. Bacterial sphingomyelinase recognizes sphingomyelinase solely and therefore its action can quantitatively release the choline labeled phosphocholine head group from sphingomyelin in the lipid extract. Structural determination and composition of sphingolipids (including sphingomyelin and ceramide) may be achieved by derivatization, HPLC analysis, or GC mass spectroscopy.

It is obvious that as this field progresses, more quantitative tools are needed. More importantly, additional tools are needed with higher sensitivity to detect species that may exist in much lower levels and yet may constitute important players in signal transduction pathways. (In analogy with the eicosanoid field, many of the biologically active eicosanoids exist at very low levels and require specific methods for detection, such as radio immunoassays and electrospray mass spectroscopy.)

Defining the key regulatory enzymes and their mechanisms of regulation

As shown in Figure 1.1, the incoming signal, which is deciphered as the input in these pathways of signal transduction, must somehow cause regulation of specific enzymes that recognize their sphingolipid substrates. This is still an area in its relative infancy in the field of sphingolipid signal transduction. In analogy with glycerolipid-mediated signaling, this has proven to be a complex area of study that has required intensive effort over decades of research. For example, intensive investigation on the PI cycle eventually resulted in the cloning of different PI-specific phospholipases C. Each of the isoenzymes of phospholipase C is activated in a specific mechanism. For example, phospholipase Cγ is activated by tyrosine phosphorylation, whereas phospholipase Cβ is activated by G protein. PI-3 kinases exist as dimers with a regulatory subunit which, in turn, is directly activated by tyrosine phosphorylation. Little is currently known about the direct regulation of the sphingomyelinases, ceramidases, sphingosine kinases, or other enzymes intimately related to regulation of signal transduction through sphingolipids. However, intensive investigation is yielding important clues, and these studies are illustrated in the ensuing chapters. Unfortunately, none of these enzymes has been cloned. Therefore, while this area of research is handicapped by a relative paucity of a biochemical infrastructure, it promises to be an area of significant growth, both in the fields biochemistry and cell biology.

Defining specific targets for sphingolipid-derived second messengers and signals

The initiation of the output from signal transduction pathway requires the recognition of the second messenger product of these pathways by specific targets, which are usually proteins. These targets include receptors (for example, the receptors for IP_3 and the various eicosanoids) and protein kinases (for example, protein kinase A for cyclic AMP). The general area of signaling through lipids has been particularly handicapped by the difficulty in determining and establishing specific targets for the putative lipid second messengers. Even in the better-established glycerolipid field of signal transduction, many of the targets are still yet to be determined (for example, specific targets for phosphatidic acid or for PI-3P). With lipids, these difficulties may arise primarily from the user-unfriendly physical properties of lipid molecules, which usually make them difficult to solubilize and to deliver to the internal compartments of the cell. Also, many of these lipid second messengers are active at intermediate concentrations in the low micromolar range, thus rendering classical binding studies of little use. This also makes the interpretation of interactions of lipids with their protein targets more difficult. In contradistinction, for lipid products that are active at much lower concentrations (for example, eiconasoids that are active in the nanomolar range), it has been easier to ascertain the targets for these molecules. Not unexpectedly, these targets usually turn out to be transmembrane receptors with high affinity for their ligands.

To compound the difficulties in determining specific targets for specific second messengers, the simplest paradigm may not always operate. It is quite possible that specific second messengers may interact with several targets, or even with a whole class of targets. For example, DAG interacts not only with PKC, but also with the protein unc-13,[51] which does not have a protein kinase domain. Nitric oxide (NO) is thought to interact with a whole variety of sulfhydryl-containing proteins and molecules, yet it plays important roles in signal transduction and cell regulation.

In sphingolipid signal transduction, this has proven to be a particularly recalcitrant area of study. Nonetheless, a number of targets have already been identified for sphingosine, ceramide, and sphingosine-1-phosphate. These are discussed in subsequent chapters in detail.

Manipulation of the levels of the putative sphingolipid second messengers

The ability to manipulate the cellular levels of the suspected sphingolipid second messengers is of paramount importance to establish two parameters:

i. Is this lipid second messenger sufficient to activate a specific pathway of signal transduction

For example, if ceramide is shown to activate a protein phosphatase in vitro, then it becomes essential to show that an increase in ceramide levels in the cell results in activation of that phosphatase. This also becomes a requirement to demonstrate any physiologic or biologic function for a ceramide-generated signal. This objective may be achieved by either delivering to the cell functional analogs of the putative lipids (such as cell-permeable analogs of ceramide), or by manipulating enzymes of lipid metabolism (such as activation of enzymes that generate ceramide or inhibition of enzymes that metabolize ceramide). The effects of such manipulations on the levels of the lipids and on subsequent events are then monitored. This then can be used to determine if the particular lipid is *sufficient* to activate its target and to inducc a certain biochemical function.

ii. To determine if the putative lipid second messenger is necessary for the proposed cellular response

Inhibitors of the key regulated enzymes that generate the particular lipid second messenger would allow investigators to determine if this lipid is necessary for initiating the signal transduction pathway and propagating a cellular response. In sphingolipid signal transduction, inhibitors exist for ceramide synthase, ceramidase, cerebroside synthase, and sphingosine kinase (see chapters by McKay, Merrill et al, and Shayman and Radin). At this point, there are no known powerful inhibitors of sphingomyelinases. Also, DNA tools are not available for most of these enzymes, with the exception of acid sphingomyelinase. The cloning of the genes responsible for the regulated metabolism of sphingolipids has become a high priority in this area of research. It is anticipated that the availability of such clones would provide significant additional and complementary tools for elucidating the regulation and significance of these pathways. For example, overexpression of relevant enzymes may activate key signaling pathways, whereas dominant negative or gene knockout would provide a complementary approach to "inhibiting" a specific sphingolipid signaling pathway.

Hazards and Caution in the Study of Lipid-Mediated Signal Transduction and Cell Regulation

Although the study of signaling through sphingolipids (and through lipids in general) promises great rewards, this remains a very complex area of research fraught with difficulties and hazards. Some of these difficulties that require caution and innovation include:

Lack of a well-developed biochemical infrastructure for the study of sphingolipid metabolism

As indicated in the previous section, very few of the key enzymes involved in known sphingolipid signaling have been purified, characterized, or cloned. This has deprived this area of research from important and necessary tools that could provide insight into the biochemical regulation of these pathways and their biologic roles. It appears that this has now moved to the forefront of this area of research, and it promises great returns in terms of novel tools to be developed and novel insights to be gained.

"Uncooperative" physical and chemical properties of sphingolipids and their derived products

Most cell biologists are used to working with water-soluble molecules and analogs that are easy to deliver to cells in order to elicit appropriate responses. At least two distinct physicochemical properties of sphingolipids (and glycerolipids) have precluded such direct assessments. (1) Most of these lipids and their derived products are significantly hydrophobic with low or negligible solubility in aqueous solutions. This precludes their easy delivery to cells in tissue culture, which is a mainstay of modern investigation into cell biology. These properties may, at times, be overcome through the development of analogs of these lipids that are more water soluble. For example, short chain ceramides and short chain diacylglycerols have been developed and shown to be efficacious in the study of cell biology. (2) Many lipids contain bulky, hydrophilic, and/or charged head groups. Even if such lipids are soluble, they probably will not gain access to intracellular compartments due to the presence of the plasma membrane barrier unless specific mechanisms of transport and uptake exist. This may represent a very rate-limiting step in the examination of biologic functions of such lipids when they act in the intracellular compartments. For example, ceramide phosphate and sphingosine phosphate are probably very poorly taken up by cells, and therefore addressing their intracellular mechanisms of action is difficult and may require much higher extracellular concentrations to achieve low intracellular levels. In some cases, these lipid molecules act on plasma membrane receptors, thus negating this problem of access. However, with other molecules, such as phosphatidic acid or ceramide phosphate, this problem may be a main reason why such molecules have not been studied further in cell biology.

Lack of potency for many candidate lipid second messengers

While many lipid-derived molecules, such as eicosanoids, act in low nanomolar concentrations by binding to plasma membrane receptors, many candidate lipid second messengers, such as ceramide and diacylglycerol, appear to act at higher concentrations which are usually in the low micromolar range. This lower level of potency invokes anxiety and a higher level of discomfort among cell biologists. This concern is quite reasonable, since lipids in the micromolar range may have nonspecific effects on cell membranes which may complicate interpretation of biologic activity. Several approaches have been employed to evaluate the significance of action of such lipids. For example, with DAG it has been very fortunate that phorbol esters are much more potent analogs of this molecule that act in the nanomolar range, and therefore their actions support the roles of DAG in the cell. Also, with both DAG and ceramide, the specificity of action of these molecules has negated the operation of nonspecific effects. Finally, mechanistic insight, developed from the action of these molecules, supports specific mechanisms of action, rather than nonspecific activities. However, it should be recognized that these are significant concerns that should be dealt with in almost every single case of defining novel actions for lipids that act in the low micromolar range.

Competing functions for sphingolipids

Most lipids have important functions in membrane formation and, probably, in membrane function. For example, sphingomyelin is postulated to regulate the level of rigidity of plasma membranes. Other sphingolipids may be important in delineating specific cell types or perhaps even specific compartments within an individual cell. Recently, sphingomyelin and sphingolipids have been shown to localize preferentially to specialized structures of the plasma membrane, termed caveoli. Many components of signal transduction pathways appear to localize to caveoli, which have been proposed as a specialized site for signal transduction events and sphingolipids may be important in the organization of these structures. These various functions for sphingolipids may, at times, "compete" with proposed signaling functions for these molecules.

Problems in defining specificity of action

The pathways and second messengers that are "easier to accept" are those that are linear and highly specific. For example, cyclic AMP, in mammalian cells, activates exclusively cyclic AMP-dependent protein kinases. With some lipids that bind specific membrane receptors, this one-to-one paradigm also applies. However, with some other candidate lipid second messengers, these paradigms may not apply, and therefore their acceptance is usually delayed until more mechanistic insight is available. For example, at least two, if not three, different targets for ceramide have been proposed, and sphingosine has been shown to regulate a number of enzymes and signaling proteins in vitro and in cells. Although DAG is now known to activate more than one target, it seems to have escaped this stigma because its best-studied target, PKC, was identified prior to the realization that DAG may possess second messenger function.

SUMMARY AND FUTURE DIRECTIONS

It is becoming very clear that sphingolipids and their derived products serve important functions in signal transduction and cell regulation. Information is accruing at an increasing rate, and has served to bring together heretofore disparate areas of research encompassing sphingolipid metabolism, signal transduction, and key areas of cell biology (such as apoptosis, cell cycle regulation, cell senescence, and tumor invasion and metastasis). Further investigation of these areas of research promises great rewards in deciphering precise biochemical mechanisms of action and specific pathways that regulate these biological processes and that utilize sphingolipids as key components. It should be noted that the elucidation of these pathways would be seriously deficient without invoking sphingolipid metabolism and sphingolipid-derived second messengers and bioeffector molecules. It is anticipated that future research will be directed at developing more tools aimed at quantitating and analyzing the sphingolipids and sphingolipid derivatives with higher sensitivity, biochemical characterization of enzymes involved in sphingolipid signal transduction, cloning of genes for these enzymes and developing DNA-based molecular tools for probing these systems. The greater task is to develop a more complete understanding of how these regulated pathways of sphingolipid metabolism fit into these exciting areas of growth regulation and cell biology.

Acknowledgments

The work in the author's laboratory is supported by grants from the NIH (GM-43825) and from the DoD (AIBS-516). I also would like to thank Dr. Lina Obeid for careful reading of the manuscript, Dr. Alicja Bielawska for preparing Figure 1.2, Heather Hayter for preparing Figure 1.1, and Don Garrett for expert secretarial assistance.

References

1. Hokin MR, Hokin LE. Enzyme secretion and the incorporation of ^{32}P into phospholipids of pancreas slices. J Biol Chem 1953; 203:967-977.
2. Nishizuka Y. Intracellular signaling by hydrolysis of phospholipids and activation of protein kinase C. Science 1992; 258:607-614.
3. Majerus PW, Bansal VS, Lips DL et al. The phosphatidylinositol pathway of platelets and vascular cells. Ann NY Acad Sci 1991; 614:44-50.
4. Berridge MJ. Calcium oscillations. J Biol Chem 1990; 265:9583-9586.
5. Rhee SG, Suh PG, Ryu SH et al. Studies of inositol phospholipid-specific phospholipase C. Science 1989; 244:546-550.
6. Bishop WR, Bell RM. Functions of diacylglycerol in glycerolipid metabolism, signal transduction and cellular transformation. Oncogene Res 1988; 2:205-218.
7. Dennis EA, Rhee SG, Billah MM et al. Role of phospholipases in generating lipid second messengers in signal transduction. FASEB J 1991; 5:2068-2077.
8. Divecha N, Irvine RF. Phospholipid signaling. Cell 1995; 80:269-278.
9. Exton JH. Phosphatidylcholine breakdown and signal transduction. Biochem Biophys Acta Lipids Lipid Metab 1994; 1212:26-42.
10. Cockcroft S. ARF-regulated phospholipase D: A potential role in membrane traffic. Chem Phys Lipids 1996; 80:59-80.
11. Olson SC, Lambeth JD. Biochemistry and cell biology of phospholipase D in human neutrophils. Chem Phys Lipids 1996; 80:3-19.
12. Lapetina EG, Crouch MF. The relationship between phospholipases A_2 and C in signal transduction. Ann NY Acad Sci 1989; 559:153-157.
13. Khan WA, Blobe GC, Hannun YA. Arachidonic acid and free fatty acids as second messengers and the role of protein kinase C. Cell Signal 1995; 7:171-184.
14. Graber R, Sumida C, Nunez EA. Fatty acids and cell signal transduction. J Lipid Mediat 1994; 9:91-116.
15. Serhan CN. Lipoxin biosynthesis and its impact in inflammatory and vascular events. Biochem Biophys Acta 1994; 1212:1-25.
16. Toker A, Meyer M, Reddy KK et al. Activation of protein kinase C family members by the novel polyphosphoinositides PtdIns-3,4-P_2 and PtdIns-3,4,5-P_3. J Biol Chem 1994; 269:32358-32367.
17. Hakomori S. Glycosphingolipids in cellular interaction, differentiation, and oncogenesis. Annu Rev Biochem 1981; 50:733-764.
18. Hannun YA, Bell RM. Functions of sphingolipids and sphingolipid breakdown products in cellular regulation. Science 1989; 243:500-507.
19. Wiegandt H. Gangliosides. In: Wiegandt H, ed. Glycolipids. New York: Elsevier, 1985:199-259.
20. Hannun YA, Loomis CR, Merrill AH Jr et al. Sphingosine inhibition of protein kinase C activity and of phorbol dibutyrate binding in vitro and human platelets. J Biol Chem 1986; 261:12604-12609.
21. Hakomori S. Bifunctional role of glycosphingolipids. J Biol Chem 1990; 265: 18713-18716.
22. Spiegel S, Milstien S. Sphingolipid metabolites: Members of a new class of lipid second messengers. J Membr Biol 1995; 146:225-237.

23. Liscovitch M, Cantley LC. Lipid Second Messengers. Cell 1994; 77:329-334.
24. Michell RH, Wakelam MJO. Sphingolipid signalling. Curr Biol 1994; 4: 370-373.
25. Merrill AH Jr, Wang E, Gilchrist DG et al. Fumonisins and other inhibitors of de novo sphingolipid biosynthesis. Adv Lipid Res 1993; 26:215-234.
26. Shayman JA. Sphingolipids: Their role in intracellular signaling and renal growth. J Am Soc Nephrol 1996; 7:171-182.
27. Mathias S, Kolesnick R. Ceramide: A novel second messenger. Adv Lipid Res 1993; 25:65-88.
28. Ballou LR, Laulederkind SJF, Rosloniec EF et al. Ceramide signalling and the immune response. Biochim Biophys Acta Lipids Lipid Metab 1996; 1301:273-287.
29. Obeid LM, Hannun YA. Ceramide: A stress signal and mediator of growth suppression and apoptosis. J Cell Biochem 1995; 58:191-198.
30. Merrill AH Jr, Wang E. Enzymes of ceramide biosynthesis. Methods Enzymol 1992; 209:427-437.
31. Hannun YA. Ths Sphingomyelin cycle and the second messenger function of ceramide. J Biol Chem 1994; 269: 3125-3128.
32. Hannun YA. Sphingolipid metabolism and biology. Encyclopedia of Human Biology 1991; 7:179-189.
33. Hannun YA. Sphingolipid-derived breakdown products: role in cell regulation. In: Krogsgaard-Larsen P, Christensen SB, Kofod H, eds. Alfred Benzon Symposium 33: New Leads and Targets in Drug Research. Munksgaard, Copenhagen, 1992: 257-269.
34. Lee TC, Ou MC, Shinozaki K et al. Biosynthesis of N-acetylsphingosine by platelet-activating factor: Sphingosine CoA-independent transacetylase in HL-60 cells. J Biol Chem 1996; 271: 209-217.
35. Abe A, Shayman JA, Radin NS. A novel enzyme that catalyzes the esterification of N-acetylsphingosine—Metabolism of C_2-ceramides. J Biol Chem 1996; 271: 14383-14389.
36. Lester RL, Wells GB, Oxford G et al. Mutant strains of *Saccharomyces cerevisiae* lacking sphingolipids synthesize novel inositol glycerophospholipids that mimic sphingolipid structures. J Biol Chem 1993; 268:845-856.
37. Van Veldhoven PP, Mannaerts GP. Sphingosine-phosphate lyase. Adv Lipid Res 1993; 26:69-98.
38. Ferlinz K, Hurwitz R, Vielhaber G et al. Occurrence of two molecular forms of human acid sphingomyelinase. Biochem J 1994; 301:855-862.
39. Ichikawa S, Sakiyama H, Suzuki G et al. Expression cloning of a cDNA for human ceramide glucosyltransferase that catalyzes the first glycosylation step of glycosphingolipid synthesis. Proc Natl Acad Sci USA 1996; 93:4638-4643.
40. Bernardo K, Hurwitz R, Zenk T et al. Purification, characterization, and biosynthesis of human acid ceramidase. J Biol Chem 1995; 270:11098-11102.
41. Yada Y, Higuchi K, Imokawa G. Purification and biochemical characterization of membrane-bound epidermal ceramidases from guinea pig skin. J Biol Chem 1995; 270:12677-12684.
42. Hanada K, Nishijima M, Akamatsu Y. A temperature-sensitive mammalian cell mutant with thermolabile serine palmitoyltransferase for the sphingolipid biolsynthesis. J Biol Chem 1990; 265: 22137-22142.
43. Hanada K, Nishijima M, Kiso M et al. Sphingolipids are essential for the growth of chinese hamster ovary cells: restoration of the growth of a mutant defective in sphingoid base biosynthesis by exogenous sphingolipids. J Biol Chem 1992; 267:23527-23533.
44. Pinto WJ, Srinivasan B, Shepherd S et al. Sphinoglipid long-chain-base auxotrophs of *Saccharomyces cerevisiae*: genetics, physiology, and a method for their selection. J Bacteriol 1992; 174:2565-2574.
45. Hannun YA, Linardic CM. Sphingolipid breakdown products: anti-proliferative and tumor-suppressor lipids. Biochem Biophys Acta Bio-Membr 1993; 1154: 223-236.
46. Preiss J, Loomis CR, Bishop WR et al. Quantitative measurement of *sn*-1,2-diacylglycerols present in platelets, hepatocytes, and *ras*- and *sis*-transformed normal rat kidney cells. J Biol Chem 1986; 261: 8597-8600.

47. Okazaki T, Bielawska A, Bell RM et al. Role of ceramide as a lipid mediator of 1a,25-dihydroxyvitamin D_3-induced HL-60 cell differentiation. J Biol Chem 1990; 265:15823-15831.
48. Merrill AH Jr, Wang E, Mullins RE et al. Quantitation of free sphingosine by high performance liqiud chromatography. Anal Biochem 1988; 171:373-381.
49. Olivera A, Rosenthal J, Spiegel S. Sphingosine kinase from Swiss 3T3 fibroblasts: A convenient assay for the measurement of intracellular levels of free sphingoid bases. Anal Biochem 1994; 223:306-312.
50. Jayadev S, Linardic CM, Hannun YA. Identification of arachidonic acid as a mediator of sphingomyelin hydrolysis in response to tumor necrosis factor α. J Biol Chem 1994; 269:5757-5763.
51. Ahmed S, Maruyama IN, Kozma R et al. The *Caenorhabditis elegans unc-13* gene product is a phospholipid-dependent high-affinity phorbol ester receptor. Biochem J 1992; 287:995-999.

CHAPTER 2

Ceramide: A Stress Response Mediator Involved in Growth Suppression

Ghassan Dbaibo and Yusuf A. Hannun

I. INTRODUCTION

Cell membranes are composed of a lipid bilayer comprised mostly of phospholipids. The majority of these lipids are glyceride-based and are termed gylcerolipids. Sphingolipids comprise less than 3% of cell membrane lipids and utilize ceramide as a backbone. This group of membrane lipids exhibits wide diversity, mostly due to the complexity of head groups at the carbon 1 (C1) position which exhibit both species and tissue specificity (Fig. 2.1). For a long time, much of the research on sphingolipids focused on the glycosphingolipids and the role of the glycosyl head group in biologic processes.[1] A number of functions were associated with the glycosphingolipids, including a role in cellular interaction, differentiation, and oncogenic transformation.[2] Additionally, complex glycosphingolipids were shown to serve as antigens important in the immune response, as receptors for viruses or bacteria, and as tumor markers.[1,3]

The discovery of inhibition of protein kinase C by the sphingolipid breakdown product, sphingosine, shifted the attention towards the lipid moeity of sphingolipids.[4-6] The early findings with sphingosine raised the possibility that sphingolipids may function in a manner analogous to the glycerolipids as reservoirs for bioactive molecules.[7,8] Thus, specific stimuli may result in the hydrolysis of membrane sphingolipids to yield breakdown products with defined biologic activities. These sphingolipid-derived molecules would be the counterparts of diacylglycerol, phosphatidic acid, platelet-activating factor, arachidonic acid, and other glycerolipid-derived molecules which are produced after the specific action of a variety of phospholipases.[9,10] Indeed, the discovery of the sphingomyelin cycle in 1989 further supported this expanding role of sphingolipids[11] and focused attention on another sphingolipid breakdown product, namely ceramide, which differs structurally from sphingosine by the addition of a fatty acyl group through an amide linkage at the 2-amino position.

Sphingolipid-Mediated Signal Transduction, edited by Yusuf A. Hannun.

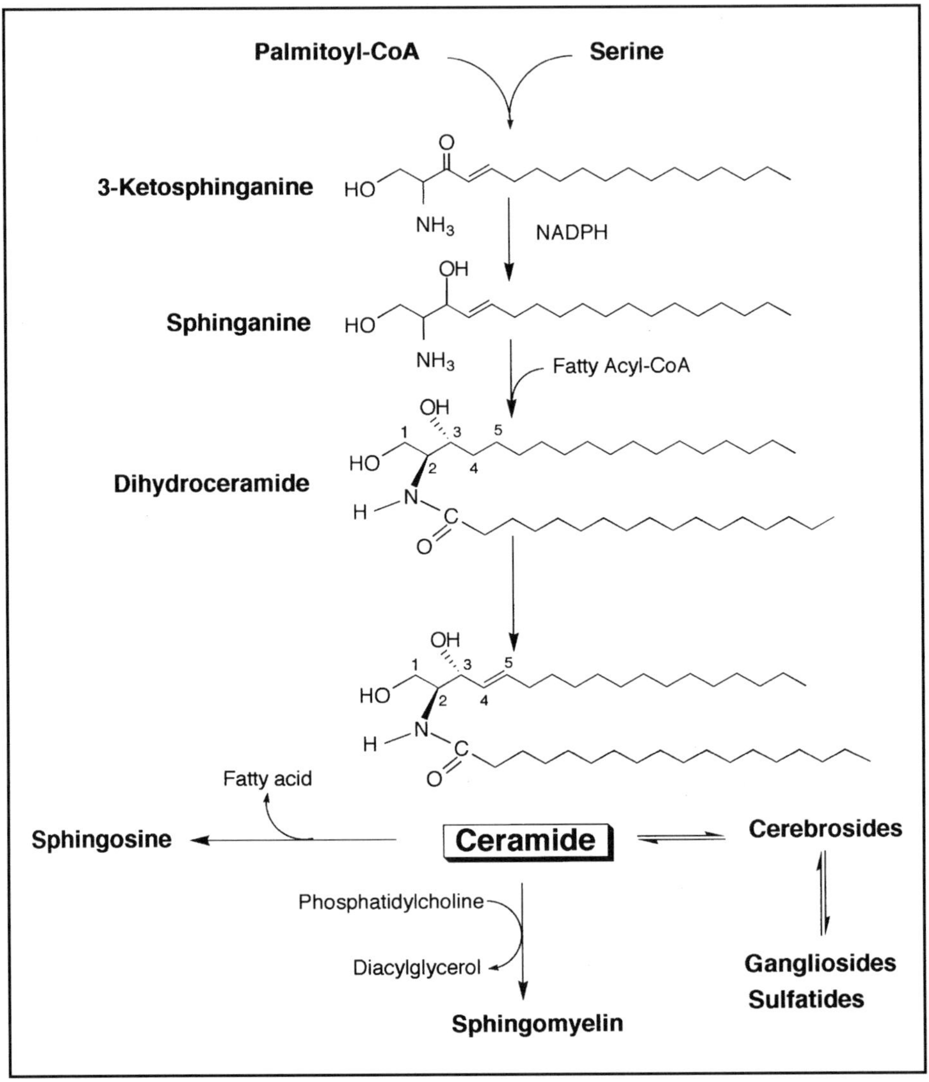

Fig. 2.1. Metabolic pathways regulating ceramide. These pathways and the involved enzymes are discussed in the text. The sphingolipid double bond is shown here to be inserted at the level of dihydroceramide, although current studies do not rule out introduction of the double bond at a later step.

OVERVIEW OF THE SPHINGOMYELIN CYCLE AND CERAMIDE BIOLOGY AND METABOLISM

Overview

In the sphingomyelin cycle, the action of a stimulus, such as 1α,25-dihydroxyvitamin D_3 (vitamin D_3), tumor necrosis factor α (TNFα), interleukin-1β, nerve growth factor (NGF), γ-interferon, chemotherapeutic agents, or other inducers of stress results in the hydrolysis of sphingomyelin and the accumulation of cellular ceramide and phosphorylcholine. These events occur within minutes and the changes peak at 30 minutes to 2 hours depending on the stimulus.[11-13]

Elevated ceramide is thought to function as a second messenger mediating a number of biologic functions.[14-16] Subsequently, sphingomyelin is regenerated and ceramide levels return to baseline. In contrast, ceramide accumulation occurring over several hours and reaching cellular levels which are several-fold more than those previously described was recently reported (refs. 17, 18; Dbaibo et al, submitted). These prolonged kinetics of ceramide accumulation were seen in response to stressful conditions, including serum deprivation, chemotherapeutic agent exposure, and treatment with TNFα. Thus, a new role for ceramide in the cellular stress response emerged. This role, and the related biologic functions of ceramide will be the focus of this chapter.

Metabolic Pathways Regulating Ceramide

Ceramide is at the center of metabolic pathways regulating the synthesis of various sphingolipids. Synthesis of ceramide commences with the condensation of serine and palmitoyl-CoA which results in the formation of 3-ketosphinganine. Dihydrosphingosine (sphinganine) is produced after the reduction of 3-ketosphinganine and functions as a substrate for ceramide synthase (sphinganine N-acyl transferase) which catalyzes the addition of a fatty acid (from fatty acyl CoA) to the 2-amino group of dihydrosphingosine.

The endogenous levels of ceramide may be regulated by the balance of the rate of its de novo synthesis, the rate of its breakdown (to sphingosine), the rate of its incorporation into more complex sphingolipids, and the rate of its generation from the hydrolysis of these compounds. The discovery of the sphingomyelin cycle focused attention on sphingomyelinases, which are a specialized form of phospholipase C that cleave sphingomyelin to yield ceramide and phosphorylcholine. These enzymes can have optimal pHs which are either acidic, alkaline, or neutral and will vary in topographical distribution within the cell.[21-23] These and other enzymes that regulate cellular ceramide levels are the focus of intense research due to the emerging role of ceramide in cell regulation.

THE STRESS RESPONSE: AN EMERGING ROLE FOR CERAMIDE

The response of organisms to stressful conditions generally constitutes a predictable and genetically determined set of events. The specific stress responses generally provide a survival advantage and have been preserved and optimized throughout evolution. The response may be relatively simple in lower organisms and multidimensional in more complex organisms. For example, nutritional starvation of bacteria may result in downregulation of a set of enzymes involved in normal metabolism and upregulation of enzymes which control alternate metabolic pathways. In contrast, infection of a mammal by a pathogenic virus results in a complicated immune response consisting of an early inflammatory reaction with secretion of a wide array of cytokines and chemokines. The latter will help recruit lymphocytes, neutrophils, and macrophages to the site of infection which, in turn, secrete more cytokines. Infected cells are eliminated by natural killer cells and a specific antibody response is elaborated to help prevent further infection with the same virus. Thus, the type of stressful stimulus and the level of the organism on the evolutionary ladder will determine which response is activated in a particular organism.

While the driving force behind an elaborate stress response is usually survival of the organism and genetic preservation through production of intact progeny, the actual biochemical mediators of the various stress responses are not completely understood. In recent years, ceramide emerged as a candidate regulator of the stress response. This hypothesized role for ceramide is supported by its ability to produce a number of biologic effects seen in response to stress such as growth suppression, cell cycle arrest, differentiation, apoptosis (programmed cell death), and modulation of inflammation. Furthermore, agents which result in the elevation of cellular ceramide levels are ei-

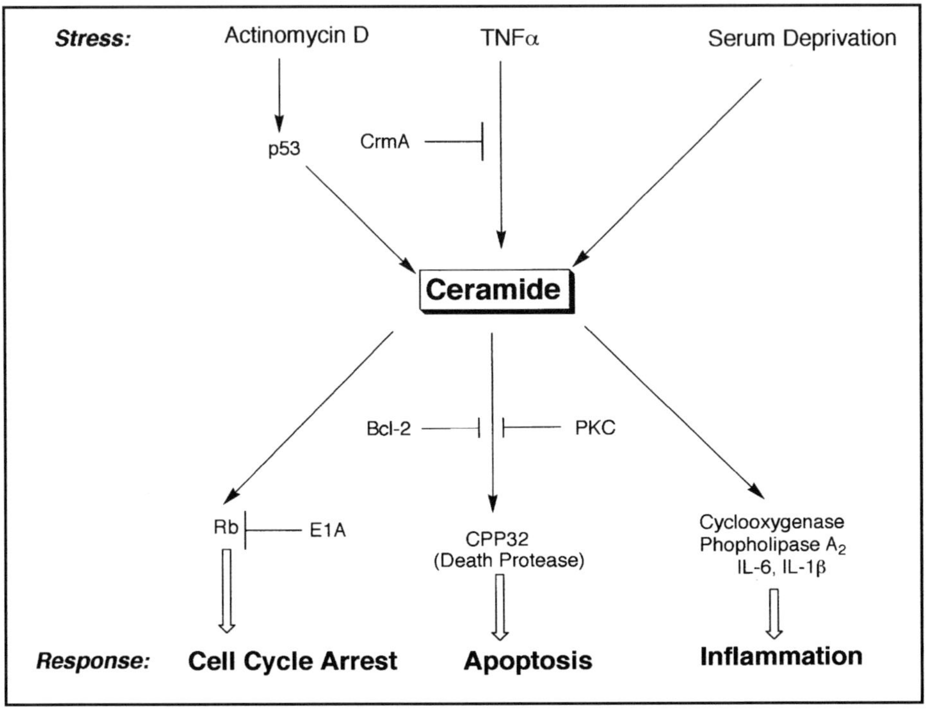

Fig. 2.2. Schematic representation for the role of ceramide in the stress response. Generation of ceramide due to the action of one of several stimuli of stress (such as TNF, chemotherapeutic and cytotoxic agents, and serum withdrawal) activates one or more pathways involved in the response to stress (such as cell cycle arrest, apoptosis, or inflammation). Modulating factors (such as PKC or Bcl-2) may influence or determine the ultimate fate of the cell.

ther inducers of apoptosis (TNFα, Fas ligand, dexamethasone), inducers of differentiation (vitamin D_3, TNFα, NGF), promoters of cellular damage (chemotherapeutic agents, brefeldin A, ultraviolet light, ionizing irradiation), or modulators of the inflammatory response (γ-interferon, interleukin 1β, TNFα). Thus, the emerging role of ceramide in the stress response may be analogous to the role of the tumor suppressors p53 and the retinoblastoma protein (Rb) which have been shown to modulate cell growth, differentiation, and apoptosis and to play a particularly important role in the stress response generated following genotoxic damage.[24] Not surprisingly, ceramide appears to have significant interactions with these protein tumor suppressors. These interactions, as well as other biologic effects of ceramide which are potentially related to the stress response, will be summarized in the sections which follow.

Ceramide and Differentiation

One of the earliest biologic effects attributed to ceramide was its role in inducing cell differentiation. These experiments were done as part of the early studies on vitamin D_3-induced activation of the sphingomyelin cycle.[12] In these experiments, vitamin D_3 was shown to activate a neutral sphingomyelinase which hydrolyzed membrane sphingomyelin to generate ceramide and phosphorylcholine. Since vitamin D_3 and TNFα induce monocytic differentiation in the HL-60 myelocytic leukemia cell line, it was important to show whether ceramide could mimic the effects of vitamin D_3 and TNFα on differentiation. Indeed, treatment of HL-60 cells for as short as two hours with a cell-permeable analog of ceramide, C_2-ceramide (N-acetylsphingosine), at micromolar concentrations resulted in monocytic differentiation. When submicromolar concentrations of C_2-ceramide were used, differentiation was only obtained if subthreshold doses of vitamin D_3 or TNFα were used concurrently. Similar results were obtained using the macrolide brefeldin A which activates the sphingomyelin cycle and induces differentiation of HL-60 cells.[25] Inducers of granulocytic (retinoic acid, dibutyryl cAMP) or macrophage-like (phorbol ester) differentiation did not induce sphingomyelin hydrolysis or ceramide generation, indicating that the effects of ceramide are specific to the monocytic differentiation pathway.[13]

More recently, similar results were obtained in rat T9 anaplastic glioblastoma cells.[26] Following treatment with NGF, these cells undergo astrocytic differentiation due to activation of the p75 neurotrophin receptor, a member of the TNF family of receptors.[27] Treatment with NGF resulted in sphingomyelin hydrolysis and accumulation of cellular ceramide within minutes. In turn, low micromolar concentrations of C_2-ceramide were shown to induce differentiation of these cells to an astrocyte-like phenotype. These findings support a general role for ceramide in the induction of differentiation. However, the biochemical mechanisms by which ceramide induces differentiation have not been delineated, although Rb activation (see below) is likely to be important due to the critical role that Rb plays in differentiation.[28,29]

Ceramide as a Cell Cycle Inhibitor

Normal cell growth and division is regulated by a complex network of cyclins and cyclin-dependent kinases (cdks) which controls the progression of the cell through the different phases of the cell cycle. Rb is a tumor suppressor which plays a central growth-inhibitory role in this process and is targeted by the different cdks which phosphorylate it at multiple sites. Loss of both alleles of the *RB1* gene results in the development of a variety of cancers including retinoblastoma of childhood, osteosarcoma, as well as bladder and breast carcinoma.[30] The phosphorylation state of Rb is critical to its function and varies during the cell cycle in a phase-specific manner.[31,32] Rb is phosphorylated in S, G_2 and M phases of the cell cycle where it is inactive and does not appear to suppress growth. In G_0 and G_1, Rb is predominantly in the dephosphorylated, i.e., active, form where it induces cell cycle arrest.[33] The mechanism of action of Rb appears to be its ability to bind, while in the dephosphorylated form, to multiple transcription factors necessary for cell cycle progression and inhibit their activities.[34,35] Some DNA viral proteins specifically bind the dephosphorylated product of Rb and inactivate it resulting in the release of transcription factors, which allows viral replication. These include E1A of adenovirus, E7 of human papillomavirus, and large T antigen of simian virus-40 (SV-40).[36-40] This underscores the importance of Rb in the regulation of the cell growth and proliferation.

Inhibition of the cell cycle is a major mechanism for growth control which is activated during the stress response to genotoxic damage. Presumably, arrest of the cell in G_1 or G_2 allows DNA repair prior to replication or mitosis, respectively, and prevents the propagation of damaged DNA with its deleterious consequences.[41-42] Several protein inhibitors of the cell cycle have been described which function by inhibiting cdks resulting in the predominance of hypophosphorylated Rb in the cell.[35] The ability of ceramide to inhibit cell growth led to a closer examination of the mechanisms by which this is accomplished and the effects on Rb in particular.

Early studies with ceramide revealed its potent antiproliferative effects. Treatment of HL-60 cells with C_2-ceramide at concentrations of 1-10 μM resulted in significant dose-dependent growth inhibition.[12,43] Similar results were obtained in other lymphocytic cell lines including Jurkat, Molt-4, and U937 as well as the T9 glioblastoma cell line.[26,44,45] Additionally, the growth of epithelial mink lung cells and rat fibroblasts was inhibited by low micromolar concentrations of C_6-ceramide.[45,46] These concentrations of exogenous ceramides are within an order of magnitude of levels of endogenous ceramide seen in response to TNF and other inducers of ceramide accumulation. Moreover, the effects of exogenous ceramides are specific in that many related molecules do not mimic the effects of ceramide. Most notably, dihydroceramide, which differs from ceramide in lacking the 4-5 trans double bond of the sphingoid backbone, is inactive in biologic assays although its uptake and metabolism is similar to that of ceramide.

Additional evidence supporting a role for ceramide in growth inhibition was obtained from experiments utilizing primary T lymphocytes. These cells were stimulated to proliferate with interleukin-2 (IL-2) or a combination of phorbol ester and calcium ionophore and their ceramide content was measured. Compared to nonstimulated cells, ceramide levels were significantly lower in proliferating cells at 48 and 72 hours.[47] In another approach, cells were treated with the UDP-glucosyl:ceramide transferase inhibitor, D,L-*threo*-1-phenyl-2decanoylamino-3-morpholino-1-propanol (PDMP), which results in significant accumulation of cellular ceramide. Thymidine incorporation into DNA was measured at 72 hours and found to be decreased 10-fold while ceramide levels increased 4-fold when compared to nontreated cells.[48] These studies clearly support a role for ceramide in growth suppression although they do not shed light on the mechanisms involved.

In order to examine the mechanism by which ceramide induces growth inhibition, studies were performed to evaluate its effects on the cell cycle. Serum deprivation of Molt-4 cells was used as a model. These cells, like many other cell lines grown in vitro, are dependent on various growth factors present in fetal bovine serum (FBS) with which they are supplemented. Removal of FBS from their culture medium results in a specific growth arrest in G_0/G_1 starting at 24 hours and becoming maximal by 48 hours following deprivation.[17] Measurement of cellular ceramide following serum deprivation revealed that ceramide levels increased dramatically in a time-dependent manner. While levels in control, i.e., serum-supplemented, cells were relatively stable, levels in serum-deprived cells increased by 3-, 8-, and 15-fold following 24, 48, and 96 hours of serum deprivation, respectively. These increases correlated with the development of a specific cell cycle arrest in G_0/G_1 as mentioned above. In order to test whether ceramide was inducing this specific arrest, Molt-4 cells were treated with increasing concentrations of C_6-ceramide and cell cycle analysis was performed. At a concentration of 15 μM, C_6-ceramide induced cell cycle arrest which was identical to that produced by serum deprivation starting at 16 hours following treatment. These effects became more dramatic at higher concentrations and were specific to C_6-ceramide in that dihydro-C_6-ceramide, which only differs by the lack of a double bond between carbons 4 and 5 of the sphingoid backbone, was not effective.[17]

The effects of ceramide on the cell cycle and the known role of Rb in regulating the cell cycle prompted the investigation of a possible relationship between the two growth suppressor molecules. Using the same model of serum deprivation of Molt-4 cells, the status of Rb phosphorylation was examined following an increasing period of serum deprivation.[45] Rb was found to become gradually dephosphorylated starting at 24 hours of serum deprivation reflected by the appearance of a rapidly migrating band on gel-electrophoresis. At 48 and 72 hours, several rapidly migrating bands appeared and their intensity increased, indicating the predominance of hypophosphorylated species of Rb. By 96 hours, dephosphorylation appeared to be complete.[45] Therefore, serum deprivation induced Rb dephosphorylation which correlated with cell cycle arrest and elevation of cellular ceramide.

Next, it was important to examine whether elevation of cellular ceramide was mediating the cell cycle effects of serum deprivation through activation, i.e., dephosphorylation, of Rb. In Molt-4 cells treated with increasing concentrations of C_6-ceramide for 4 hours, Rb was found to be dephosphorylated starting at a concentration of 10 μM. This effect was time-dependent in that longer treatments resulted in more pronounced dephosphorylation of Rb. The specificity of these effects was ascertained by using other putative lipid mediators, as well as dihydro-C_6-ceramide, none of which had any effects on Rb.[45] Sphingosine, which is metabolically and structurally related to ceramide, was shown earlier to induce Rb dephosphorylation.[49] However, a role for sphingosine in serum deprivation was shown to be unlikely based on two findings: (1) sphingosine levels did not change from baseline with serum deprivation and (2) exogenous ceramide was not metabolized to sphingosine intracellularly. These findings indicated that ceramide, and not sphingosine, may be playing a physiologic role in mediating cell cycle arrest induced by serum deprivation through its ability to induce Rb dephosphorylation.

Similar results were obtained using another approach.[50] NIH 3T3 fibroblasts overexpressing the insulin-like growth factor-1 (IGF-1) receptor were treated with PDMP, which increases cellular ceramide levels by inhibiting glucosylceramide synthase (see above), in the presence of serum or IGF-1 stimulation. PDMP produced a reversible cell cycle arrest at the G_1/S and G_2/M transition points which was maximal at 12-24 hours. This was accompanied by a 3- to 4-fold increase in cellular ceramide. This correlated with a decrease in the activity of two cyclin-dependent kinases, $p34^{cdc2}$ kinase and cdk2 kinase raising the possibility that these kinases are inhibited by ceramide.

Whether Rb plays a role in mediating the growth inhibitory effects of ceramide on various cell types was then tested. Different cell lines either containing or lacking functional Rb were examined.[45] In general, cells containing functional Rb were sensitive to ceramide-induced growth suppression whereas cells in which Rb was disabled (due to genetic mutation or deletion, e.g., WERI-Rb-1 retinoblastoma cells, or due to transfection of large T antigen which binds to and inhibits Rb) were resistant to ceramide. Similarly, when Molt-4 cells were transfected with the adenoviral E1A gene they showed no cell cycle arrest following ceramide treatment.[45]

Therefore, the role of ceramide as a cell cycle inhibitor in response to stress appears to hinge, in part, on the presence of functional Rb. These studies support a role for ceramide as a "lipid tumor suppressor" which may be involved in orchestrating the cellular response to stressful conditions.

Role for Ceramide in Regulating Apoptosis

Early experiments with ceramide analogs consistently showed evidence of cytotoxicity which accompanied the growth suppressive effects.[44] These effects were later shown to be the result of specific activation of programmed cell death (apoptosis) by ceramides.[51-53] Initial evidence implicating ceramide in the regulation of apoptosis came from studies utilizing U937 cells.[51] These cells undergo apoptosis, characterized by pathognomonic DNA fragmentation, in response to TNFα. Treatment with low micromolar concentrations of C_2-ceramide produced characteristic DNA fragmentation within 1-3 hours which was indistinguishable from that produced by TNFα. Additionally, the effects were specific to ceramide where dihydroceramide and other amphiphilic lipids were ineffective. These results were later confirmed and extended to other cell lines.[17,52,54] Ceramide is now suggested as a mediator of apoptosis following TNFα, Fas ligation, ionizing radiation, chemotherapeutic agents, and serum deprivation.[17,18,54-57] The common denominator among these stimuli is that they either participate in the stress response or that they are themselves stressful to the cell.

The mechanism by which ceramide drives the apoptotic pathway remains unclear. Recent studies showed that the family of cysteine proteases related to the Interleukin-1β converting enzyme (ICE), are involved in the execution phase of apoptosis.[58]

These proteases may work sequentially, in a cascade analogous to that of the complement and clotting pathways, to promote cell death.[59,60] These proteases became possible targets for ceramide action. One substrate for these proteases is the enzyme poly(ADP-ribose) polymerase (PARP), which is cleaved by some, but not all, members of this family of proteases.[61,62] PARP, a 116 kDa protein involved in DNA repair, becomes cleaved during apoptosis to an 85 kDa fragment. The protease with greatest activity against PARP was recently identified as CPP32 (prICE/Yama/apopain).[63-65] Treatment of Molt-4 cells with C_2- or C_6-ceramide was found to induce PARP cleavage within 3-4 hours yielding the specific apoptotic fragment.[66] These effects were specific to ceramide and were not seen after treatment with dihydroceramide or dioctanoylglycerol.

Further insight into the role of ceramide in apoptosis was obtained utilizing the antiapoptotic molecule Bcl-2. When overexpressed, Bcl-2 confers resistance to apoptosis when induced by a variety of stimuli.[67] Overexpression of Bcl-2 in Molt-4 cells, as well as other cell types, rendered them resistant to the apoptotic effects of cell permeable ceramide analogs (refs. 18, 68, 69; Dbaibo et al, submitted). Importantly, Bcl-2 did not prevent cellular accumulation of ceramide in response to the apoptotic stimulus, e.g., the chemotherapeutic agent vincristine or TNFα, rather it appeared to work at a point further downstream by inhibiting an apoptotic effector molecule that is probably induced by ceramide.

The cowpox cytokine response modifier A (CrmA) gene product is another anti-apoptotic molecule that was utilized to gain better understanding about the regulation of ceramide in apoptosis. CrmA is a serine protease inhibitor that was found to be a potent inhibitor of ICE and related family members.[70] Not only does CrmA inhibit mature IL-1β formation, but it also protects from TNFα-induced apoptosis presumably by inhibiting ICE-related proteases involved in the apoptotic pathway. When CrmA-overexpressing MCF-7 breast carcinoma cells were treated with C_6-ceramide, they were not protected from its apoptotic effects although they were resistant to TNFα-induced apoptosis (Dbaibo et al, submitted). Additionally, when these cells were treated with TNFα there was no accumulation of cellular ceramide compared to vector cells which had a 4- to 5-fold increase in ceramide. Since, at very high concentrations, CrmA had been shown to inhibit CPP32 in a cell free system,[65] it became important to examine the effects of CrmA expression on PARP cleavage. When cells overexpressing CrmA were treated with TNFα, no PARP cleavage was observed. However, treatment with ceramide induced characteristic PARP cleavage. Therefore, these studies implicate the presence of at least two ICE-like proteases, one upstream and the other downstream of ceramide generation. The upstream protease is more sensitive to CrmA inhibition and is possibly involved in ceramide synthesis or sphingomyelinase activation while the downstream protease is more intimately involved with the execution phase of apoptosis. Indeed, recent studies have shown the existence of a TNFα receptor and Fas linked ICE-like protease termed MACH or FLICE.[71,72] Thus, ceramide may function as a "sensor" that gauges the degree of stimulation of proteases or other molecules involved in the *signaling* phase of apoptosis which in turn may reflect the degree of stress or damage to which the cell is being exposed. If the damage is repairable or the stress tolerable, the cell will be driven towards cell cycle arrest to allow recovery. However, upon reaching a certain intrinsically defined threshold, perhaps reflecting irreversible damage, ceramide can activate the downstream proteases involved in the *execution* phase of apoptosis.

Support for the hypothesized role of ceramide in the stress response comes from recent evidence showing a relationship between p53 induction and the cellular accumulation of ceramide (Dbaibo et al, submitted). p53 has been proposed as the "guardian of the genome" because of its demonstrated role in the response to DNA-damaging agents and its requirement for cell cycle arrest and/or apoptosis in response to these agents.[73] Genotoxic insults such as chemotherapeutic agents, gamma irradiation, or ultraviolet irradiation, have been shown to induce p53 expression. The role of p53 appears to be at the "decision fork" where it either drives the cell towards apoptosis or it induces cell cycle arrest, allowing the repair of damaged DNA. The similar functions of p53 and ceramide led to the examination of their mutual relationship. When a p53-dependent pathway was examined, e.g., induction of apoptosis of Molt-4 cells by low concentrations of actinomycin D, it was found that induction of p53 was followed by a dramatic rise in endogenous ceramide occurring over several hours (Dbaibo et al, submitted). At time points occurring later than the onset of ceramide increase, early evidence of apoptosis or cell cycle arrest was seen. When Molt-4 cells were treated with exogenous cell-permeable ceramide analogs, effects similar to those of actinomycin D on apoptosis were observed. However, this occurred in the absence of induction of p53 indicating that the effects of ceramide are independent of p53, and that p53 functions upstream of ceramide on the apoptotic pathway stimulated by low concentrations of actinomycin D. These results indicate that ceramide may mediate some of the effects of p53 in the response to genotoxic damage. Additionally, there was evidence for the presence of p53-independent pathways of apoptosis and cell cycle arrest which are mediated by ceramide, e.g., TNFα, indicating that ceramide accumulation may be a more ubiquitous response to stress compared to p53.

Role of Ceramide in the Inflammatory Response

As discussed above, inflammation is one form of the stress response which is particularly relevant to infections with microorganisms. A number of inducers of ceramide accumulation play an important role in the inflammatory response. In turn, ceramide appears to mediate specific biologic events seen in the inflammatory response. These effects include the stimulation of prostaglandin E_2 formation by enhancing gene expression of cyclooxygenase in human fibroblasts[74] as well as enhancing the transcription of cytosolic phospholipase A_2 and cyclooxygenase 2 in murine fibrosarcoma cells (L929).[75] Also, ceramide was shown to induce the transcription of IL-6 in human fibroblasts[76] and to augment the IL-1β-induced secretion of IL-2 by murine thymocytes.[77] A role for ceramide in the activation of NF-κB was also suggested although the balance of evidence does not substantiate such a role.[16,44,78-83] These, and other, inflammatory effects of ceramide are discussed in more detail in chapter 3 by L.R. Ballou (this volume).

MOLECULAR TARGETS OF CERAMIDE ACTION

The mechanism by which ceramide exerts its effects remains unclear and is subject to intense research.

Direct (In Vitro) Targets

Several candidate enzymes have been described and implicated as direct targets in the ceramide-activated pathways. These include the ceramide-activated, proline directed, protein kinase[84,85] which does not appear to be directly activated by ceramide since it loses its ceramide responsiveness upon renaturation from SDS gels.[86] Also, protein kinase C ζ (PKCζ) has been suggested as a candidate mediator of the effects of ceramide, particularly on NF-κB[87,88] although its significance in ceramide-activated pathways awaits further study. Recently, ceramide was shown to bind and activate protein kinase c-Raf;[89] however, appropriate lipid controls were not used which makes the interpretation of these results difficult.

A ceramide-activated protein phosphatase (CAPP) has been described.[90] It appears to be a serine/threonine phosphatase which is inhibitable by low concentrations of okadaic acid. CAPP is hypothesized to be a subtype of heterotrimeric protein phosphatase 2A (PP2A) where the B (regulatory) subunit is necessary for ceramide-induced activation whereas the C (catalytic) and A (structural) subunits, are not sufficient for activation by ceramide when present alone or in combination.[91] These results are compatible with the hypothesized role of the B subunit in regulating the catalytic activity and substrate specificity of PP2A. However, other studies showed that a dimeric form of PP2A lacking the B subunit can also be activated by ceramide, which suggested that the C subunit may contain the ceramide-binding domain.[92] CAPP is activated at concentrations of synthetic ceramides which are similar to those that produce growth suppression. Additionally, partially purified CAPP is activated by natural ceramides but not by sphingosine, dihydroceramide, or other lipids. Moreover, CAPP activity was identified in the yeast *Saccharomyces cerevisiae*, whose growth is strongly inhibited by ceramide, indicating that a ceramide-related pathway is conserved in lower eukaryotes.[93] These findings were further supported by the demonstration of ceramide-related biologic effects that are reversed by okadaic acid which, as mentioned above, can inhibit CAPP at low concentrations.[94] More recently, studies in yeast have shown that yeast CAPP is composed of the Sit4 catalytic subunit, the A subunit encoded by Tpd3, and the regulatory subunit encoded by CDC55.[95] Mutations in these subunits render *S. cerevisiae* resistant to the growth suppressor effects of ceramide. Thus, emerging evidence is beginning to support a role for CAPP as a mediator of the effects of ceramide on growth regulation.

Cellular Targets for Ceramide

The addition of ceramides to cells or the manipulation of endogenous levels of ceramide has been shown to modulate the activity of a number of signaling molecules. As discussed previously, ceramide causes activation of Rb, activation of proteases, down regulation of c-myc, and regulation of phosphatases and kinases. Recently, another target for ceramide action was described. The stress-activated protein kinases (SAPKs), such as c-Jun N-terminal protein kinase (JNK), are presumably involved in the cellular response to intra- and extracellular stress such as TNFα, protein synthesis inhibition, or exposure to ultraviolet light. These kinases appear to function by phosphorlyating and activating transcription factors exemplified by the phosphorylation of c-Jun and ATF2 by JNK.[96] The emerging role of ceramide in mediating the effects of stressful responses such as those that activate JNK led to the examination of its effects on JNK. Treatment of HL-60 cells with C_2-ceramide or sphingomyelinase, but not dihydroceramide, resulted in the activation of JNK within 5 minutes.[97] Bacterial sphingomyelinase was also found to reproduce the effects of TNFα on JNK activation in HepG2 cells.[96] The functional significance of this activation was evident when cells were protected from apoptosis induced by C_2-ceramide upon inhibition of JNK, which was accomplished by expressing dominant negative mutants of the upstream kinases responsible for phosphorylating and activating it.[98] Additionally, the inhibition of JNK activation did not interfere with ceramide accumulation following TNFα. These studies indicated that activation of the JNK pathway was potentially required for the ceramide-mediated stress response although there is no evidence, so far, that ceramide directly interacts with and activates these enzymes.

CONCLUSIONS

Ceramide is emerging as a candidate regulator of the stress response. Mounting evidence suggests that ceramide plays a central role in growth suppression by inducing differentiation, cell cycle arrest, or apoptosis. The conditions which lead to ceramide accumulation, which include the actions of cytokines, growth factor deprivation, chemotherapeutic agents, or genotoxic damage, represent different forms of stress to which the cell appears to respond in a uniform manner. Accumulation of ceramide may be a well-conserved and basic form of a "warning sign" which alerts the cell to impending danger. Ceramide may also participate in the response to this perceived danger by driving the cell towards apoptosis or cell cycle arrest. The fate of the cell is not determined solely by the level of endogenous ceramide since other cellular pathways appear to be involved in modulating the cellular response to elevated ceramide. Thus, if the diacylglycerol/PKC pathway is concomitantly activated, the cell does not undergo apoptosis but is rather driven towards cell cycle arrest. Similarly, if Bcl-2 is overexpressed, increased levels of ceramide cause Rb dephosphorylation and cell cycle arrest but not apoptosis. Other pathways may also be involved in modulating the cellular response to elevated ceramide. Hence, the cellular response to stress and subsequent accumulation of ceramide is determined by a complex interplay of signals which could be specific to the tissue or organism involved.

Future studies will aim at discovering the role of the different biochemical pathways that lead to ceramide accumulation and the mechanisms of their regulation. Thus, the role of the different sphingomyelinases, ceramidases, and ceramide synthase needs to be clarified. Also, the targets of ceramide function need to be verified and their respective substrates examined. The elucidation of these pathways will help us understand the basic biochemical pathways involved in the stress response and may offer possible targets for manipulation or intervention in clinical conditions where this may be necessary.

Acknowledgments

The studies in the authors' laboratory have been supported by NIH grant GM 43825 and DoD grant AIBS 516.

References

1. Hakomori S. Glycosphingolipids in cellular interaction, differentiation, and oncogenesis. Annu Rev Biochem 1981; 50:733-764.
2. Hakomori S. Bifunctional role of glycosphingolipids. J Biol Chem 1990; 265: 18713-18716.
3. Karlsson KA. Animal glycosphingolipids as membrane attachment sites for bacteria. Annu Rev Biochem 1989; 58:309-350.
4. Hannun YA, Loomis CR, Merrill AH Jr, Bell RM. Sphingosine inhibition of protein kinase C activity and of phorbol dibutyrate binding in vitro and human platelets. J Biol Chem 1986; 261:12604-12609.
5. Hannun YA, Bell RM. Lysosphingolipid inhibition of protein kinase C: implications in the pathogenesis of the sphingolipidoses. Science 1987; 235:670-674.
6. Hannun YA, Greenberg CS, Bell RM. Sphingosine inhibition of agonist dependent secretion and activation of human platelets implies that protein kinase C is a necessary and common event of the signal transduction pathway. J Biol Chem 1987; 262:13620-13626.
7. Hannun YA, Bell RM. Functions of sphingolipids and sphingolipid breakdown products in cellular regulation. Science 1989; 243:500-507.
8. Divecha N, Irvine RF. Phospholipid Signaling. Cell 1995; 80:269-278.
9. Dennis EA, Rhee SG, Billah MM, Hannun YA. Role of phospholipases in generating lipid second messengers in signal transduction. FASEB J 1991; 5:2068-2077.
10. Nishizuka Y. Intracellular signaling by hydrolysis of phospholipids and activation of protein kinase C. Science 1992; 258:607-614.
11. Okazaki T, Bell RM, Hannun YA. Sphingomyelin turnover induced by vitamin D_3 in HL-60 cells. Role in cell differentiation. J Biol Chem 1989; 264:19076-19080.
12. Okazaki T, Bielawska A, Bell RM, Hannun YA. Role of ceramide as a lipid mediator of 1α,25-dihydroxyvitamin D_3-induced HL-60 cell differentiation. J Biol Chem 1990; 265:15823-15831.
13. Kim MY, Linardic C, Obeid L, Hannun Y. Identification of sphingomyelin turnover as an effector mechanism for the action of tumor necrosis factor alpha and gamma-interferon. Specific role in cell differentiation. J Biol Chem 1991; 266:484-489.
14. Hannun YA. The Sphingomyelin cycle and the second messenger function of ceramide. J Biol Chem 1994; 269:3125-3128.
15. Kolesnick R, Golde DW. The sphingomyelin pathway in tumor necrosis factor and interleukin-1 signaling. Cell 1994; 77:325-328.

16. Hannun YA, Obeid LM, Dbaibo GS. Ceramide: a novel second messenger and lipid mediator. Lipid Second Messengers 1996; 8:177-204.
17. Jayadev S, Liu B, Bielawska AE, Lee JY, Nazaire F, Pushkareva MYU, Obeid LM, Hannun YA. Role for ceramide in cell cycle arrest. J Biol Chem 1995; 270:2047-2052.
18. Zhang J, Alter N, Reed JC, Borner C, Obeid LM, Hannun YA. Bcl-2 interrupts the ceramide-mediated pathway of cell death. Proc Natl Acad Sci USA 1996; 93:5325-5328.
19. Merrill AH Jr, Jones DD. An update of the enzymology and regulation of sphingomyelin metabolism. Biochim Biophys Acta 1990; 1044:1-12.
20. Rother J, Van Echten G, Schwarzmann G, Sandhoff K. Biosynthesis of sphingolipids: Dihydroceramide and not sphinganine is desaturated by cultured cells. Biochem Biophys Res Commun 1992; 189:14-20.
21. Spence MW. Sphingomyelinases. Adv Lipid Res 1993; 26:3-23.
22. Chatterjee S. Neutral sphingomyelinase. Adv Lipid Res 1993; 26:25-47.
23. Okazaki T, Bielawska A, Domae N, Bell RM, Hannun YA. Characteristics and partial purification of a novel cytosolic, magnesium-independent, neutral sphingomyelinase activated in the early signal transduction of 1α,25-dihydroxyvitamin D_3-induced HL-60 cell differentiation. J Biol Chem 1994; 269:4070-4077.
24. Canman CE, Chen CY, Lee MH, Kastan MB. DNA damage responses: p53 induction, cell cycle perturbations, and apoptosis. Cold Spring Harb Symp Quant Biol 1994; 59:277-286.
25. Linardic CM, Jayadev S, Hannun YA. Activation of the sphingomyelin cycle by brefeldin A: Effects of brefeldin A on differentiation and implications for a role for ceramide in regulation of protein trafficking. Cell Growth Diff 1996; 7:765-774.
26. Dobrowsky RT, Werner MH, Castellino AM, Chao MV, Hannun YA. Activation of the sphingomyelin cycle through the low-affinity neurotrophin receptor. Science 1994; 265:1596-1599.
27. Smith CA, Farrah T, Goodwin RG. The TNF receptor superfamily of cellular and viral proteins: activation, costimulation, and death. Cell 1994; 76:959-962.
28. Whyte P, Eisenman RN. Dephosphorylation of the retinoblastoma protein during differentiation of HL60 cells. Biochem Cell Biol 1992; 70:1380-1384.
29. Lee EY-HP, Hu N, Yuan S-SF, Cox LA, Bradley A, Lee W-H, Herrup K. Dual roles of the retinoblastoma protein in cell cycle regulation and neuron differentiation. Genes & Dev 1994; 8:2008-2021.
30. Horowitz JM, Park S-H, Bogenmann E, Cheng J-C, Yandell DW, Kaye FJ, Minna JD, Dryja TP, Weinberg RA. Frequent inactivation of the retinoblastoma antioncogene is restricted to a subset of human tumor cells. Proc Natl Acad Sci USA 1990; 87:2775-2779.
31. Buchkovich K, Duffy LA, Harlow E. The retinoblastoma protein is phosphorylated during specific phases of the cell cycle. Cell 1989; 58:1097-1105.
32. DeCaprio JA, Ludlow JW, Lynch D, Furukawa Y, Griffin J, Piwnica-Worms H, Huang C-M, Livingston DM. The product of the retinoblastoma susceptibility gene has properties of a cell cycle regulatory element. Cell 1989; 58:1085-1095.
33. Goodrich DW, Wang NP, Qian Y-W, Lee EY-HP, Lee W-H. The retinoblastoma gene product regulates progression through the G1 phase of the cell cycle. Cell 1991; 67:293-302.
34. Wang JY, Knudsen ES, Welch PJ. The retinoblastoma tumor suppressor protein. Adv Cancer Res 1994; 64:25-85.
35. Weinberg RA. The retinoblastoma protein and cell cycle control. Cell 1995; 81:323-330.
36. Nevins JR. E2F: a link between the Rb tumor suppressor protein and viral oncoproteins. Science 1992; 258:424-429.
37. Bagchi S, Raychaudhuri P, Nevins JR. Adenovirus E1A proteins can dissociate heteromeric complexes involving the E2F transcription factor: a novel mechanism for E1A trans-activation. Cell 1990; 62:659-669.
38. Chellappan S, Kraus VB, Kroger B, Munger K, Howley PM, Phelps WC, Nevins JR. Adenovirus E1A, simian virus 40 tumor antigen, and human papil-

lomavirus E7 protein share the capacity to disrupt the interaction between transcription factor E2F and the retinoblastoma gene product. Proc Natl Acad Sci USA 1992; 89:4549-4553.
39. Ludlow JW. Interactions between SV40 large-tumor antigen and the growth suppressor proteins pRB and p53. FASEB J 1993; 7:866-871.
40. Vousden K. Interactions of human papillomavirus transforming proteins with the products of tumor suppressor genes. FASEB J 1993; 7:872-879.
41. Hartwell LH, Kastan MB. Cell cycle control and cancer. Science 1994; 266:1821-1828.
42. Hartwell LH, Weinert TA. Checkpoints: controls that ensure the order of cell cycle events. Science 1989; 246:629-634.
43. Bielawska A, Crane HM, Liotta D, Obeid LM, Hannun YA. Selectivity of ceramide-mediated biology: lack of activity of erythro-dihydroceramide. J Biol Chem 1993; 268:26226-26232.
44. Dbaibo GS, Obeid LM, Hannun YA. TNFα signal transduction through ceramide: dissociation of growth inhibitory effects of TNFα from activation of NF-κB. J Biol Chem 1993; 268:17762-17766.
45. Dbaibo GS, Pushkareva MY, Jayadev S, Schwarz JK, Horowitz JM, Obeid LM, Hannun YA. Retinoblastoma gene product as a downstream target for a ceramide-dependent pathway of growth arrest. Proc Natl Acad Sci USA 1995; 92:1347-1351.
46. Gomez-Muñoz A, Martin A, O'Brien L, Brindley DN. Cell-permeable ceramides inhibit the stimulation of DNA synthesis and phospholipase D activity by phosphatidate and lysophosphatidate in rat fibroblasts. J Biol Chem 1994; 269:8937-8943.
47. Borchardt RA, Lee WT, Kalen A, Buckley RH, Peters C, Schiff S, Bell RM. Growth-dependent regulation of cellular ceramides in human T-cells. Biochim Biophys Acta 1994; 1212:327-336.
48. Inokuchi J, Momosaki K, Shimeno H, Nagamatsu A, Radin NS. Effects of D-threo-PDMP, an inhibitor of glucosylceramide synthetase, on expression of cell surface glycolipid antigen and binding to adhesive proteins by B16 melanoma cells. J Cell Physiol 1989; 141:573-583.
49. Chao R, Khan W, Hannun YA. Retinoblastoma protein dephosphorylation induced by D-erythro-sphingosine. J Biol Chem 1992; 267:23459-23462.
50. Rani CS, Abe A, Chang Y, Rosenzweig N, Saltiel AR, Radin NS, Shayman JA. Cell cycle arrest induced by an inhibitor of glucosylceramide synthase: Correlation with cyclin-dependent kinases. J Biol Chem 1995; 270:2859-2867.
51. Obeid LM, Linardic CM, Karolak LA, Hannun YA. Programmed cell death induced by ceramide. Science 1993; 259: 1769-1771.
52. Jarvis WD, Kolesnick RN, Fornari FA, Traylor RS, Gewirtz DA, Grant S. Induction of apoptotic DNA damage and cell death by activation of the sphingomyelin pathway. Proc Natl Acad Sci USA 1994; 91:73-77.
53. Ji L, Zhang G, Uematsu S, Akahori Y, Hirabayashi Y. Induction of apoptotic DNA fragmentation and cell death by natural ceramide. FEBS Lett 1995; 358: 211-214.
54. Haimovitz-Friedman A, Kan CC, Ehleiter D, Persaud RS, McLoughlin M, Fuks Z, Kolesnick RN. Ionizing radiation acts on cellular membranes to generate ceramide and initiate apoptosis. J Exp Med 1994; 180:525-535.
55. Cifone MG, De Maria R, Roncaioli P, Rippo MR, Azuma M, Lanier LL, Santoni A, Testi R. Apoptotic signaling through CD95 (Fas/Apo-1) activates an acidic sphingomyelinase. J Exp Med 1994; 180:1547-1552.
56. Tepper CG, Jayadev S, Liu B, Bielawska A, Wolff R, Yonehara S, Hannun YA, Seldin MF. Role of ceramide as an endogenous mediator of Fas-induced cytotoxicity. Proc Natl Acad Sci USA 1995; 92:8443-8447.
57. Bose R, Verheij M, Haimovitz-Friedman A, Scotto K, Fuks Z, Kolesnick RN. Ceramide synthase mediates daunorubicin-induced apoptosis: an alternative mechanism for generating death signals. Cell 1996; 82:405-414.
58. Martin SJ, Green DR. Protease activation

during apoptosis: death by a thousand cuts? [Review]. Cell 1995; 82:349-352.

59. Chinnaiyan AM, Tepper CG, Seldin MF, O'Rourke K, Kischkel FC, Hellbardt S, Krammer PH, Peter ME, Dixit VM. FADD/MORT1 is a common mediator of CD 95 (Fas/APO-1) and tumor necrosis factor receptor-induced apoptosis. J Biol Chem 1996; 271:4961-4965.
60. Enari M, Talanian RV, Wong WW, Nagata S. Sequential activation of ICE-like and CPP32-like proteases during Fas-mediated apoptosis. Nature 1996; 380:723-726.
61. Kaufmann SH, Desnoyers S, Ottaviano Y, Davidson NE, Poirier GG. Specific proteolytic cleavage of poly(ADP-ribose) polymerase: an early marker of chemotherapy-induced apoptosis. Cancer Res 1993; 53:3976-3985.
62. Lazebnik YA, Kaufmann SH, Desnoyers S, Poirier GG, Earnshaw WC. Cleavage of poly(ADP-ribose) polymerase by a proteinase with properties like ICE. Nature 1994; 371:346-347.
63. Fernandes-Alnemri T, Litwack G, Alnemri ES. CPP32, a novel human apoptotic protein with homology to Caenorhabditis elegans cell death protein Ced-3 and mammalian interleukin-1 beta-converting enzyme. J Biol Chem 1994; 269:30761-30764.
64. Tewari M, Quan LT, O'Rourke K, Desnoyers S, Zeng Z, Beidler DR, Poirier GG, Salvesen GS, Dixit VM. Yama/CPP32 beta, a mammalian homolog of CED-3, is a CrmA-inhibitable protease that cleaves the death substrate poly(ADP-ribose) polymerase. Cell 1995; 81:801-809.
65. Nicholson DW, Ali A, Thornberry NA, Vaillancourt JP, Ding CK, Gallant M, Gareau Y, Griffin PR, Labelle M, Lazebnik YA, Munday NA, Raju SM, Smulson ME, Yamin T, Yu VL, Miller DK. Identification and inhibition of the ICE/CED-3 protease necessary for mammalian apoptosis. Nature 1995; 376:37-43.
66. Smyth MJ, Perry DK, Zhang J, Poirier GG, Hannun YA, Obeid LM. prICE: a downstream target for ceramide-induced apoptosis and for the inhibitory action of Bcl-2. Biochem J 1996; 316:25-28.
67. Oltvai ZN, Korsmeyer SJ. Checkpoints of dueling dimers foil death wishes [comment]. [Review]. Cell 1994; 79:189-192.
68. Martin SJ, Takayama S, McGahon AJ, Miyashita T, Corbeil J, Kolesnick RN, Reed JC, Green DR. Inhibition of ceramide-induced apoptosis by Bcl-2. Cell Death Differ 1995; 2:253-257.
69. Karasavvas N, Erukulla RK, Bittman R, Lockshin R, Hockenbery D, Zakeri Z. BCL-2 suppresses ceramide-induced cell killing. Cell Death Differ 1996; 3:149-151.
70. Ray CA, Black RA, Kronheim SR, Greenstreet TA, Sleath PR, Salvesen GS, Pickup DJ. Viral inhibition of inflammation: cowpox virus encodes an inhibitor of the interleukin-1β converting enzyme. Cell 1992; 69:597-606.
71. Boldin MP, Goncharov TM, Goltsev YV, Wallach D. Involvement of MACH, a novel MORT1/FADD-interacting protease, in Fas/APO1- and TNF receptor-induced cell death. Cell 1996; 85:803-815.
72. Muzio M, Chinnaiyan AM, Kischkel FC, O'Rourke K, Shevchenko A, Ni J, Scaffidi C, Bretz JD, Zhang M, Gentz R, Mann M, Krammer PH, Peter ME, Dixit VM. FLICE, a novel FADD-homologous ICE/CED-3-like protease, is recruited to the CD95 (Fas/Apo-1) death-inducing signaling complex. Cell 1996; 85:817-827.
73. Lane DP. p53, guardian of the genome. Nature 1992; 358:15-16.
74. Ballou LR, Chao CP, Holness MA, Barker SC, Raghow R. Interleukin-1-mediated PGE_2 production and sphingomyelin metabolism. Evidence for the regulation of cyclooxygenase gene expression by sphingosine and ceramide. J Biol Chem 1992; 267:20044-20050.
75. Hayakawa M, Jayadev S, Tsujimoto M, Hannun YA, Ito F. Role of ceramide in stimulation of the transcription of cytosolic phospholipse A_2 and cyclooxygenase 2. Biochem Biophs Res Comm 1996; 220:681-686.
76. Laulederkind SJF, Bielawska A, Raghow R, Hannun YA, Ballou LR. Ceramide induces interleukin 6 gene expression in human fibroblasts. J Exp Med 1995; 182:599-604.
77. Mathias S, Younes A, Kan C-C, Orlow I,

Joseph C, Kolesnick RN. Activation of the sphingomyelin signaling pathway in intact EL4 cells and in a cell-free system by IL-1β. Science 1993; 259:519-522.

78. Schütze S, Potthoff K, Machleidt T, Berkovic D, Wiegmann K, Krönke M. TNF activates NF-κB by phosphatidylcholine-specific phospholipase C-induced "acidic" sphingomyelin breakdown. Cell 1992; 71:765-776.

79. Wiegmann K, Schütze S, Machleidt T, Witte D, Krönke M. Functional dichotomy of neutral and acidic sphingomyelinases in tumor necrosis factor signaling. Cell 1994; 78:1005-1015.

80. Betts JC, Agranoff AB, Nabel GJ, Shayman JA. Dissociation of endogenous cellular ceramide from NF-κB activation. J Biol Chem 1994; 269:8455-8458.

81. Johns LD, Sarr T, Ranges GE. Inhibition of ceramide pathway does not affect ability of TNF-α to activate nuclear factor-κB. J Immunol 1994; 152:5877-5882.

82. Kuno K, Sukegawa K, Ishikawa Y, Orii T, Matsushima K. Acid sphingomyelinase is not essential for the IL-1 and tumor necrosis factor receptor signaling pathway leading to NFκB activation. Int Immunol 1994; 6:1269-1272.

83. Yang Z, Costanzo M, Golde DW, Kolesnick RN. Tumor necrosis factor activation of the sphingomyelin pathway signals nuclear factor κB translocation in intact HL-60 cells. J Biol Chem 1993; 268:20520-20523.

84. Mathias S, Dressler KA, Kolesnick RN. Characterization of a ceramide-activated protein kinase: Stimulation by tumor necrosis factor α. Proc Natl Acad Sci USA 1991; 88:10009-10013.

85. Joseph CK, Byun H-S, Bittman R, Kolesnick RN. Substrate recognition by ceramide-activated protein kinase: Evidence that kinase activity is proline-directed. J Biol Chem 1993; 268:20002-20006.

86. Liu J, Mathias S, Yang Z, Kolesnick RN. Renaturation and tumor necrosis factor-alpha stimulation of a 97-kDa ceramide-activated protein kinase. J Biol Chem 1994; 269:3047-3052.

87. Diaz-Meco MT, Berra E, Municio MM, Sanz L, Lozano J, Dominguez I, Diaz-Golpe V, De Lera MTL, Alcamí J, Payá CV, Arenzana-Seisdedos F, Virelizier J-L, Moscat J. A dominant negative protein kinase C ζ subspecies blocks NF-κB activation. Mol Cell Biol 1993; 13:4770-4775.

88. Lozano J, Berra E, Municio MM, Diaz-Meco MT, Dominguez I, Sanz L, Moscat J. Protein kinase C ζ isoform is critical for κB-dependent promoter activation by sphingomyelinase. J Biol Chem 1994; 269:19200-19202.

89. Huwiler A, Brunner J, Hummel R, Vervoordeldonk M, Stabel S, Bosch HVD, Pfeilschifter J. Ceramide-binding and activation defines protein kinse c-Raf as a ceramide-activated protein kinase. Proc Natl Acad Sci USA 1996; 93:6959-6963.

90. Dobrowsky RT, Hannun YA. Ceramide stimulates a cytosolic protein phosphatase. J Biol Chem 1992; 267:5048-5051.

91. Dobrowsky RT, Kamibayashi C, Mumby MC, Hannun YA. Caramide activates heterotrimeric protein phosphatase 2A. J Biol Chem 1993; 268:15523-15530.

92. Law B, Rossie S. The dimeric and catalytic foms of protein phosphatase 2A from rat brain are stimulated by C_2-ceramide. J Biol Chem 1995; 270:12808-12813.

93. Fishbein JD, Dobrowsky RT, Bielawska A, Garrett S, Hannun YA. Ceramide-mediated biology and CAPP are conserved in *Saccharomyces cerevisiae.* J Biol Chem 1993; 268:9255-9261.

94. Wolff RA, Dobrowsky RT, Bielawska A, Obeid LM, Hannun YA. Role of ceramide-activated protein phosphatase in ceramide-mediated signal transduction. J Biol Chem 1994; 269:19605-19609.

95. Nickels JT, Broach JR. A ceramide-activated protein phosphatase mediates ceramide-induced G1 arrest of Saccharomyces cerevisiae. Genes Dev 1996; 10:382-394.

96. Kyriakis JM, Banerjee P, Nikolakaki E, Dai T, Rubie EA, Ahmad MF, Avruch J, Woodgett JR. The stress-activated protein kinase subfamily of c-Jun kinases. Nature 1994; 369:156-160.

97. Westwick JK, Bielawska AE, Dbaibo G, Hannun YA, Brenner DA. Ceramide activates the stress-activated protein kinases. J Biol Chem 1995; 270:22689-22692.

98. Verheij M, Bose R, Lin XH, Yao B, Jarvis WD, Grant S, Birrer MJ, Szabo E, Zon LI, Kyriakis JM, Haimovitz-Friedman A, Fuks Z, Kolesnick RN. Requirement for ceramide-initiated SAPK/JNK signalling in stress-induced apoptosis. Nature 1996; 380:75-79.

CHAPTER 3

Ceramide and Inflammation

Leslie R. Ballou

INTRODUCTION

Inflammation is a general term used to describe an extraordinarily complex set of biochemical events mounted by living tissues either in response to physical injury or as part of the host-defense mechanism. The pain, swelling, heat, and redness characteristic of inflammation involves an extensive network of cellular and molecular interactions. These complex interactions are characterized by the infiltration of inflammatory cells (e.g., neutrophils, mononuclear leukocytes and macrophages) to the site of injury and/or invasion and the generation of an almost unimaginable array of biologically active molecules by both infiltrating and resident cells, which function as the primary biochemical mediators of the inflammatory response; (please see the scheme depicting the interactions of a variety of different types of cells involved in inflammation). These inflammatory mediators include a wide variety of soluble molecules such as vaso- and neuroactive peptides, lipid mediators such as the eicosanoids, and growth factors and cytokines which are generated in response to specific inflammatory stimuli. Adding to the overall complexity of this soluble mediator signaling network is the fact that several of the proinflammatory cytokines may elicit differential biologic effects on different cell types, induce overlapping responses, have synergistic or antagonistic effects on one another, or affect the synthesis of other cytokines. Thus, in addition to their respective roles in the inflammatory process, many of these soluble mediators play important roles in the regulation of other physiologically important processes such as modulation of the immune response, cell-cell interaction and adhesion, apoptosis, cell growth and differentiation, to mention only a few.

Two prototypic proinflammatory cytokines, interleukin-1β (IL-1) and tumor necrosis factor (TNF) α, are produced and secreted by monocytes and other cell types in response to a variety of stimuli.[1] These cytokines in particular have been implicated as playing a central role in inflammatory responses because monoclonal antibodies directed against the cytokine itself or disruption of receptor binding by specific receptor antagonists appear to abrogate at least some of the inflammatory responses seen in animal models of disease. Furthermore, both cytokines are potent inducers of arachidonic acid mobilization and its metabolism to prostaglandin, leukotrienes, lipoxins, hydroxyeicosatetraenoic acids, thromboxanes, and other eicosanoids, a hallmark of the inflammatory response.[2-6] Eicosanoids function as both

Sphingolipid-Mediated Signal Transduction, edited by Yusuf A. Hannun.

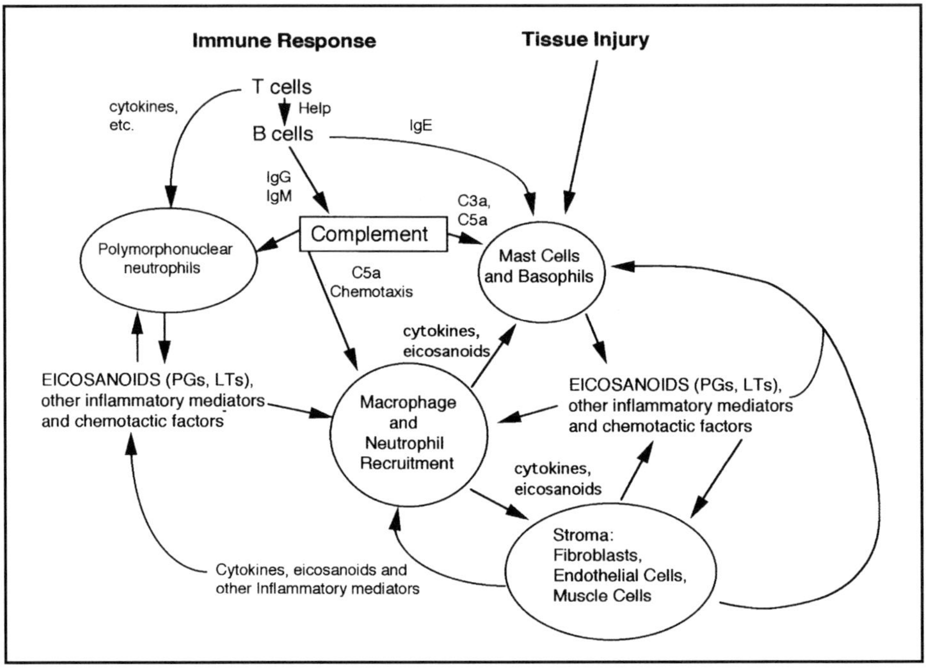

Scheme 3.1. The inflammatory response involves complex interactions between activation of the immune system, the production of inflammatory mediators (e.g., C3a, C5a, histamine, bradykinin, chemotactic factors, eicosanoids) by a variety of cell types (primarily mast cells/basophils) which control blood flow, vascular permeability and the migration and adherence of inflammatory cells (macrophages, neutrophils, leukocytes). At the site of inflammation, both immune and stromal cells produce cytokines which are involved in perpetuating inflammatory reactions. A major question in deciphering the complex interactions among the various cell types involved in inflammation concerns the cytokine-induced signal transduction pathways utilized by each type of cell.

paracrine and autocrine hormones which interact with specific eicosanoid receptors and subsequently affect a variety of signal transduction pathways associated with the inflammatory response.[7,8] The central role of the eicosanoids as mediators of inflammation is dramatically underscored by virtue of the prevalent use of nonsteroidal anti-inflammatory drugs (NSAIDs) such as aspirin which target the inhibition of prostaglandin production as a primary treatment of inflammation.[9,10]

The nature of the primary intracellular signals delivered in response to IL-1 and TNF receptor stimulation are poorly understood and remain controversial. Therefore, it is well beyond the scope of this chapter to discuss the many aspects of IL-1 and TNF signaling which are currently under active investigation. It is clear however, that in addition to being potent inflammatory mediators and inducers of eicosanoid biosynthesis, IL-1 and TNF also activate the sphingomyelin signaling pathway resulting in the generation of ceramide in a variety of cell types. Furthermore, ceramide functions as an important signal transduction molecule in its own right, by exerting its influence on a number of key signal transduction pathways.[11-27]

We have had a long-standing interest in the signaling pathways involved in the regulation of IL-1 and TNF-induced eicosanoid biosynthesis. In retrospect, because IL-1 and TNF induce both ceramide generation and eicosanoid biosynthesis, ceramide becomes a reasonable candidate for participation in the signaling pathway regulating eicosanoid biosynthesis, but we took a much more circuitous route to the identification of ceramide as an inflammatory mediator. In this chapter I will discuss our work leading to our present understanding of the biochemical mechanism(s) by which ceramide modulates cytokine-induced eicosanoid biosynthesis, and thus plays an important role as an inflammatory mediator.

CYTOKINE-INDUCED CERAMIDE GENERATION, PROSTAGLANDIN PRODUCTION, AND THEIR PHYSIOLOGIC ROLE IN INFLAMMATION

The release of proinflammatory cytokines such as IL-1 and TNF from infiltrating immune cells is a hallmark of the early host-defense reactions characteristic of inflammation and once released, both have the ability to stimulate the production and release of other proinflammatory agents. A classic example of one such group of secondarily produced downstream proinflammatory agents are arachidonic acid metabolites, the eicosanoids; the regulation of their biosynthesis in response to cytokines has long been a major focus of my laboratory.[8] Clearly, many different inflammatory mediators in addition to cytokines are involved in the modulation of prostaglandin production (and the metabolism of other biologically active eicosanoid metabolites) in a wide variety of cell types and the potential influence of ceramide relative to the biosynthesis of other arachidonic acid metabolites has yet to be investigated. For the purposes of this discussion I will focus attention on the relationship between cytokine signaling, ceramide generation, and the regulation of prostanoid production.

The Regulation of Prostaglandin Biosynthesis

Regardless of the stimulus, the essential signal transduction pathways regulating the synthesis of prostaglandins are essentially the same in all eicosanoid producing cells. Collectively, all compounds which are derived from C_{20} fatty acids (primarily arachidonic acid) are referred to as eicosanoids and generally include all of those arachidonic acid metabolites produced as a result of cellular activation. The eicosanoids are further subdivided into two major categories, the leukotrienes which are formed from free arachidonic acid via the 5'-lipoxygenase pathway (5'-LO) and the prostanoids (including the prostaglandins) which are produced from free arachidonic acid via the prostaglandin synthase/cyclooxygenase (COX) pathway (Fig. 3.1). We will focus our attention on the arm of the eicosanoid biosynthetic pathway leading to the synthesis of prostaglandins because all of our work in the area has been using cells (human foreskin fibroblasts) which produce a single prostanoid product, prostaglandin E_2 (PGE_2). We plan to examine the effects of sphingolipids on the synthesis of other arachidonic acid metabolites using appropriate cell types.

Figure 3.1 shows a generalized schematic pathway for the regulation of eicosanoid biosynthesis. Initially, the ligation of a cell surface receptor leads to the activation of phospholipase A_2 (PLA_2), an enzyme (actually, a group of enzymes) which catalyze the release of arachidonic acid from membrane phospholipids (for review see ref. 28). Once released, free arachidonic acid can be converted to the leukotriene A_4 (LTA_4) via the action of 5'-LO. LTA_4 may be subsequently converted to other leukotriene metabolites by the action of other lipoxygenases.[29] Free arachidonic acid is converted to prostanoids, by its sequential conversion to an unstable intermediate, prostaglandin G_2 (PGG_2) and then to prostaglandin H_2 (PGH_2) which is the sole precursor for synthesis of all of the prostanoids. Interestingly, COX performs two separate and distinct catalytic reactions, one is its cyclooxygenase activity which converts arachidonate to PGG_2 and the second is its hydroperoxide activity which converts PGG_2 to PGH_2.[30] The specific prostanoid product(s) produced by each type of cell is dependent upon the presence or absence of

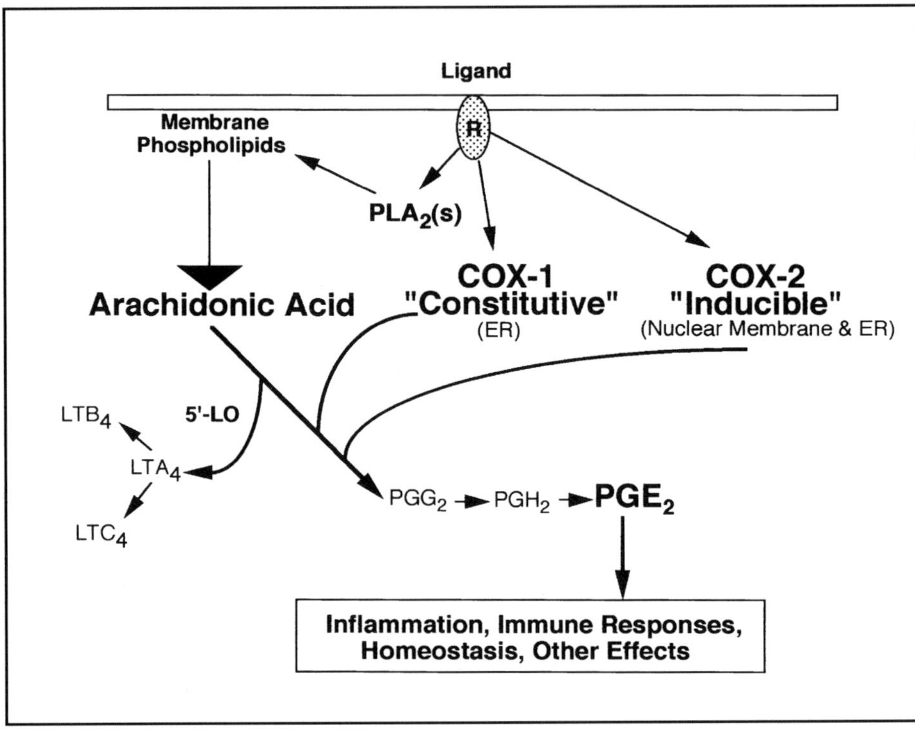

Fig. 3.1. The prostaglandin and leukotriene biosynthetic pathway. COX, cyclooxygenase; 5'-LO, 5'-lipoxygenase; PG, prostaglandin; LT, leukotriene; R, receptor; ER, endoplasmic reticulum; PLA_2, phospholipase A_2.

those enzymes which convert PGH_2 to the vast array of potential prostanoid products.

One very exciting and rapidly evolving area of research into the regulation of prostanoid biosynthesis has emerged as a result of the discovery of two cyclooxygenase isoforms, COX-1 and COX-2, both of which catalyze the same reaction, the conversion of arachidonic acid to PGH_2, but appear to be differentially regulated at the transcriptional level and perform different biological functions.[31,32] In general, COX-1 is constitutively expressed in many tissues but its expression can be affected under certain circumstances. Conversely, COX-2 expression is often undetectable in most tissues but its expression is rapidly induced by a variety of proinflammatory agents. The precise respective roles of COX-1 and COX-2 in normal physiology and in pathophysiology remain somewhat controversial.[33,34] Both IL-1 and ceramide induce the transcription of both COX isoforms and work is currently in progress to determine the precise role of each isozyme in regulation of PGE_2 production in our model system.

Establishing a Relationship Between Cytokine Signaling, Ceramide, and the Regulation of Prostanoid Biosynthesis

As mentioned in the Introduction, I first became interested in the potential role of ceramide as an inflammatory mediator quite by chance. My first exposure to sphingolipids as signal transduction molecules was when I read a review in *Science* by Hannun and Bell published in 1989 regarding the effects of sphingolipids on cell function.[35] What specifically attracted my attention was the observation that sphingosine, a metabolite of ceramide, was a potent inhibitor of protein kinase C (PKC) activation. At the time I read the article, we were testing the effects of various PKC inhibitors (and other pharmacological agents) on IL-1 signal transduction, with respect to their effects on PGE_2 synthesis. Since sphingosine is a naturally occurring compound we decided to test its effect on IL-1-induced PGE_2 production. Thus, as a direct result of a series of experiments testing the effects of sphingosine on IL-1 and TNF-induced PGE_2 production in cultured human fibroblasts we eventually became interested in the potential role of ceramide as a modulator of PGE_2 synthesis.[36,37] In our early experiments testing the effect of sphingosine on cytokine induced PGE_2 production, we consistently observed a dramatic increase in the amount of PGE_2 produced in response to IL-1 or TNF in the presence of exogenously added sphingosine; interestingly, sphingosine added by itself had little or no effect on PGE_2 production.[36,37] Because all of our recent work has focused on IL-1 signal transduction and the regulation of PGE_2 synthesis I will limit the following discussion to IL-1 signaling, ceramide generation, and the regulation of PGE_2 production in human fibroblasts; very similar mechanisms are likely to be involved with TNF signaling. In order to examine the mechanism(s) by which sphingosine potentiated the ability of IL-1 to stimulate PGE_2 production and the role that PKC may play in the effects of sphingosine, we compared the effect of sphingosine on IL-1-induced PGE_2 production in normal and PKC deficient cells.[36] As expected, IL-1 significantly stimulated PGE_2 production over controls but unexpectedly, PKC downregulation similarly enhanced IL-1-induced PGE_2 production, strongly suggesting that PKC activation is not required in the signaling pathway for PGE_2 production. The effect of sphingosine on IL-1-induced PGE_2 production was not significantly diminished in PKC-deficient nor did it affect PMA-induced PGE_2 production. Further H-7, a synthetic PKC inhibitor, did not inhibit IL-1-induced PGE_2 production which also supports the concept that the IL-1 signal transduction pathway leading to prostanoid synthesis does not necessarily involve PKC activation. That sphingosine and not H-7 stimulates IL-1-mediated PGE_2 production suggests that sphingosine affects PGE_2 synthesis independently of its effects on PKC. These findings led us to examine the effects of sphingosine on the enzymes most directly involved in regulating PGE_2 synthesis, PLA_2 and COX.

To examine the effect of IL-1 and sphingosine on the expression and activation of PLA_2 in human fibroblasts we labeled cells to equilibrium with [^{14}C]arachidonic acid and then treated them with sphingosine, IL-1, or a combination of IL-1 and sphingosine for 24 hours; the amount of labeled arachidonate released into the media was determined. Compared to untreated control wells (948 ± 77 cpm), sphingosine had little effect upon arachidonic acid release (1072 ± 110 cpm) while IL-1 exposure resulted in increased release of arachidonic acid (2002 ± 144 cpm). The combination of sphingosine and IL-1 resulted in approximately a two-fold increase in arachidonic acid release (4314 ± 152 cpm) compared to IL-1 alone suggesting that sphingosine potentiates the IL-1-mediated release of arachidonic acid from membrane phsopholipids.[36] To directly assess the effects of sphingosine and IL-1 on PLA_2 activity we prepared a soluble fibroblast fraction (containing $> 95\%$ of the total PLA_2 activity) from untreated cells and cells pretreated overnight with IL-1, sphingosine or the combination of both. In vitro PLA_2 activity was measured using phosphatidylcholine (PtdCho) labeled at the *sn*-2-position with [^{14}C]arachidonic acid as substrate.[36] In short, PLA_2 activity was not affected by either pretreating with sphingosine or by the addition of sphingosine to the cell extracts. However, when cells were pretreated with IL-1, PLA_2 activity increased three-fold over untreated controls and when sphingosine was added to extracts from IL-1 pretreated cells, PLA_2 activity increased another 2-fold when compared to IL-1 levels alone. Treatment of cells with cycloheximide during the IL-1-preincubation period completely abrogated the stimulatory effects of IL-1 on PLA_2 activity suggesting that IL-1 induces the de novo synthesis of PLA_2 and that sphingosine stimulates this activity in vitro; sphingosine does not appear to affect the basal PLA_2 activity in cells not exposed to IL-1. Cycloheximide added at the time of IL-1 addition completely blocked the increase in PLA_2 activity, suggesting a requirement for the synthesis of new protein. While these results clearly indicate that sphingosine stimulates IL-1-induced PLA_2 activity and this may account for its ability to enhance PGE_2 production in response to IL-1, sphingosine is not able to deliver all of the necessary signals required to induce PGE_2 production in the absence of added cytokine.

At the time we began to consider the possible mechanisms involved in the sphingosine-mediated enhancement of cytokine-induced PGE_2 production (e.g., effects on PLA_2 and COX expression) we started becoming more and more interested in the possibility that ceramide might function as a modulator of PGE_2 production not only because of its structural relationship to sphingosine but also because there was a great deal of information coming out regarding the central role of ceramide as a lipid second messenger in a number of cytokine signal transduction pathways. Most notably, the proinflammatory cytokine TNF induced sphingomyelin hydrolysis and ceramide accumulation in several different cell types suggesting a possible connection between proinflammatory cytokine signaling pathways, ceramide and PGE_2 production. Another idea that piqued our interest in ceramide as an active participant in regulating PGE_2 synthesis was the possibility that exogenously added sphingosine could be converted to ceramide. We found that indeed, ceramide levels did increase significantly in cells treated with sphingosine. We also checked to see whether the addition of a cell permeable analog of ceramide, C2-ceramide, could stimulate cytokine-induced PGE_2 production like sphingosine. Indeed, C2-ceramide was as effective as sphingosine in stimulating IL-1-induced PGE_2 production (Fig. 3.2) clearly demonstrating that ceramide also modulates cytokine-induced PGE_2 synthesis.

To compare the effect of endogenously produced and exogenously added C2-ceramide in regulating PGE_2 production, we treated cells with bacterial sphingomyelinase, alone or in combination with IL-1, and measured PGE_2 synthesis. As shown in Figure 3.3, sphingomyelinase treatment enhances IL-1 mediated PGE_2 production in a dose-dependent manner. Treatment of cells with sphingomyelinase alone had no effect on PGE_2 production at any concentration tested. These results strongly suggest that endogenous sphingomyelin metabolites such as ceramide and sphingosine, generated via the hydrolysis of sphingomyelin, plays an important role in modulating IL-1-mediated-PGE_2 production.

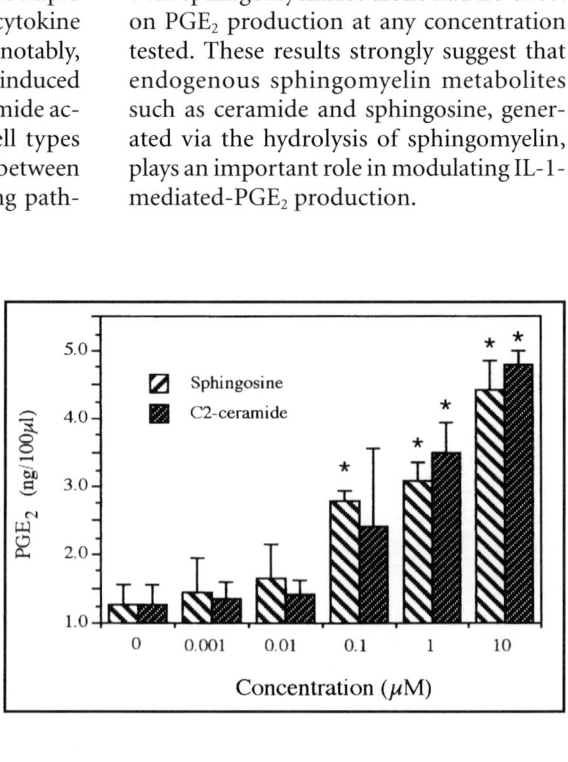

Fig. 3.2. Effect of sphingosine and C2-ceramide on IL-1-mediated PGE_2 production. Confluent dermal fibroblast cultures were treated with IL-1 (0.05 ng/ml) and increasing concentrations of sphingosine or C2-ceramide. After 24 hours PGE_2 was measured by RIA (* indicates a *p* value ≤ 0.05). Reprinted with permission from Ballou LR, Chao CP, Holness MA et al, J Biol Chem 1992; 267: 20044-20050.

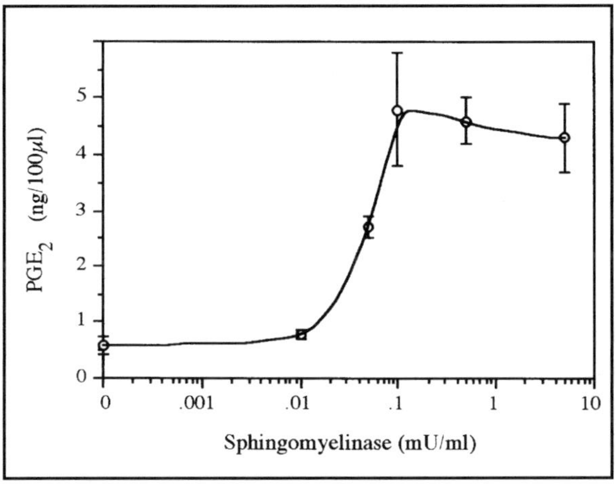

Fig. 3.3. Effect of exogenous sphingomyelinase treatment on IL-1-induced PGE_2 production. Confluent cultures were treated with IL-1 (0.05 ng/ml) along with increasing concentrations of bacterial sphingomyelinase as indicated. After 24 hours PGE_2 was measured by RIA. Results are expressed as ng of PGE_2/100 μl medium and represent the mean ± S.D. of duplicate assays from two different dermal fibroblast cell lines. Reprinted with permission from Ballou LR et al, J Biol Chem 1992; 267:20044-20050.

Another key issue to be examined to make a case for the involvement of ceramide in the IL-1 signaling cascade leading to the induction of PGE_2 synthesis was to determine whether IL-1, like TNF, stimulated sphingomyelin turnover and the generation of ceramide. Our reasoning was based upon the observation that a number of different cytokines stimulated sphingomyelin hydrolysis, via the activation of sphingomyelinase(s), and the accumulation of ceramide.[11,15-17,21,22] In a historical sense, TNF was one of the earliest cytokines to be identified as a stimulator of sphingomyelin turnover. Even though we had only just begun our work with ceramide and other sphingolipids with respect to their ability to modulate IL-1 signal transduction, we were intrigued by the possibility that ceramide may actually be a component of the IL-1 signal transduction pathway. Because TNF is a prototypical proinflammatory cytokine with the well established ability to induce prostanoid synthesis in a wide variety of cells and given the fact that TNF and IL-1 elicit many of the same responses in a wide variety of cell types (including the induction of prostanoid synthesis), it seemed likely that there could be a parallel between TNF and IL-1 signaling with respect to the activation of sphingomyelin hydrolysis and ceramide accumulation.

After searching the literature and finding that there was no information regarding the effects of IL-1 on sphingomyelin turnover or ceramide generation we decided to examine this in human foreskin fibroblasts. As shown in Figure 3.4, IL-1 indeed induced a significant hydrolysis of the labeled sphingomyelin pool which reached maximal levels after ≈2 hours.[38] In a similar experiment we used [^{3}H]serine-labeled cells in an effort to determine whether levels of cellular sphingosine or ceramide increased as a result of IL-1-induced sphingomyelin turnover. We were unable to detect an increase in labeled sphingosine in response to IL-1 treatment, although labeled ceramide was readily detectable. We also failed to measure an increase in cellular sphingosine when serine-labeled cells were treated with sphingomyelinase even though a significant amount of labeled ceramide was produced indicating the preferential accumulation of ceramide in these cells. In other experiments, we measured the effect of IL-1 on ceramide accumulation using the diacylglycerol kinase assay. As shown in Figure 3.5, cellular ceramide levels increased as a function of incubation time in the presence of IL-1. Maximal ceramide levels were measured ≈4 hours after the addition of IL-1. Thereafter, cellular ceramide levels decreased steadily, although even 24 hours following IL-1 addition, ceramide levels

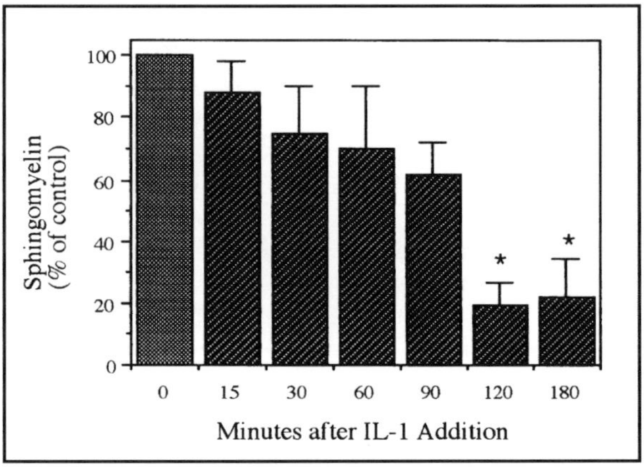

Fig. 3.4. Effect of IL-1 on sphingomyelin turnover. Fibroblasts were labeled with [^{3}H]choline for 72 hours. Prelabeled cells were treated with IL-1 (0.05 ng/ml) and excess cold choline for the indicated times. Lipids were extracted and sphingomyelin separated from phosphatidylcholine by thin-layer chromatography and quantified by means of a Bioscan 2000 imaging system. Results are expressed as a (mean) per cent ± S.E.M. compared with time-matched controls which did not receive IL-1 treatment (* represents *p* values 0.05). R eprinted with permission from Ballou LR et al, J Biol Chem 1992; 267:20044-20050.

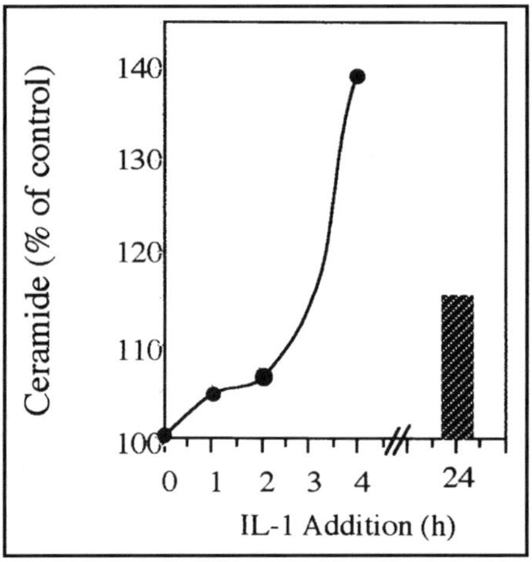

Fig. 3.5. Effect of IL-1 on intracellular ceramide accumulation. Confluent cultures were treated with IL-1 (0.05 ng/ml) for the times indicated and ceramide was measured using the diacylglycerol kinase assay. Reprinted with permission from Ballou LR et al, J Biol Chem 1992; 267:20044-20050.

remained higher than untreated controls. These data are consistent with the conclusion that sphingomyelin hydrolysis is indeed stimulated by IL-1, along with the accumulation of intracellular ceramide, and that the activation of this pathway is a significant signal transduction event mediating at least some of the many actions of IL-1 in vivo. Shortly after our initial report,[38] IL-1 was shown to stimulate sphingomyelin hydrolysis in both intact EL4 cells and in a cell free system.[39]

The finding that IL-1 activates the sphingomyelin signaling pathway together with our observations regarding the effects of exogenously added ceramide on IL-1-induced PGE_2 synthesis establish a connection between proinflammatory cytokine (IL-1 and TNF) signal transduction and ceramide as a lipid mediator in the signaling pathway(s) regulating PGE_2 synthesis. The mechanism by which ceramide enhances cytokine-induced PGE_2 production, while at the same time is unable to stimulate PGE_2 synthesis in the absence of cytokine, has been somewhat of a paradox and has been the focus of much of our recent work. In our original studies we clearly showed that sphingosine and ceramide played a role in regulating PGE_2 production via their ability to induce COX transcription and translation.[38] Before discussing this and our more recent work related to ceramide and the regulation of PGE_2 production I would like to point out caveats in the original studies which have since been addressed. First, we have now examined the effects of ceramide on an 85 kDa, cytosolic PLA_2 ($cPLA_2$), rather than the 14 kDa pancreatic type used in the original study, because $cPLA_2$ is now known to be upregulated by a variety of cytokines, and is far more likely to participate in the signaling pathway involved in cytokine/ceramide-mediated PGE_2 synthesis than the 14 kDa enzyme.[40-42] Second, we have now analyzed COX-1 and COX-2 mRNA by Northern blot instead of PCR, which allows for much more selectivity in differentiating between the COX-1 and COX-2 isoforms and for quantitative analysis of mRNA levels. In any event these original studies led us to the conclusion that ceramide induces COX expression and that this is a primary mechanism by which ceramide stimulates PGE_2 production. In the following section I will focus on recent studies regarding the kinetics of the transcription and translation of the COX-1 and COX-2 isoforms in response to IL-1 and ceramide as well as the transcription and translation of a cytosolic form of PLA_2 ($cPLA_2$) in an effort to elucidate the biochemical mechanism(s) underlying the ability of ceramide to modulate cytokine-induced PGE_2 production.

THE BIOCHEMICAL BASIS BY WHICH CERAMIDE REGULATES PROSTAGLANDIN BIOSYNTHESIS: THE INDUCTION OF COX TRANSCRIPTION AND TRANSLATION

We have clearly shown that both sphingosine and ceramide upregulate cytokine induced PGE_2 synthesis and that the most likely biochemical mechanism(s) by which these molecules affect PGE_2 synthesis are via effects on PLA_2 and COX expression and/or activation. However, even though sphingosine appears to enhance PLA_2 activity both in intact cells and in vitro, ceramide does not seem to have a similar affect on PLA_2 activity in our cells. This is clearly an important distinction between the potential mechanism(s) of action for sphingosine and ceramide as modulators of PGE_2 production. Further, this apparent lack of an effect of ceramide on PLA_2 activity served to focus our attention toward the potential effects of ceramide on COX gene expression/activation as a mechanism for its ability to stimulate PGE_2 synthesis. In the next sections I will briefly describe the original studies identifying COX as a target for ceramide action and focus on a discussion of more recent, kinetic analyses of the effects of ceramide on the expression of the COX isoforms and $cPLA_2$.

Cross-Talk Between Cytokines, the Sphingomyelin Pathway and the Prostaglandin Biosynthetic Pathway

Because cytokine-induced PGE_2 biosynthesis is dependent on the sequential activation of PLA_2 and COX, and because ceramide did not appear to affect arachidonic acid mobilization in human fibroblasts, we focused our early attention of the effect of ceramide on COX gene expression. Also, we initially examined the effects of ceramide on the expression of a Group I, pancreatic type, 14 kDa PLA_2. Using PCR to measure changes in COX and 14 kDa PLA_2 mRNA we found that ceramide rapidly induced the expression of COX mRNA with maximal mRNA levels (8-fold over control) detected 1 hour following treatment and returned to near basal levels by 24 hours. The effect of ceramide appeared to be highly selective since it had no effect on the expression of 14 kDa PLA_2 or on the steady-state levels of either GAPDH or ProαI (III) collagen gene transcripts. Structurally similar amphiphilic lipids such as stearylamine and N-stearoyl sphingosine, did not have any effect on IL-1-mediated PGE_2 production or on COX/PLA_2 expression. In response to IL-1, we detected little change in the steady-state levels of COX expression at early time points; however, after 24 hours significant increases in the levels of COX mRNA were present and interestingly, the effects of IL-1 and ceramide on COX gene expression were additive.[38]

To summarize these results, our signaling scheme shows that IL-1 induces PGE_2 synthesis by enhancing PLA_2 and Cox expression/activity (Fig. 3.6). IL-1 also induces sphingomyelin turnover and ceramide generation, and then ceramide interacts with other arms of the IL-1 pathway to modulate PGE_2 production. Treatment of cells with sphingomyelinase or the addition

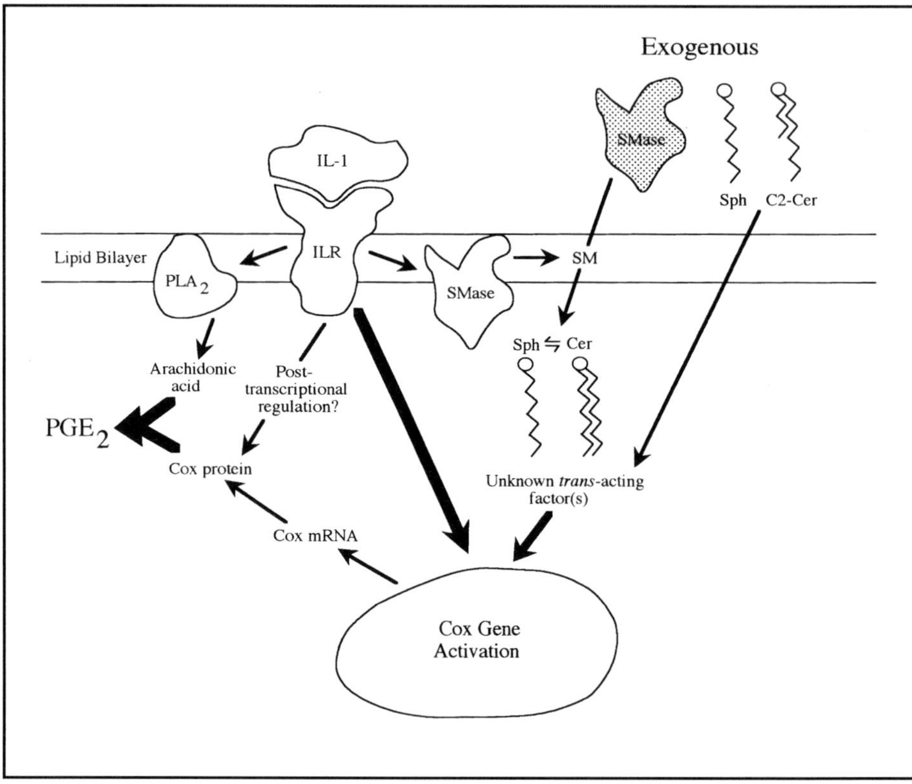

Fig. 3.6. Proposed relationship between IL-1-induced PGE_2 synthesis and sphingomyelin metabolism. IL-1 induces PGE_2 synthesis by enhancing PLA_2 and Cox expression/activity. IL-1 also induces sphingomyelin turnover. Treatment of cells with sphingomyelinase or the addition of exogenous sphingosine or C2-ceramide enhances IL-1-mediated PGE_2 production. Both sphingosine and C2-ceramide induce Cox gene expression. Thus, the generation of sphingomyelin metabolites may modulate IL-1-induced PGE_2 production at the level of Cox gene expression. While IL-1 alone can induce PGE_2 production, neither added sphingosine nor C2-ceramide can induce PGE_2 synthesis in the absence of IL-1, suggesting that the effects of IL-1 on PGE_2 production are not solely mediated by sphingosine or C2-ceramide. At present, the biochemical steps linking sphingomyelin turnover and induction of Cox expression are unknown. Abbreviations used are: ILR, interleukin-1 receptor; SM, sphingomyelin; SMase, sphingomyelinase; Cox, cyclooxygenase; PLA_2, phospholipase A_2; Sph, sphingosine; Cer, ceramide. Reprinted with permission from Ballou LR et al, J Biol Chem 1992; 267:20044-20050.

of exogenous ceramide enhances IL-1-mediated PGE_2 production but does not induce PGE_2 synthesis in the absence of cytokine because even though ceramide enhances COX expression, it does not induce arachidonic acid mobilization, making substrate rate-limiting. When cytokine is present, both PLA_2 and COX are induced and in the presence of high intracellular levels of ceramide, COX expression is additively increased resulting in enhanced rates of PGE_2 synthesis. The effects of IL-1 and ceramide on COX expression suggest that they affect COX expression via different mechanisms. Using this as a basic model to illustrate our cumulative findings we next conducted a series of experiments to answer the following specific questions about the role of ceramide in regulating PGE_2 production: (1) What are the specific effects of IL-1 and ceramide on COX-1 and COX-2 transcription and translation? (2) Do IL-1 and ceramide affect $cPLA_2$ expression/activation and can the addition of exogenous arachidonic acid provide the missing signal required for ceramide-induced PGE_2 synthesis? In order to answer these questions we performed a kinetic analysis of the effects of IL-1 and ceramide on COX-1, COX-2, and $cPLA_2$ transcription and translation compared with the kinetics of PGE_2 production (manuscript submitted).

The Effects of IL-1 and Ceramide on COX and $cPLA_2$ Expression

We initially measured the effects of ceramide and IL-1 on PGE_2 accumulation as a means of assessing the temporal relationship between the expression of the rate-limiting genes COX-1, COX-2, and $cPLA_2$ and the induction of PGE_2 biosynthesis. Treatment of human foreskin fibroblasts with IL-1 induced PGE_2 production in a time-dependent manner, but only after a delay of ≈8 hours. Treatment of cells with ceramide alone did not induce PGE_2 production but when added along with IL-1, significantly increased PGE_2 synthesis (Fig. 3.7).

In order to determine the mechanism(s) involved in the apparent cross-talk between ceramide and IL-1 in enhancing PGE_2 synthesis we first examined their effects on the quantitative pattern of COX-2 expression because it is highly inducible and its expression is specifically associated with inflammation.[43] The induction of COX-2 expression by IL-1 and ceramide is specific since the steady-state levels of mRNA of a housekeeping gene, GAPDH, remain unaltered in response to any of the treatments tested. General diagrams showing the temporal expression of COX-1, COX-2 and $cPLA_2$ are depicted in Figure 3.8A, B and C. In untreated fibroblasts COX-2 expression is essentially undetectable. IL-1 rapidly increases COX-2 mRNA levels ≈20-fold by 80 minutes and even though COX-2 mRNA levels began to decrease after 4 hours, steady-state levels remain ≈10-fold higher than control levels for at least 24 hours (Fig. 3.8A). Ceramide also rapidly induces COX-2 mRNA expression to a lesser extent (≈5-fold) and in much more transient manner. Unlike IL-1 or IL-1/ceramide treatment, steady-state levels of COX-2 mRNA return to basal levels by 4 hours and no PGE_2 is produced in response to ceramide in the absence of IL-1 (Fig. 3.8B). While there is no significant difference in COX-2 transcription when IL-1 and ceramide are added together (Fig. 3.8C) COX-2 protein levels in ceramide and IL-1 treated cells are significantly higher than in cells treated with IL-1 alone and remain higher for at least 24 hours; no COX-2 protein is detectable in untreated or ceramide treated cells (not shown). Thus, the additive effects of IL-1 and ceramide on COX-2 protein expression may, at least in part, account for the increase in PGE2 production observed when cells are treated with IL-1 and ceramide together.

Although COX-2 is generally regarded as the inducible COX isoform, COX-1 expression is also increased in response to a variety of stimuli.[44] As expected, COX-1 mRNA is constitutively expressed in un-

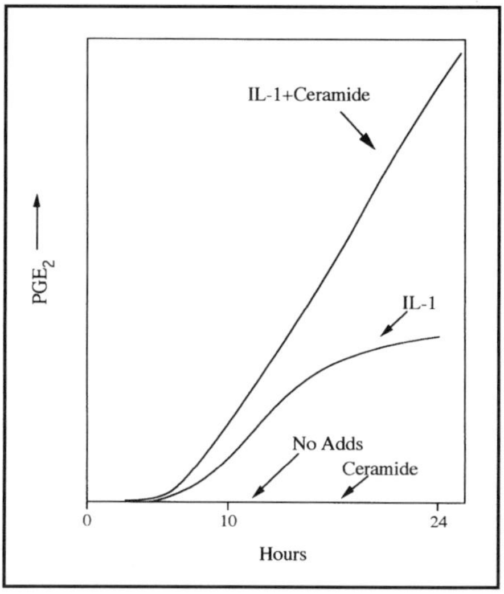

Fig. 3.7. Schematic illustration of the kinetics of IL-1 and ceramide induced PGE_2 production.

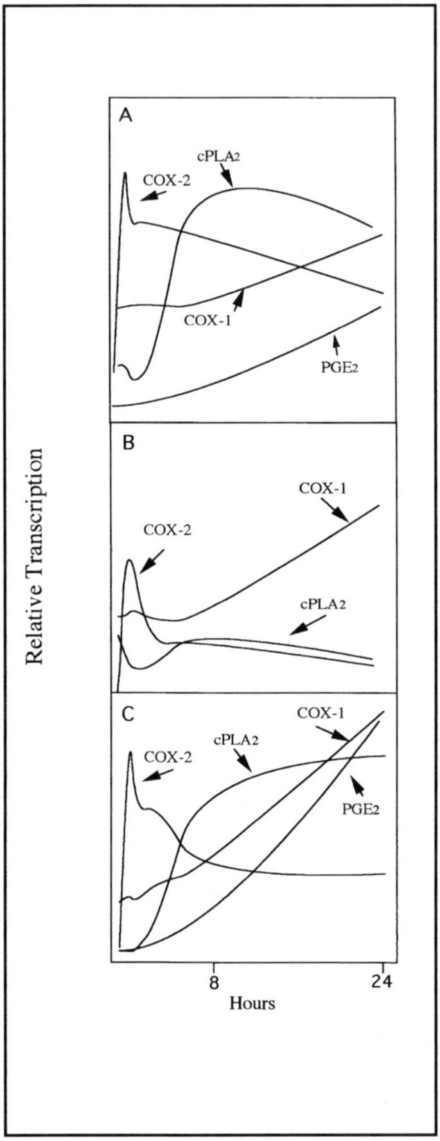

Fig. 3.8. Schematic illustration of the kinetics of IL-1 and ceramide mediated COX-1, COX-2 and $cPLA_2$ transcription. (A) Time course of transcription in response to IL-1 only. (B) Time course of transcription in response to ceramide only. (C) Time course of transcription in response to IL-1 and ceramide added together.

treated cells but its expression increases in response to either IL-1 or ceramide. In contrast to the rapid induction of COX-2 expression, IL-1 and ceramide induce a much slower, dramatic increase in COX-1 mRNA levels at later time points (see Fig. 3.8). The expression of COX-1 is characterized by the presence of a 2.7 kb COX-1 transcript as well as the dramatic induction of a larger 5.5 kb, alternatively spliced form of COX-1 mRNA (not shown). The two major COX-1 mRNAs expressed by 24 hours are most likely the result of alternative splicing of exon 9 of the human COX-1 gene;[45] while both mRNAs are enhanced by IL-1 and ceramide, the 5.5 kb COX-1 species is not evident between 80 minutes and 2 hours, but is dramatically induced compared with the constitutively expressed 2.7 kb species compared with time matched control levels at 24 hours. Steady-state COX-1 protein levels increased in proportion to mRNA levels and it appears that the expression pattern for COX-1 more closely follows the time-course for PGE2 production. Thus, ceramide increases both COX-1 and COX-2 protein expression in the presence of IL-1 but has little or no effect on COX-2 protein expression in the absence of IL-1, suggesting that the increase in PGE2 when ceramide is present may be the result of the ability of ceramide to enhance both the amount and stability of COX-1 and COX-2 protein.

Because ceramide affects COX-1 and COX-2 transcription without affecting PGE2 production, we suspected that COX expression was not the limiting factor accounting for the inability of ceramide to induce PGE_2 production. To test the prediction that arachidonic acid availability, regulated by $cPLA_2$ expression, may account for the lack of PGE_2 production, we examined the effect of ceramide and IL-1 on $cPLA_2$ gene expression. No changes in $cPLA_2$ were detected during the first 4 hours of treatment with IL-1, ceramide, or a combination of both, but $cPLA_2$ mRNA increased dramatically after 8 hours only in response to IL-1 (see Figs. 3.8A, B and C). Immunoprecipitable, radiolabeled $cPLA_2$ protein also increased ≈2- and 1.5-fold over their respective controls at 8 and 24 hours and increases in levels of the phosphorylated, activated form of the enzyme could be clearly identified. C2-ceramide treatment had no effect on either $cPLA_2$ mRNA or protein levels at any time point tested. These results are consistent with the hypothesis that the inability of ceramide to induce PGE_2 production, despite its ability to induce the expression of both COX isoforms, may be due to the fact that ceramide by itself does not stimulate $cPLA_2$ expression and arachidonic acid mobilization, and thus no PGE_2 is produced.

Next we tested whether substrate (arachidonic acid) availability or COX activity is the rate-limiting step in IL-1 and ceramide-induced PGE_2 production. Cells were either treated with IL-1, ceramide, or with a mixture of both, after which the cells were washed and arachidonic acid added in serum free media; PGE_2 production was then analyzed as a measure of total COX activity. As illustrated in Figure 3.9, the kinetics of PGE_2 production in cells treated with either IL-1 or ceramide is essentially identical in cells supplemented with arachidonic acid, supporting the hypothesis that a lack of substrate likely accounts for the inability of ceramide to induce PGE_2 production even though it enhances COX expression. At early time points, basal PGE_2 production is significantly increased in the presence of added arachidonic acid, indicating that the constitutively expressed COX-1 was enzymatically active. However, no measurable increase in PGE_2 production is observed in IL-1 treated cells at early time points even though COX-2 mRNA/protein is significantly elevated.

Together, these data support the generalized pathway shown in Figure 3.6 in which ceramide induces COX transcription and also enhances cytokine-induced COX expression. Now we know that ceramide affects the transcription of both COX-1 and COX-2 but it appears that ceramide does not induce $cPLA_2$ expression in human fibroblasts. Much more work will be required to determine the precise roles of COX-1 and COX-2 in the regulation of IL-1 and ceramide-induced prostanoid synthesis and perhaps more important, their respective functions in normal homeostasis versus inflammatory and immune responses.

SUMMARY AND FUTURE DIRECTIONS

Over the past few years we have compiled a significant amount of evidence indicating that ceramide plays a direct role in modulating prostaglandin production via its

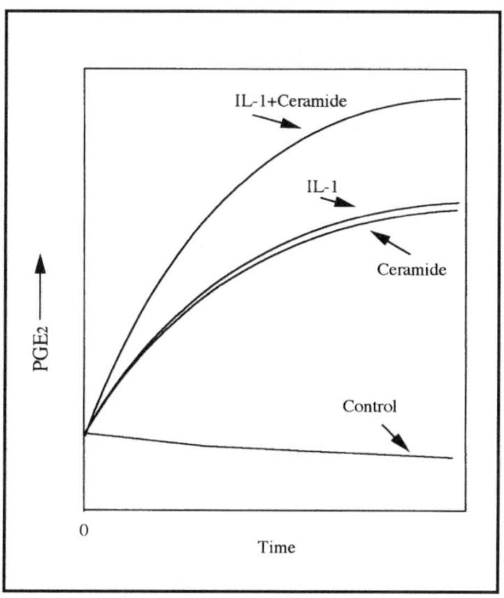

Fig. 3.9. Arachidonic acid is rate-limiting in ceramide-mediated PGE_2 production. Confluent fibroblasts were left untreated, treated with IL-1, C2-ceramide or a combination of IL-1 and C2-ceramide for increasing lengths of time (up to 24 hours) in order to induce the expression of COX-1 and COX-2 as determined in previous experiments (see Fig. 3.8). After COX induction, culture media was removed, the cells washed to remove any accumulated PGE_2, and fresh serum-free media containing arachidonic acid (50 μM) was added. After an additional 15 minute incubation in the presence of added arachidonic acid, PGE_2 was measured by RIA.

ability to induce the expression of COX-1 and COX-2. Ceramide induces COX-1 and COX-2 expression by itself but ceramide requires another signal(s), at least in human foreskin fibroblasts, to mobilize arachidonic acid for conversion to PGE_2. Our studies indicate that the inability of ceramide to mobilize arachidonic acid is responsible for the lack of prostaglandin production even though COX-1 and COX-2 expression is increased by ceramide. While we have not yet extensively examined the effects of ceramide on COX and $cPLA_2$ expression in other cell types, others have recently reported that ceramide induces both COX-2 and $cPLA_2$ expression in a mouse fibrosarcoma cell line (L929).[46] In this case, ceramide by itself is capable of inducing prostaglandin production in these cells, suggesting that there may be somewhat different mechanism(s) for ceramide action in different types of cells. These observations also indicate that a very complicated interaction between the signals generated by ceramide and those induced by cytokines (in addition to ceramide generation) are involved in the regulation of prostaglandin biosynthesis.

Certainly, it seems more than coincidental that cytokines which stimulate the synthesis of prostanoids (e.g., IL-1 and TNF) also induce ceramide generation. This correlation may in part be due to the recent observation that arachidonic acid stimulates sphingomyelin hydrolysis and ceramide accumulation.[47] Thus, a positive feedback loop may exist between ceramide generation and arachidonic acid which is mobilized in response to IL-1/TNF, or even ceramide itself, in the case of L929 cells in which ceramide also activates $cPLA_2$. The ability of ceramide to mobilize arachidonic acid (at least in some cells) and to enhance prostanoid production clearly establishes it as an inflammatory mediator given the many proinflammatory functions attributed to prostaglandins and other biologically active arachidonic acid metabolites.

Clearly, there is a significant amount of cross-talk between ceramide-mediated signaling pathways and those pathways regulating arachidonic acid metabolism. The ability of ceramide to stimulate cytokine-induced COX and $cPLA_2$ expression along with concomitant production of biologically active arachidonic acid metabolites adds yet another layer of complexity to the ceramide signal transduction cascade because of the many complex biological actions of arachidonic acid metabolites. Further, the effects of ceramide on prostaglandin production not only implicate a regulatory role for ceramide in inflammation, but because of its effects on COX expression, suggests other novel mechanism(s) by which ceramide may mediate its effects on such processes as apoptosis or the mediation of autoimmune disease. For example, COX appears to bind directly with autoimmunity- and apoptosis-associated nucleobindin (Nuc) and as a result plays a role in maintaining the intracellular retention and function of Nuc,[48] a protein which is commonly found bound to nucleosomal-laddered DNA characteristic of apoptosis. Such interactions between COX and Nuc suggest a degree of cross-talk between proinflammatory and apoptotic pathways and, given the role of ceramide in apoptosis, the upregulation of COX and its binding to Nuc may represent another factor involved in ceramide-mediated inflammation and/or apoptosis. Another candidate for cross-talk between ceramide signaling pathways and inflammation is the ability of ceramide to activate the mitogen-activated protein (MAP) kinase pathway which is intimately involved in the upregulation of those genes involved in mediating the acute inflammatory response.[49-51] MAP kinase is also known to phosphorylate and thereby activate $cPLA_2$, which is directly involved in providing substrate for conversion to biologically active eicosanoids, suggesting another crossover point between ceramide and other proinflammatory signaling pathways.[52]

Our plans for the immediate future are to examine the precise regulatory role of ceramide on the transcription of COX-1, COX-2 and $cPLA_2$ using specific promoter/

luciferase constructs transfected into fibroblasts as well as other types of cells. Using promoter deletion mutants, DNA footprinting, and other techniques, we hope to identify those regions of each promoter which are involved in regulating ceramide-induced transcription as well as aid in the identification of those transcription factor(s) which are involved in regulating COX-1, COX-2 and $cPLA_2$ expression. Studies are also currently underway to examine the possible role of ceramide in the stabilization of COX/$cPLA_2$ transcripts and/or protein as another possible mechanism by which ceramide enhances PGE_2 production. Regardless of the mechanism(s) involved, the ability of ceramide to enhance prostanoid production alone or in conjunction with cytokine stimulation, illustrates the importance of ceramide as a lipid mediator of inflammation.

References

1. Dinarello CA. Inflammatory cytokines: interleukin-1 and tumor necrosis factor as effector molecules in autoimmune disease. Curr Opin Immunol 1994; 3:941-948.
2. Vane JR, Botting RM. Regulatory mechanisms of the vascular endothelium: an update. Pol J Pharmacol 1994; 46:499-421.
3. Vane JR, Botting RM. New insights into the mode of action of anti-inflammatory drugs. Inflamm Res 1995; 44:1-10.
4. Seibert K, Masferrer JL. Role of inducible cyclooxygenase (COX-2) in inflammation. Receptor 1994; 4:17-23.
5. Seibert K, Zhang Y, Leahy K et al. Pharmacological and biochemical demonstration of the role of cyclooxygenase 2 in inflammation and pain. Proc Natl Acad Sci USA 1994; 91:12013-12017.
6. Seibert K, Masferrer J, Zhang Y et al. Mediation of inflammation by cyclooxygenase-2. Agents Actions Suppl 1995; 46:41-50.
7. Goetzl EJ, An S, Smith WL. Specificity of expression and effects of eicosanoid mediators in normal physiology and human diseases. FASEB J 1995; 9:1051-1058.
8. Isakson P, Hauser S, Zhang Y et al. Cytokine regulation of eicosanoid generation. Ann NY Acad Sci 1994; 744:181-183.
9. Smith WL, Meade EA, De Witt DL. Pharmacology of prostaglandin endoperoxide synthase isozymes-1 and -2. Ann NY Acad Sci 1994; 714:136-142.
10. Barnett J, Chow J, Ives D et al. Purification, characterization and selective inhibition of human prostaglandin G/H synthase 1 and 2 expressed in the baculovirus system. Biochim Biophys Acta 1994; 1209:130-139.
11. Ballou LR, Laulederkind SJF, Rosloneic EF et al. Ceramide signalling and the immune response. Biochimica et Biophysica Acta 1996; 1301:273-287.
12. Bell RM, Hannun YA, Merril AH Jr. Advances in Lipid Research, Sphingolipids Part A: Functions and Breakdown Products. Advances in Lipid Research, vol 25. San Diego, Ca: Academic Press, 1993; 336.
13. Bell RM, Hannun YA, Merrill AH Jr. Advances in Lipid Research, Sphingolipids Part B: Regulation and Function of Metabolism. Advances in Lipid Research, vol 26. San Diego: Academic Press, 1993; 380.
14. Hannun YA, Linardic CM. Sphingolipid breakdown products: anti-proliferative and tumor-suppressor lipids. Biochem Biophys Acta 1993; 1154:223-236.
15. Hannun YA, Obeid LM, Wolff RA. The novel second messenger ceramide: identification, mechanism of action, and cellular activity. Adv Lipid Res 1993; 25:43-64.
16. Hannun YA, Bell RM. The sphingomyelin cycle: a prototypic sphingolipid signaling pathway. Adv Lipid Res 1993; 25:27-41.
17. Hannun YA. The sphingomyelin cycle and the second messenger function of ceramide. J Biol Chem 1994; 269: 3125-3128.
18. Hannun YA, Obeid LM. Ceramide: an intracellular signal for apoptosis. TIBS 1995; 20:73-78.
19. Heller RA, Krönke M. Tumor necrosis factor receptor-mediated signaling pathways. J Cell Biol 1994; 126:5-9.
20. Kolesnick RN. Sphingomyelin and derivatives as cellular signals. Prog Lipid

Res 1991; 30:1-38.
21. Kolesnick R. Signal transduction through the sphingomyelin pathway. Mol Chem Neuropathol 1994; 21:287-297.
22. Kolesnick R, Golde DW. The sphingomyelin pathway in tumor necrosis factor and interleukin-1 signaling. Cell 1994; 77:325-328.
23. Kolesnick R, Fuks Z. Ceramide: A signal for apoptosis or Mitogenesis? J Exp Med 1995; 181:1949-1952.
24. Mathias S, Kolesnick R. Ceramide: a novel second messenger. Adv Lipid Res 1993; 25:65-90.
25. Merrill AH. Ceramide: a new lipid "second messenger"? Nutr Rev 1992; 50:78-80.
26. Michell RH, Wakelam MJO. Sphingolipid signalling. Current Biol 1994; 4:370-373.
27. Schütze S, Machleidt T, Krönke M. The role of diacylglycerol and ceramide in tumor necrosis factor and interleukin-1 signal transduction. J Leukocyte Biol 1994; 56:533-541.
28. Dennis EA. Diversity of group types, regulation, and function of phospholipase A_2. J Biol Chem 1994; 269:13057-13060.
29. Samuelsson B. Leukotrienes: mediators of immediate hypersensitivity reactions and inflammation. Science 1983; 220:568-575.
30. Samuelsson B, Goldyne M, Granstrom E et al. Prostaglandins and thromboxanes. Annu Rev Biochem 1978; 47:997-1029.
31. Herschman H. Prostaglandin synthase 2. Biochim Biophys Acta 1996; 1299: 125-140.
32. Seibert K, Masferrer J, Zhang Y et al. Expression and selective inhibition of constitutive and inducible forms of cyclooxygenase. Adv Prostaglandin Thromboxane Leukot Res 1995; 23:125-127.
33. Langenbach R, Morham SG, Tiano HF et al. Prostaglandin synthase 1 gene disruption in mice reduces arachidonic acid-induced inflammation and indomethacin-induced gastric ulceration. Cell 1995; 83:483-492.
34. Morham SG, Langenbach R, Loftin CD et al. Prostaglandin synthase 2 gene disruption causes severe renal pathology in the mouse. Cell 1995; 83:473-482.
35. Hannun YA, Bell RM. Functions of sphingolipids and sphingolipid breakdown products in cellular regulation. Science 1989; 243:500-507.
36. Ballou LR, Barker SC, Postlethwaite AE et al. Sphingosine potentiates IL-1-mediated prostaglandin E_2 production in human fibroblasts. J Immunol 1990; 145:4245-4251.
37. Candela M, Barker SC, Ballou LR. Sphingosine synergistically stimulates tumor necrosis factor α–induced prostaglandin E_2 production in human fibroblasts. J Exptl Med 1991; 174:1363-1369.
38. Ballou LR, Chao CP, Holness MA et al. Interleukin-1-mediated PGE_2 production and sphingomyelin metabolism. Evidence for the regulation of cyclooxygenase gene expression by sphingosine and ceramide. J Biol Chem 1992; 267:20044-20050.
39. Mathias S, Younes A, Kan CC et al. Activation of the sphingomyelin signaling pathway in intact EL4 cells and in a cell-free system by IL-1β. Science 1993; 259:519-522.
40. Clark JD, Lin LL, Kriz RW et al. A novel arachidonic acid-selective cytosolic PLA_2 contains a $Ca^{(2+)}$-dependent translocation domain with homology to PKC and GAP. Cell 1991; 65:1043-1051.
41. Lin LL, Lin AY, Knopf JL. Cytosolic phospholipase A_2 is coupled to hormonally regulated release of arachidonic acid. Proc Natl Acad Sci USA 1992; 89:6147-6151.
42. Lin LL, Lin AY, De Witt DL. Interleukin-1α induces the accumulation of cytosolic phospholipase A_2 and the release of prostaglandin E_2 in human fibroblasts. J Biol Chem 1992; 267:23451-23454.
43. Mitchell JA, Larkin S, Williams TJ. Cyclooxygenase-2: regulation and relevance in inflammation. Biochem Pharmacol 1995; 50:1535-1542.
44. Murakami M, Matsumoto R, Urade Y et al. c-kit ligand mediates increased expression of cytosolic phospholipase A_2, prostaglandin endoperoxide synthase-1, and hematopoietic prostaglandin D_2 synthase and increased IgE-dependent prostaglandin D_2 generation in immature mouse mast cells. J Biol Chem 1995; 270: 3239-3246.
45. Diaz A, Reginato AM, Jimenez SA. Alternative splicing of human prostaglandin G/

H synthase mRNA and evidence of differential regulation of the resulting transcripts by transforming growth factor β, interleukin-1β, and tumor-necrosis factor α. J Biol Chem 1992; 267:10816-10822.
46. Hayakawa M, Jayadev S, Tsujimoto M et al. Role of ceramide in stimulation of the transcription of cytosolic phospholipase A_2 and cyclooxygenase 2. Biochem Biophys Res Commun 1996; 220:681-686.
47. Jayadev S, Linardic CM, Hannun YA. Identification of arachidonic acid as a mediator of sphingomyelin hydrolysis in response to tumor necrosis factor α. J Biol Chem 1994; 269:5757-5763.
48. Ballif B, Mincek N, Barratt J et al. Interaction of cyclooxygenase with an apoptosis- and autoimmunity-associated protein. Proc Natl Acad Sci USA 1996; 93:5544-5549.
49. Raines MA, Kolesnick RN, Golde DW. Sphingomyelinase and ceramide activate mitogen-activated protein kinase in myeloid HL-60 cells. J Biol Chem 1993; 268:14572-14575.
50. Huwiler A, Brunner J, Hummel R et al. Ceramide-binding and activation defines protein kinase c-Raf as a ceramide-activated protein kinase. Proc Natl Acad Sci USA 1996; 93:6959-6963.
51. Davis RJ. The mitogen-activated protein kinase signal transduction pathway. J Biol Chem 1993; 268:14553-14556.
52. Lin LL, Wartmann M, Lin AY et al. $cPLA_2$ is phosphorylated and activated by MAP kinase. Cell 1993; 72:269-278.

CHAPTER 4

A Role for Ceramide in Meiosis

Jay C. Strum, Katherine I. Swenson and Robert M. Bell

INTRODUCTION

Meiotic maturation of *Xenopus laevis* oocytes is a widely used model for the study of cell cycle regulation. Progesterone, the physiological regulator of maturation, triggers cell cycle progression of prophase arrested oocytes through meiosis I and into meiosis II where the mature oocyte remains arrested in metaphase II until fertilization. Although extensively studied, the mechanism of reinitiation remains poorly understood. Progesterone addition to oocytes caused within minutes a transient decrease in cAMP levels which was necessary for maturation; agents which elevate cAMP inhibit progesterone-induced maturation.[1] The molecular mechanism(s) by which progesterone elicits these changes is unclear. However, inhibition of adenylate cyclase[2] in a G-protein dependent manner[3] appears to be involved. Subsequent microinjection experiments demonstrated that the catalytic subunit of cAMP-dependent protein kinase (PKA) inhibited maturation,[4] while the regulatory subunit induced maturation,[5] providing evidence that PKA is a negative regulator of oocyte maturation. While the role of PKA in oocyte maturation is now well established, the initial event(s) following the addition of progesterone is undefined. A transmembrane signaling mechanism is required since, unlike the classical steroid hormone receptor for transcriptional activation, the progesterone receptor is localized to the plasma membrane.[6,7] Numerous studies have proposed a phospholipase-mediated transmembrane signaling mechanism.[8-11] While progesterone has been observed to stimulate the turnover of the major glycerophospholipids with the generation of potential second messengers, these are not sufficient to induce maturation in the absence of hormone. In contrast, treatment of oocytes with exogenous sphingomyelinase (*Staphylococcus aureus*) is a potent inducer of maturation.[12,13] These investigators additionally showed that microinjection of sphingosine was capable of inducing maturation in the absence of hormone. These observations are consistent with the emerging role of sphingolipids as regulators of critical cellular processes.[14] In particular, Okazaki et al[15] originally described a sphingomyelin cycle where ceramide, generated by sphingomyelinase hydrolysis of membrane sphingomyelin, acts as a lipid second messenger functioning as a modulator of monocytic differentiation,[15] cell cycle regulation,[16,17] and apoptosis.[18,19] At the time of the original report by Varnold and Smith, sphingosine had recently been reported as a potent inhibitor of protein kinase C (PKC)[20] and ceramide was

Sphingolipid-Mediated Signal Transduction, edited by Yusuf A. Hannun.

not known to act as a lipid second messenger. Thus, these investigators assumed that sphingosine levels were increased by sphingomyelinase treatment of oocytes leading to inhibition of PKC and subsequently, reinitiation of oocyte maturation. Since ceramide is the more proximal metabolite of sphingomyelinase hydrolysis and sphingosine could be acylated to form ceramide, we further investigated the role of sphingolipids in the regulation of oocyte maturation. From these studies, we have defined a role for ceramide in the maturation pathway which suggests it may be functionally important in mediating the reinitiation of the meiotic cell cycle triggered by progesterone.[21] The involvement of sphingolipid turnover in progesterone-induced meiosis has been strengthened by recently reported observations showing that agonists which increase the level of ceramide in cells trigger a normal process of maturation and that at suboptimal concentrations, these agonists can potentiate progesterone-induced maturation.[22,23] This chapter details the observations which support a role for ceramide in meiosis and presents the current hypothesis on the function of sphingolipid turnover in reinitiation of oocyte maturation.

EXOGENOUS SPHINGOMYELINASES INDUCE NORMAL MATURATION

To investigate the role of lipid second messengers in oocyte maturation, we initially tested the ability of various phospholipases to influence progesterone- or insulin-induced maturation by following germinal vesicle breakdown (GVBD) as an indicator of cell cycle progression. Phospholipase C (*Bacillus cereus*) and cabbage phospholipase D had little or no effect on the time course of progesterone or insulin induced GVBD. However, as previously reported by Varnold and Smith,[12,13] a brief, 5 minute treatment of oocytes with sphingomyelinase (*Staphylococcus aureus*), 0.25 U/ml, was capable of inducing maturation in the absence of hormone. Enzyme inactivated by heat treatment was not effective in stimulating maturation. To prevent deleterious effects to the cells, pretreatment was not extended past 5 minutes and an extensive washing step with bovine serum albumin (1 mg/ml) included in the buffer was necessary to remove enzyme bound to the oocytes.

The morphological appearance of GVBD was indistinguishable from oocytes treated with progesterone, suggesting that sphingomyelinase treatment induces a normal maturation process. However, it has previously been shown that oocytes can undergo a degenerative process, which is incorrectly identified as GVBD, in response to some treatments.[24] Therefore, a time course of H1 kinase activity following sphingomyelinase treatment was determined. H1 kinase activity represents the complex of cdc2 kinase and cyclin B. In immature oocytes this activity is low and peaks at the time of GVBD. Following GVBD, the levels of H1 kinase activity decreases and as the cells enter into meiosis II, the activity increases and remains elevated as the cells arrest in metaphase II. We found that levels of H1 kinase activity in oocytes treated with progesterone or sphingomyelinase oscillated in a similar manner, indicating that a normal process of maturation was initiated by enzyme treatment. These observations were subsequently confirmed by De Smedt et al[22] who reported that sphingomyelinase from *Streptomyces* was also effective in activating cdc2 kinase and eliciting a normal maturation process.

We additionally found that the kinetics of sphingomyelinase-induced GVBD could be accelerated by increasing the amount of enzyme during treatment indicating that the concentration of products from sphingomyelinase hydrolysis were important in the timing of maturation. Since the primary products of sphingomyelinase hydrolysis are ceramide and phosphocholine and the former product has been widely implicated as a potential second messenger with biological activity, we measured the amount of ceramide, using *E. coli* DG kinase,[25] which was produced under the conditions of our assays. We found that the levels of ceramide increased by about 25% during the initial enzyme treatment. However, the levels continued to slowly increase for several hours after the cells had been extensively washed. The concentration of ceramide never increased more than 2-fold over basal levels, which varied in different stage VI oocytes and batches of cells from different females from 250-350 pmoles/oocyte. This suggests that relatively small changes in ceramide can induce a normal maturation process.

SPHINGOMYELINASE-INDUCED GVBD REQUIRES PROTEIN SYNTHESIS

It is well known that progesterone-induced oocyte maturation requires protein synthesis.[26] To examine whether protein synthesis is required for sphingomyelinase-induced GVBD, oocytes were pretreated with 100 μM cycloheximide for one hour and subsequently treated with sphingomyelinase (*Staphylococcus aureus*) or progesterone. Cycloheximide treatment completely blocked both sphingomyelinase and progesterone-induced GVBD indicating that protein synthesis is necessary for sphingomyelinase-induced GVBD.

MOS IS REQUIRED FOR SPHINGOMYELINASE-INDUCED GVBD

In immature oocytes, the levels of the protooncogene *c-mos* are very low.[27] Following progesterone treatment c-mos is translated from preexisting mRNA. It is believed that the inhibition of *c-mos* synthesis by cycloheximide is the reason for cycloheximide inhibition of progesterone-induced maturation.[27] In support of this, it has been shown using antisense *mos* oligonucleotides that synthesis of *c-mos* is required for progesterone-induced maturation.[28] Since we had found that protein synthesis was necessary for sphingomyelinase-induced GVBD, we next investigated if mos synthesis was required. Microinjection of antisense mos oligonucleotides prior to sphingomyelinase treatment blocked GVBD, demonstrating that mos synthesis is necessary for sphingomyelinase-induced GVBD. Microinjection of sense oligonucleotides of mos had no effect on sphingomyelinase-induced GVBD. This suggests that there are some commonalities in the pathway through which sphingomyelinase and progesterone induce maturation.

PDMP INDUCES NORMAL MEIOTIC MATURATION

Recently, De Smedt et al[22] have investigated the role of sphingolipids in oocyte maturation and reported that DL-threo-1-phenyl-2-decanoylamino-3-morpholino-1-propanol (PDMP), an inhibitor of glucosylceramide synthase,[29] activates cdc2 kinase and induces a normal maturation process in a concentration dependent manner. Additionally, they found that cycloheximide blocked PDMP-induced GVBD indicating that protein synthesis was necessary for maturation and agents which elevate cAMP levels blocked PDMP-induced GVBD. Most interestingly, these investigators showed that PDMP could potentiate progesterone-induced maturation suggesting common elements in the signaling pathways used by both agonists.

This study provides additional evidence that an increase in cellular ceramide, generated indirectly by inhibition of a ceramide metabolizing enzyme, triggers a normal pathway of maturation.

BREFELDIN A INDUCES NORMAL MEIOTIC MATURATION

Brefeldin A is a fungal metabolite which has been shown to disrupt protein traffic through the golgi apparatus.[30] Recent observations reported by Mulner-Lorillon et al,[23] show that brefeldin A is capable of inducing the activation of cdc2 kinase and normal meiotic maturation. Brefeldin A-induced GVBD was sensitive to cycloheximide and cAMP modulating agents. When brefeldin A was used in combination with a subthreshold concentration of progesterone, a significant decrease in the time course of brefeldin A-induced GVBD was observed. These investigators proposed that the levels of c-mos in immature oocytes may be sufficient to trigger maturation but the protein is sequestered. Upon progesterone treatment the protein may be transported to a location within the cell where it is activated and stimulates maturation. However, this hypothesis is difficult to reconcile with antisense experiments previously discussed. The recent work of Linardic et al[31] suggests an alternative explanation. These studies showed that brefeldin A activates the sphingomyelin cycle in HL-60 cells resulting in an increase in ceramide levels. Therefore, brefeldin A may stimulate the formation of ceramide in oocytes and thus induce GVBD.

MICROINJECTED SPHINGOLIPIDS INDUCE GVBD

Varnold and Smith[12,13] were the first to demonstrate that microinjected sphingosine, complexed to BSA, was capable of inducing GVBD. We repeated and extended upon these initial experiments by microinjecting short-chain analogs of ceramide, the more proximal metabolite of sphingomyelin hydrolysis. Attempts were made to exogenously add sphingolipids solubilized in ethanol; however, they were not able to penetrate the vitelline envelope and thus were not taken up by the oocytes. Sphingolipids were therefore complexed to BSA and 40 pmoles were microinjected into oocytes, resulting in a final concentration of 40 μM. Higher concentrations of the short-chain ceramides could not be attained due to insolubility. Since both sphingosine and short-chain ceramides induce GVBD and can be rapidly interconverted, it was necessary to determine a time course of metabolism for each using radiolabeled substrates to determine which sphingolipid was the biologically active species. When these were microinjected into oocytes as BSA complexes, we found that sphingosine was readily acylated to form ceramide and was subsequently recovered in sphingomyelin. Surprisingly, we found that the short-chain ceramides were rapidly metabolized to long-chain ceramides with little or no accumulation of sphingosine. In order to further identify which sphingolipid was responsible for triggering maturation, the fungal metabolite fumonisin B_1 was used. Fumonisin B_1 shares structural homology with sphingosine and thus is a competitive inhibitor of sphingosine N-acyltransferase.[32] We found that microinjection of 50 μM fumonisin B_1 prior to sphingosine blocked its conversion to ceramide and its ability to induce GVBD. Taken together, these results suggest that ceramide is the metabolite of sphingomyelin hydrolysis responsible for triggering GVBD.

PROGESTERONE STIMULATES CERAMIDE FORMATION THROUGH ACTIVATION OF A MG^{2+}-DEPENDENT NEUTRAL SPHINGOMYELINASE

We next investigated the possibility that progesterone stimulated the formation of ceramide. Oocytes were treated with progesterone for various times and the mass of ceramide was determined using *E. coli* DG kinase.[25] A significant increase was seen as early as 2-5 minutes following addition of progesterone. Within the first hour the mass of ceramide had approximately doubled and increased thereafter until the time of GVBD. Ceramide levels were not determined in cells following GVBD. The mass of sphingomyelin was similarly determined and found to decrease by 40% of control in 3 hours. Since the decrease in the mass of sphingomyelin was greater than the increase in the mass of ceramide this indicates that a significant amount of the ceramide is being further metabolized.

When using the DG kinase assay on nonderivatized total lipid extracts, the mass of ceramide and diacylglycerol in a particular cellular extract can simultaneously be quantitated. We found that diacylglycerol mass did not change until just prior to GVBD, at which time it increased by about 2-fold. The source of this diacylglycerol and its significance is presently unknown. Whether it functions as an intermediate in de novo biosynthesis of glycerolipids or as a signaling molecule will be the subject of future studies.

Experiments were conducted to determine the mechanism of progesterone stimulated ceramide formation. We found that the specific activity of a Mg^{2+}-dependent neutral sphingomyelinase increased within 2-5 minutes following progesterone treatment of cells and peaked at 3- to 4-fold above untreated cells. The activity of this sphingomyelinase decreased after 1 hour of progesterone treatment but remained significantly above basal levels. However, as previously noted the level of ceramide continued to increase through 3 hours following treatment. It is possible that a sphingomyelinase with an acidic pH optimum is also activated by progesterone treatment and is responsible for a portion of the increase in ceramide. This would be consistent with some reports in the literature which suggests that both enzymes are activated following stimulation of mammalian cells with TNFα.[33] Further studies are necessary in order to characterize the sphingomyelinases in *Xenopus* oocytes.

DISCUSSION

The observations discussed here provide evidence that ceramide is involved in progesterone-induced reinitiation of meiotic maturation in *Xenopus* oocytes. Definitive experiments which would demonstrate whether ceramide formation is required for progesterone-induced maturation cannot presently be performed because of the lack of the necessary molecular tools. This will require the cloning of a member of the gene family of Mg^{2+}-dependent neutral sphingomyelinases and an understanding of the regulation of this enzyme. Presently, we can speculate on the following questions. Why does the time course of GVBD vary with different ceramide-inducing agents? Where do sphingomyelin hydrolysis and ceramide formation fit into the pathway of oocyte maturation and how does ceramide trigger GVBD?

Ceramide is an intermediate in the de novo pathway for the biosynthesis of complex sphingolipids. Thus, it occupies a similar position in intermediary metabolism to diacylglycerol and appears to function in an analogous, yet opposing, manner to diacylglycerol.[14] The majority of each of these neutral lipids in cells is involved in the de novo biosynthetic pathway of more complex lipids and thus does not serve a signaling function. The dichotomous nature of these lipids dictates that they are separated into synthesis and signaling pools. This prevents the uncontrolled activation of target enzymes by lipid second messengers. Therefore, the location of ceramide formation is likely important in order to elicit a particular biological response. In the case of exogenous sphingomyelinase-induced GVBD, changes in ceramide levels occur at the plasma membrane. Likewise, since the progesterone receptor is located in the plasma membrane, it can be assumed that ceramide may be generated in this membrane and thus may explain why the time course of progesterone and sphingomyelinase-induced GVBD were very similar. When ceramide is increased in cells by PDMP or brefeldin A treatment or microinjection of ceramide:BSA complexes, the time course is delayed relative to progesterone-induced GVBD. When mammalian cells are treated with PDMP, ceramide increases primarily in the golgi apparatus.[29] Also, it has previously been shown that exogenously added ceramides accumulate in the golgi.[34] Therefore, microinjected ceramide or ceramide generated indirectly by PDMP treatment may have to redistribute to a more proximal location to exert its effect on maturation. Thus, sphingomyelinase mimics progesterone-induced maturation better than other treatments. It is also possible that the time course of maturation, induced by microinjected short-chain ceramide, is delayed because the analogs may have to be converted to long-chain ceramide to serve as an effector molecule. Therefore, not only does the lipid have to relocate, but a threshold concentration may have to be achieved to trigger maturation.

From our data, it is clear that ceramide formation is an early event following progesterone addition to oocytes and is likely generated in the plasma membrane. We also know that ceramide formation is prior to mos synthesis. Further studies are needed to determine the temporal relationship of ceramide formation and the decrease in cAMP following progesterone stimulation.

In preliminary studies, our results are contradictory. We found that cholera toxin blocks progesterone but not sphingomyelinase-induced GVBD, suggesting that ceramide formation is downstream of cAMP-dependent protein kinase. This would imply that oocyte sphingomyelinase may be negatively regulated by PKA. However, pretreatment with 100 μM forskolin blocks both progesterone and sphingomyelinase-induced GVBD. This discrepancy will be the subject of future studies.

There are several targets for ceramide which have recently been identified. The most well characterized is the serine/threonine protein phosphatase, PP2A. This enzyme was found by Dobrowsky et al[35] to be activated in vitro by ceramide. Additional studies suggested that the B subunit contained the ceramide binding site since dimeric and catalytic subunit forms were insensitive.[36] However, subsequent studies by Law and Rossie[37] suggested that the catalytic subunit contained the ceramide binding site. It has previously been reported that okadaic acid, a potent inhibitor of PP2A, induces maturation. Therefore, this is not likely the target enzyme of ceramide in oocyte maturation. A ceramide activated, proline directed protein kinase has been described.[38] To date, this enzyme has not been further characterized. Protein kinase ζ has been reported to be a target of ceramide.[39] However, studies have shown that antisense to PKC ζ or dominant negative PKC ζ blocks insulin but not progesterone-induced GVBD, suggesting that this enzyme is not the primary target for ceramide in progesterone-induced maturation.[40] Recently, ceramide was reported to bind to and activate the serine/threonine protein kinase c-raf-1.[41] C-mos is a functional homologue of raf found primarily in germ cells. It is possible that ceramide might directly bind to and activate mos, thereby leading to maturation. Our results imply that mos must be present for ceramide to induce maturation. This would suggest that ceramide either directly or indirectly triggers the translation of mos or activates mos. In the latter case, it is possible that ceramide could function to localize mos to the membrane where it is activated by additional protein kinases. Further studies will be necessary to define the target(s) of ceramide in oocytes.

References

1. Maller JL. Interaction of steroids with the cyclic nucleotide system in amphibian oocytes. Adv Cyclic Nucleotide Res 1983; 15:295-302.
2. Finidori-Lepicard J, Schorderet-Slatkine S, Hanoune J, Baulieu EE. Progesterone inhibits membrane-bound adenylate cyclase in Xenopus laevis oocytes. Nature 1981; 292:255-257.
3. Jordana X, Allende CC, Allende JE. Guanine nucleotides are required for progesterone inhibition of amphibian oocyte adenylate cyclase. Biochem Int 1981; 3:527-532.
4. Bornslaeger EA, Mattei P, Schultz RM. Involvement of cAMP-dependent protein kinase and protein phosphorylation in regulation of mouse oocyte maturation. Dev Biol 1986; 114:453-462.
5. Maller JL, Krebs EG. Progesterone-stimulated meiotic cell division of Xenopus oocytes: Induction by regulatory subunit and inhibition by catalytic subunit of adenosine 3':5'-monophosphate-dependent protein kinase. J Biol Chem 1977; 252:1712-1718.
6. Smith LD, Ecker RE. The interaction of steroids with Rana pipiens oocytes in the induction of maturation. Dev Biol 1971; 25:232-247.
7. Godeau JF, Schorderet-Slatkine S, Hubert P, Baulieu EE. Progesterone-induced meiosis in Xenopus laevis oocytes: A role for cAMP at the "maturation-promoting factor" level. Proc Natl Acad Sci USA 1978; 75:2353-2357.
8. Stith BJ, Maller JL. Induction of meiotic maturation by 12-O-tetradecanoylphorbol 13-acetate. Exp Cell Res 1987; 169: 514-523.
9. Chein EJ, Morrill GA, Kostellow AB. Progesterone-induced second messengers at the onset of meiotic maturation in the amphibian oocyte: Interrelationships between phospholipid N-methylation, calcium and diacylglycerol release, and inositol phospholipid turnover. Mol Cell

Endo 1991; 81:53-67.
10. Kostellow AB, Ma G-Y, Morrill GA. Steroid action at the plasma membrane: Progesterone stimulation of phosphatidylcholine-specific phospholipase C following release of the prophase block in amphibian oocytes. Mol Cell Endo 1993; 92:33-44.
11. Carnero A, Lacal JC. Phospholipase-induced maturation of Xenopus laevis oocytes: Mitogenic activity of generated metabolites. J Cell Biochem 1993; 52: 440-448.
12. Varnold RL, Smith LD. The role of protein kinase C in progesterone-induced maturation. In: Davidson E, Ruderman J, Posakony J, eds. Developmental Biology, UCLA Symposia on Molecular and Cellular Biology, New Series, Vol. 125. New York: Alan R Liss, 1990:1-7
13. Varnold RL, Smith LD. Protein kinase C and progesterone-induced maturation in Xenopus oocytes. Development 1990; 109:597-604.
14. Hannun YA. The sphingomyelin cycle and the second messenger function of ceramide. J Biol Chem 1994; 269: 3125-3128.
15. Okazaki T, Bell RM, Hannun YA. Sphingomyelin turnover induced by vitamin D3 in HL-60 cells: Role in cell differentiation J Biol Chem 1989; 264:19076-19080
16. Jayadev S, Liu B, Bielawska AE, Lee JY, Nazaire F, Pushkareva MYu, Obeid LM, Hannun YA. Role for ceramide in cell cycle arrest. J Biol Chem 1995; 270: 2047-2052.
17. Sheela Rani CS, Abe A, Chang Y, Rosenzweig N, Saltiel AR, Radin NS, Shayman JA. Cell cycle arrest induced by an inhibitor of glucosylceramide synthase: Correlation with cyclin-dependent kinases. J Biol Chem 1995; 270:2859-2867.
18. Obeid LM, Linardic CM, Karolak LA, Hannun YA. Programmed cell death induced by ceramide. Science 1993; 259:1769-1771.
19. Jarvis WD, Kolesnick RN, Fornari FA, Traylor RS, Gewitz DA, Grant S. Induction of apoptotic damage and cell death by activation of the sphingomyelin pathway. Proc Natl Acad Sci USA 1994; 91:73-77.
20. Hannun YA, Loomis CR, Merrill AH, Bell RM. Sphingosine inhibition of protein kinase C activity and of phorbol dibutyrate binding in vitro and in human platelets. J Biol Chem 1986; 261: 12604-12609.
21. Strum JC, Swenson KI, Turner JE, Bell RM. Ceramide triggers meiotic cell cycle progression in Xenopus ooyctes: A potential mediator of progesterone-induced maturation. J Biol Chem 1995; 270:13541-13547.
22. De Smedt V, Rime H, Jessus C, Ozon R. Inhibition of glycosphingolipid synthesis induces p34cdc2 activation in Xenopus oocyte. FEBS Lett 1995; 375:249-253.
23. Mulner-Lorillon O, Belle R, Cormier P, Drewing S, Minella O, Poulhe R, Schmalzing G. Brefeldin A provokes indirect activation of cdc2 kinase (MPF) in Xenopus oocytes, resulting in meiotic cell division. Dev Biol 1995; 170:223-229.
24. Smith LD. The induction of oocyte maturation: transmembrane signaling events and regulation of the cell cycle. Development 1989; 107:685-699.
25. Preiss JE, Loomis CR, Bell RM, Niedel JE. Quantitative measurement of sn-1,2-diacylglycerols. Methods Enzymol 1987; 141:294-300.
26. Wasserman WJ, Masui Y. Effects of cycloheximide on a cytoplasmic factor initiating meiotic maturation in Xenopus oocytes. Exp Cell Res 1975; 91:381-388.
27. Sagata N, Oskarsson M, Copeland T, Brumbaugh J, Van de Woude GF. Function of c-mos proto-oncogene product in meiotic maturation in Xenopus ooyctes. Nature 1988; 335:519-525.
28. Kanki JP, Donoghue DJ. Progression from meiosis I to meiosis II in Xenopus oocytes requires de novo translation of the mos protooncogene. Proc Nat Acad Sci USA 1991; 88:5794-5798.
29. Radin NS, Shayman JA, Inokuchi J. Metabolic effects of inhibiting glucosylceramide synthesis with PDMP and other substances. Adv Lipid Res 1993; 26:183-213.
30. Donaldson JG, Finazzi D, Klausner RD. Phospholipase-induced maturation of Xenopus laevis oocytes: Mitogenic activity of generated metabolites. Nature 1992; 360:350-352.
31. Linardic CM, Jayadev S, Hannun YA.

Activation of the sphingomyelin cycle by brefeldin A: effects of brefeldin A on differentiation and implications for a role for ceramide in regulation of protein trafficking. Cell Growth and Diff 1996; 7:765-774.

32. Schroeder JJ, Crane HM, Xia J, Liotta DC, Merrill AH Jr. Disruption of sphingolipid metabolism and stimulation of DNA synthesis by fumonisin B1: a molecular mechanism for carcinogenesis associated with Fusarium moniliforme . J Biol Chem 1994; 269:3475-3481.
33. Weigmann K, Schutze S, Machleidt T, Witte D, Kronke M. Functional dichotomy of neutral and acidic sphingomyelinases in tumor necrosis factor alpha signaling. Cell 1994; 78:1005-1015.
34. Lipsky NG, Pagano RE. A vital stain for the Golgi apparatus. Science 1985; 228:745-747.
35. Dobrowsky RT, Hannun YA. Ceramide stimulates a cytosolic protein phosphatase. J Biol Chem 1992; 267:5048-5051.
36. Dobrowsky RT, Kamibayashi C, Mumby MC, Hannun YA. Ceramide activates heterotrimeric protein phosphatase 2A. J Biol Chem 1993; 268:15523-15530.
37. Law B, Rossie S. The dimeric and catalytic subunit forms of protein phosphatase 2A from rat brain are stimulated by C2-ceramide. J Biol Chem 1995; 270:12808-12813.
38. Mathias S, Dressler KA, Kolesnick RN. Characterization of a ceramide-activated protein kinase: stimulation by tumor necrosis factor alpha. Proc Nat Acad Sci USA 1991; 88:10009-10013.
39. Lozano J, Berra E, Municio MM, Diaz-Meco MT, Dominguez I, Sanz L, Moscat J. Protein kinase C zeta isoform is critical for κB-dependent promoter activation by sphingomyelinase. J Biol Chem 1994; 269:19200-19202.
40. Berra E, Diaz-Meco MT, Dominguez I, Municio MM, Sanz L, Lozano J, Chapkin RS, Moscat J. Protein kinase C zeta isoform is critical for mitogenic signaling. Cell 1993; 74:555-563.
41. Huwiler H, Brunner J, Hummel R, Vervoordeldonk M, Stabel S, van den Bosch H, Pfeilschifter J. Ceramide-binding and activation defines protein kinase c-raf as a ceramide-activated protein kinase. Proc Natl Acad Sci USA 1996; 93:6959-6963.

CHAPTER 5

Ceramide, Aging and Cellular Senescence

Joanna Y. Lee and Lina M. Obeid

INTRODUCTION

Cellular senescence is defined as the finite life span of mammalian cells in culture and has been investigated for the last thirty years. Research employing mammalian cells as well as cells from lower organisms such as *C. elegans* and yeast has established several models to explain the altered cellular functions in senescence. One model proposes that cellular senescence is associated with accumulated DNA damage and reduced DNA repair function. The other model argues that alterations of genetic control programs could lead to cellular senescence. There is evidence to support both hypotheses, yet the underlying mechanisms of cellular senescence are still not clear.

Sphingolipids have long been considered as stable structural components of cell membranes. Hannun and colleagues[1] provided the first evidence of sphingolipids as signaling molecules by demonstrating that sphingosine, an endogenous sphingolipid metabolite, reversibly inhibits protein kinase C (PKC). Subsequent discovery of the existence of a sphingomyelin cycle and sphingolipid-mediated biology has suggested a potential role for ceramide in cell differentiation,[2] programmed cell death,[3] cell cycle arrest,[4] and cellular senescence.[5] This chapter will review the literature with emphasis on the genetic control of cellular senescence, the current understanding of the role of lipid-mediated biology involving cell growth inhibition, tumor suppressor activation and cellular senescence. This will be followed by a brief discussion on the future directions in the study of sphingolipids and cellular senescence.

CELLULAR SENESCENCE: MECHANISMS AND REGULATION

Normal cells in culture are unable to proliferate indefinitely.[6,7] The finite replicative life span of cells in culture and the limited number of cell divisions is defined by a process known as cellular senescence. Senescent cells are viable and metabolically active for a long time if maintained in culture medium with regular changes.[8] Phenotypically, these cells show increased cell size, reduced capacity to incorporate [^{3}H]thymidine, decreased saturation density, altered sensitivity to cell contact, and they resemble terminally differentiated cells.[8,9]

Sphingolipid-Mediated Signal Transduction, edited by Yusuf A. Hannun.

Senescent cells also have reduced rates of protein synthesis and degradation and increased lysosome biogenesis.[10]

Relevance to the In Vivo Process of Aging

Most studies on cellular senescence have been carried out using cell culture. An important question is whether studies on senescence in vitro have any relevance to the mechanisms of aging in animals. An enormous amount of studies have demonstrated that characteristic senescence changes in vivo are expressed in cell culture.[8,9] Cells cultured from aged donors tend to senesce more rapidly than cells from young donors.[11] Cells from short-lived species usually have a shorter replicative life span than cells from long-lived species.[12] Cells obtained from patients with Werner's syndrome, a premature aging disorder[13,14] have a reduced proliferative capacity when compared to cells from age-matched controls. Similar alterations in the regulation of some genes such as c-fos and heat shock protein 70 have been observed in culture and in vivo.[8] Recently, the expression of a neutral β-galactosidase activity was found in senescent cells but not in young proliferating cells nor in serum deprived young cells.[15] Importantly, this marker was also found to be present in aging skin in vivo. Thus, this β-galactosidase activity provides an additional tool to differentiate senescent cells from quiescent or terminally differentiated cells in tissue. Taken together, studies in vitro could provide a useful window to the understanding of senescence processes in vivo.

Genetic Control of Cellular Senescence

Senescence is a dominant process as demonstrated by cell fusion experiments.[16] The resultant hybrids of proliferating prosenescent cells and senescent cells acquire a finite replication capacity. Similarly, when immortal tumor cells are fused to senescent cells, DNA replication is inhibited.[8] These results suggest immortality can result from inactivation of senescence-related genes. Therefore, lesions in different genes of immortal cells can complement one another resulting in the production of hybrids with a finite replicative life span. Immortal cells with lesions in the same genes cannot complement, and the resulting hybrids remain immortal. Currently, four complementation groups have been assigned from fusion analysis of many different immortal human cell lines implying that senescence is controlled by multiple gene pathways.[17] Studies on introducing single genes thought to be dysregulated in senescence such as c-fos and CDC2 demonstrated that these genes are unable to induce DNA synthesis in senescent cells. These studies are consistent with the multiple gene pathway theory of senescence.[18,19]

Senescent cells characteristically lack the immediate early gene c-fos, which is a component of the AP-1 transcription factor.[20] Consequently, they are lacking in AP-1 activation.[21] The repression of other early response genes such as Id1 and Id2 which encode negative regulators of basic helix-loop-helix transcription factors[22,23] may also contribute to the failure of senescent cells to proliferate.

Molecular Control of Cellular Senescence

Biochemical characterization has revealed that senescent cells are blocked in the late G1 phase of the cell cycle and cannot enter S phase upon growth factor stimulation. A number of genes that are normally expressed in late G1 or at the G1/S boundary are not expressed in senescent cells. Examples of such genes are thymidine kinase, thymidylate synthetase, dihydrofolate reductase, and replication-dependent histones.[8] The repression of the transcription factor E2F activity in senescent cells may be partially responsible for the regulation of some of the genes mentioned above.

The expression of DNA tumor viral genes such as simian virus 40 (SV40) T antigen, or the manipulation of tumor suppressor genes p53 and Rb by antisense oligonucleotides, can delay the cell senescence process.[24] SV40 T antigen is able to induce senescent cells to enter S phase, but cells are unable to enter mitosis,[25,26] suggesting that T antigen stimulated senescent cells appear to be blocked at the G2/M boundary of the cell cycle. Thus, senescent cells fail to proliferate by arresting the cells at both G1 and G2/M phases of the cell cycle.

Senescent cells are unable to phosphorylate the retinoblastoma (Rb) protein which is a critical component in the regulation of cell cycle progression. The Rb protein is differentially phosphorylated during the cell cycle and cellular differentiation.[27-29] Immunoprecipitation of Rb protein followed by SDS-PAGE analysis reveals multiple bands which reflect different phosphorylated forms of the protein. Hypophosphorylated forms of Rb protein predominate in the G1 phase of cell cycle it becomes phosphorylated as cells enter S phase it remains phosphorylated during mitosis and returns to the hypophosphorylated form as cells exit from mitosis.

Rb function during the cell cycle progression is regulated by cyclin and cyclin-dependent kinase (CDK) complexes.[30-32] Studies have demonstrated that senescent cells underexpress cyclins A and B proteins, CDC2 and CDK2 mRNA.[19,33] Senescent cells also have decreased cyclin D- and E-dependent CDK activity.[34] Since the E2F transcription factor normally binds to Rb, p107 and p130 proteins, and phosphorylation of Rb protein releases the bound E2Fs which then trigger the expression of genes including cyclin A and CDC2,[35] the repression of cyclin A and CDC2 may reflect the alteration of E2F activity. Cyclin A and CDK2 are critical for DNA synthesis and G1-S transition,[36] and cyclin B and CDC2 are important for regulating cell cycle progression through G2.[37] Furthermore, CDK2 kinase is a positive regulator of CDC2-cyclin B complex.[38] These findings suggest that a defect in the CDK-cyclin pathway in senescence may be responsible for the subsequent Rb hypophosphorylation and cell cycle arrest. In addition, senescent cells overexpress CDK cyclin inhibitors such as p21[39] and p16 which have been postulated to bind to CDK-cyclin complexes and inactivate their kinase activity. Therefore, overexpression of p21 in senescent cells may induce dephosphorylation of Rb by inhibiting CDK activities.

An emerging hypothesis accounting for the loss of DNA synthetic capacity in senescent cells is the shortening of telomeres. Telomere length shortens with increased population doubling of cells in culture and with organismic age.[40,41] In contrast, immortal cells express the enzyme telomerase and preserve telomere length.[42] This suggests that limited replicative capacity is associated with loss of genetic information at or near the ends of chromosomes and that the lack of telomerase activity in normal cells may be responsible for the shortening of telomere length.[43] This hypothesis does not adequately explain some of the observations made in related studies. First, the existence of telomerase negative immortal cell lines has been reported.[44,45] Second, no phenotypic effect was found in immortal cells when treated with chemical inhibitors to reduce telomerase activity to levels of normal cells.[44] Third, some normal cells also express telomerase with no effect on preserving their telomere length.[46,47] Fourth, some senescent hybrids continue to express telomerase.[45] These results raise the question of whether shortening of telomere length is a primary cause or an end result of senescence. Telomere length shortening represents a characteristic of senescence in mammalian cells, since no such change was found during aging of *Saccharomyces cerevisiae*.[48] Nonetheless, the telomere shortening hypothesis remains an attractive model for aging research.

Other biochemical parameters include alterations of certain protease inhibitors and collagenase in senescent cells.[8,9] However, the specificity, function and mechanisms of these alterations in the process of senescence remain to be elucidated.

Apoptosis and Senescence

The cell cycle activation and apoptosis (or programmed cell death) appear to be two contradictory cellular responses, and yet, they have attracted a lot of attention in the past few years. Although senescent cells are arrested at G1 phase of the cell cycle, very little is known about the control of cellular senescence and regulation of programmed cell death. Wang and colleagues[49] have reported that senescent cells overexpress Bcl-2 protein which has been shown to inhibit apoptosis in response to a number of extracellular agents,[50,51] whereas the immortalized mouse 3T3 fibroblasts express low or undetectable levels of bcl-2 protein accounting for the resistance of senescent cells to apoptosis. Numerous studies relating to cell cycle have implicated the participation of cyclin A and cyclin E in apoptosis.[52-54] Premature activation of $p34^{cdc}$ kinase has been shown to be necessary in apoptosis of a transformed T cell line.[55] Since senescent cells are arrested at G1 phase of the cell cycle and they have low or undetectable levels of cyclin A and CDC2 kinase activity, the inability of senescent cells to proliferate may represent a protection mechanism from programmed cell death. This hypothesis is supported by the finding that Rb protein can overcome p53-mediated or ionizing radiation induced apoptosis.[56,57] Furthermore, a recent study has demonstrated that differentiated myotubes were resistant to apoptosis through the induction of CDK inhibitor p21.[58] These data suggest that activation of Rb and induction of p21 may serve to protect cells from programmed cell death. Consistent with these findings, senescent cells have predominantly dephosphorylated Rb protein and elevated expression of p21 which may contribute to the nonproliferating and antiapoptotic character of senescent cells.

DNA Damage and Cellular Senescence

Our knowledge on the mechanisms of DNA damage and cellular senescence is sketchy. DNA damage has been accepted as one of the possible mechanisms underlying the aging process. Oxidative DNA damage has been proposed to be the primary cause of aging.[59] This was supported by the observation that the formation of free radicals generated during oxygen metabolism is inversely correlated with life span.[60] It is highly likely that the combination of deterioration of DNA repair function and direct damage on DNA could alter the genetic control program and cause cellular dysfunction.

In an effort to determine the relationship between oxidative damage and cellular senescence, Chen et al[61] have reported that reduction of oxidative damage delays senescence. When human diploid fibroblasts were cultured under a reduced O_2 concentration (3%), they achieved 50% more population doublings during their life span as compared to those cells cultured at the normal O_2 concentration (20%). The extension of life span was associated with an increased rate of cell growth, increased saturation density, and delayed onset of senescence. PBN (α-phenyl-t-butyl nitrone), a spin-trapping agent that acts as an antioxidant, delayed senescence and extended the replicative life span of IMR90 HDF. These findings support the hypothesis that oxidative DNA damage is associated with the cellular senescence of HDF.

Other Model Systems in Cellular Senescence

Other model systems used to study cellular senescence include yeast, fungi, *C. elegans* and *Drosophila*. These species provide excellent genetic systems for dissection of aging processes and life span determination. For example, in the budding yeast, starvation-resistant strains display different sensitivity to death and have relatively longer life spans.[62] In fungi, on the other hand, the mechanisms of senescence involve mitochondrial DNA rearrangements. Perhaps they are useful for examining the mitochondrial DNA alterations which are critical for longevity and homeostasis in humans.[63]

The *C. elegans* model has been developed for genetic analysis on the functions of specific age-related genes.[64] Current efforts include isolating age mutants selectively for extended life span. One of these mutants, age-1 is associated with an increased activity of the cytoplasmic Cu/Zn superoxide dismutase and catalase[65,66] and a lower rate of accumulation of mitochondrial DNA deletions.[67] Thus, these studies support the hypothesis that oxidative damage may induce senescence and limit life span.[61]

Owing to its short life span, small body size, and ease of assay, *Drosophila* remains an attractive model system for the study of the genetics of aging.[68] Consistent with *C. elegans* studies, it has been demonstrated that *Drosophila* strains which contain extra copies of genes coding both superoxide dismutase and catalase have increased life spans.[69] Despite the differences between *Drosophila* and mammals, adult *Drosophila* show age-dependent cellular, morphological and biochemical changes also observed in mammals.[70]

In animals, the aging process includes changes in proliferative and postmitotic cells. Cell culture studies do not generate information about the senescence process of postmitotic cells. Since adult *Drosophila* and *C. elegans* are composed entirely of postmitotic cells, these model systems may provide additional understanding of the senescence process. However, the critical issue is always whether information generated from these model systems can be extrapolated to humans.

SPHINGOLIPIDS IN AGING

Despite the significant advancement in our understanding of cellular senescence, the picture of signal transduction pathways in cellular senescence is far from complete and our knowledge of lipid-mediated signaling pathways in senescence remains scarce. The majority of studies to date that have focused on sphingolipids and aging have been predominantly related to composition aspects of sphingolipid and have emphasized the changes in membrane fluidity associated with aging.

In an early study using mitochondrial membranes isolated from the myocardium of young and old male Long-Evans rats, Lewin and Timiras[71] have demonstrated marked increases in sphingomyelin, cardiolipin (diphosphatidylglycerol) and cholesterol contents in old rats. Thus, age-related decrease in membrane fluidity and energy transduction of the mitochondrial membrane may contribute to the impairment of cardiac function with aging. Using cells cultured from newborn rat heart myocytes, Yechiel and Barenholz[72] were able to show that the major difference between young cells (5 to 6-day-old) and old cells (14 to 15-day-old) was the alteration of phosphatidylcholine to sphingomyelin mole ratio. In parallel with the morphological changes and the reduction of beating rate in old cells, sphingomyelin and cholesterol levels were significantly increased while phosphatidylcholine was reduced by 15-20%. In addition, treatment of old cells with phosphatidylcholine liposomes, which increased the PC/SM ratio and decreased cholesterol levels, reversed the biological and biochemical changes associated with old cells. Therefore, modification of cellular lipid composition may contribute to the altered cellular functions seen in aging.

Similar studies have extended to other systems such as the central nervous system where regional changes in lipid composition have been observed.[73-75] In platelets, modification of lipid composition has also been implicated in the onset of atherosclerosis and other age-related diseases.[76] Alterations of lipid composition associated with aged epidermis, however, have not been conclusive. One study using aged human subjects showed a significant increase in ceramide levels in the stratum corneum of females.[77] In contrast, utilizing a senescent murine model, Ghadially and colleagues[78] found that although the total lipid content was decreased in the stratum corneum of aged mice, the content of sphingolipids including ceramide did not change significantly under basal conditions. These studies raise the possibility that the impairment of barrier integrity and barrier repair to external insults in aged skin is responsible for the altered cellular functions. Studies based on changes in lipid composition may oversimplify the complicated processes in aging. Nonetheless, these results provided the primary evidence connecting sphingolipids and aging.

The elucidation of the sphingomyelin cycle provided clear evidence that sphingolipids may function as signaling molecules in cell regulation. Recently, some studies have begun to indicate changes in components of this signaling pathway in cellular and organismal aging. One study using rat liver plasma membrane found that neutral sphingomyelinase activity was increased between 1-13 months.[79] In parallel to this change, the sphingomyelin level was reduced and phosphatidylcholine was augmented. It was not elucidated, however, what the ultimate impact of increased sphingomyelin hydrolysis was on ceramide formation and downstream targets. An earlier study done by the same group demonstrated that phospholipase A_2 activity was increased in rat liver plasma membrane between 3-16 months old.[80] Interestingly, a recent study has found that stimulation of HL-60 cells by TNFα induces phospholipase A2 activation and generates arachidonic acid, which subsequently activates a neutral magnesium-independent sphingomyelinase responsible for the loss of membrane sphingomyelin, ceramide formation, and growth inhibiton.[81] Thus, it is likely that the generation of arachidonate metabolites is responsible for the subsequent change in sphingomyelinase activity seen in aged animals.

In another study, Tamiya-Koizumi and Kojima[82] examined the effects of various phospholipids on the neutral sphingomyelinase in the dilipidized plasma membrane. Their studies on the plasma membrane of rat ascites hepatoma showed that phosphatidylserine significantly activated magnesium-dependent neutral sphingomyelinase. Both phosphatidic acid and phosphatidylinositol may also be activators of neutral sphingomyelinase, whereas phosphatidylethanolamine and phosphatidylcholine were not effective. Thus, membrane acidic phospholipids may be important in the regulation of neutral sphingomyelinase activity and the cellular function.

SPHINGOLIPIDS IN SENESCENCE

Induction of the Senescent Phenotype

In an attempt to identify the role of ceramide in cellular senescence, Venable and colleagues[5] have systematically examined the ability of ceramide to induce a senescent phenotype. It was shown that endogenous levels of ceramide increased significantly and specifically when Wi-38 human diploid fibroblasts reached the senescent stage. Concomitant to the elevation of endogenous ceramide levels in senescent cells, neutral sphingomyelinase activity was increased up to 10-fold. The changes in sphingomyelinase activity were observed only in senescent cells, but not in quiescent cells following serum withdrawal or contact inhibition. Thus, activation of sphingomyelinase and ceramide formation may represent a specific signaling pathway to cellular senescence and not to quiescence. Addition of exogenous ceramide to young Wi-38 HDF produced endogenous levels of ceramide comparable to that seen in senescent cells, as determined by metabolic labeling studies. Low micromolar concentrations of ceramide induced growth inhibition of young Wi-38 HDF as measured by thymidine incorporation. When young Wi-38 HDF were treated with ceramide, they resembled senescent cells morphologically. Moreover, ceramide induced Rb dephosphorylation and inhibited serum-induced transcription factor AP-1 activity, mimicking the biochemical phenotype in senescence.[5]

Following the initial studies of HL-60 human promyelocytic cell line, it was found that micromolar concentrations of ceramide produced similar growth inhibition and induction of differentiation in a number of cell types including Jurkat,[83] T9 glioblastoma cells,[84] U937, and Molt-4 cells.[85] This effect of ceramide mimics senescent cells which have reduced capacity for DNA synthesis and resemble terminal differentiation. These findings extended the significant role of the sphingomyelin cycle and stimulated numerous studies on ceramide in other areas of cell regulation such as cell cycle arrest and tumor suppressor activation.

Ceramide and Cell Cycle

Since senescent cells are blocked at the G1 phase of the cell cycle and ceramide inhibits cell growth, one approach to identify the role of ceramide in cellular senescence has been examining the role of ceramide in cell cycle. The progression of cell cycle is tightly regulated by multiple genes and gene products in response to proliferation and antiproliferation. Normal cells require various growth factors for continued proliferation, and perturbations such as serum withdrawal would induce cell cycle arrest and cell death. Since ceramide has been shown to inhibit cell growth and induce differentiation, serum deprivation has been utilized as a model system to understand the antiproliferative responses. Molt-4 cells were arrested in the G0/G1 phase of the cell cycle after serum withdrawal as determined by fluorescence-activated cell sorting analysis of cellular DNA content.[4] Growth arrest was observed at 12 hours with maximal effect occurring at 48 hours with more than 90% of viable cells in G0/G1. Concomitant to cell cycle arrest, serum deprivation elicited sphingomyelin hydrolysis and ceramide generation in the cells, perhaps through the activation of a particulate, magnesium-dependent, neutral sphingomyelinase. Interestingly, the effects of ceramide on cell cycle arrest was even more evident than serum deprivation. Exposure of Molt-4 cells to C_6-ceramide produced a dose dependent cell cycle arrest with an earlier onset than serum deprivation whereas dihydroceramide, a closely related lipid molecule, was not effective.

Additional studies were carried out in Wi-38 HDF. Lee and colleagues have found that micromolar concentrations of ceramide arrest young Wi-38 cells at both G1 and G2/M phases of the cell cycle (Lee and Obeid, unpublished data). Moreover, treatment of ceramide resulted in a concentration and time dependent inhibition of CDC2 and CDK2 kinase activities. The inhibition of cyclin-dependent kinase activity appeared to be contributed by the alteration of cyclin A expression, since CDK inhibitors p21 and p16 seemed not to be affected by treatment with ceramide. The change in cyclin-dependent kinase activity may lead to the subsequent inhibition of RB phosphorylation (see below). Thus, following ceramide treatment, young Wi38 HDF resembled senescent cells in cell cycle progression.

A separate study has provided further evidence for ceramide induced cell cycle arrest.[86] Exposure to NIH 3T3 cells overexpressing insulin-like growth factor-1 (IGF-1) receptors with PDMP (threo-1-phenyl-2-decanoylamino-3-morpholino-1-propanol), a glucosylceramide synthase inhibitor which leads to increased cellular ceramide concentration, resulted in an arrest of the cell cycle at both G1 and G2/M transitions. The effects of PDMP treatment was time dependent and reversible and was presumably through the increased cellular ceramide formation. These findings supported the hypothetical role of ceramide in cell cycle arrest.

Ceramide and Tumor Suppression

The tumor suppressor retinoblastoma gene product (Rb) is one of the central players in the regulation of cell cycle progression as discussed in above.[3] Senescent cells contain predominantly the dephosphorylated form of Rb and cannot undergo RB phosphorylation upon growth factor stimulation. To investigate the molecular mechanisms of ceramide on cell cycle arrest and induction of senescent phenotype, the state of Rb phosphorylation following ceramide treatment of Wi-38 human diploid fibroblast was determined.[5] Addition of ceramide resulted in a specific concentration and time dependent induction of Rb dephosphorylation in Wi-38 HDF. These findings suggested that Rb functions as a downstream target for ceramide in the regulation of cell growth.

To substantiate the effect of ceramide on cell cycle arrest, the induction of Rb dephosphorylation following ceramide treatment on Molt-4 cells was examined.[85] Addition of ceramide inhibited hyperphosphorylation and induced dephosphorylation of Rb. The specific effect of ceramide on Rb was concentration- and time-dependent, and was independent of sphingosine. To determine whether a functional Rb is necessary in mediating ceramide's effect on cell cycle, studies were performed using different cell lines expressing or lacking Rb. Cells devoid of Rb were resistant to ceramide treatment whereas growth of normal Molt-4 cells was significantly inhibited by ceramide. Moreover, cells overexpressing either E1A of adenovirus, which binds to Rb and inactivates it, or large tumor antigen of SV40 were protected from growth inhibition and cell cycle arrest mediated by ceramide. In contrast, cells transfected with the mutant large T antigen containing functional Rb was potently inhibited by ceramide.[85] These findings indicate that the effects of ceramide on Rb may represent a common mechanism in mediating growth suppression seen in senescent cells.

Effect of Ceramide on Protein Kinase C and Cellular Senescence

Protein kinase C (PKC) has been implicated as one of the critical components of signaling pathways regulating cell growth. Protein kinase C activation elicits a variety of cellular responses by phosphorylating target proteins on serine/threonine residues. Specific inhibitors of PKC block cell proliferation, supporting a role for PKC in the proliferation pathway.[87] Alterations in protein kinase C expression and activity correlating with impaired cellular functions have been reported in several senescent model systems. PKC activity was reduced in senescent blood vessels, and PMA-stimulated PKC translocation was significantly altered in senescent aorta, accounting for the reduced contractile responsiveness.[88] Following the initial studies elucidating a role of ceramide in cellular senescence, Venable and colleagues[89] have identified a defect in the DAG/PKC pathway in cellular senescence. Unlike young HDF, senescent cells failed to translocate PKC in response to serum stimulation. This result was consistent with an earlier finding that translocation of PKC activity in cell senescence is defective.[90] Further studies revealed that the DAG signal is defective in senescence, since PMA was equally active in inducing PKC translocation in young and senescent HDF.[89]

In support of the finding that the DAG signal is defective in senescence, Lipschitz and colleagues[91] showed that neutrophils from old donors generate much less DAG and inositol triphosphate (IP_3) than neutrophils from young donors in response to formyl-methionyl-leucyl-phenylalanine (FMLP). In addition, neutrophils from old donors showed significant reduction in the concentration of phosphatidylinositol (PI), phosphatidylinositol 4-monophosphate (PIP), and phosphatidylinositol 4,5-bisphosphate (PIP_2) in response to FMLP stimulation, accounting for the decrease of IP_3 and DAG levels. Therefore, the balance between the two lipid-mediated pathways of signal transduction, namely the DAG/PKC and ceramide/CAPP pathways may have a profound effect on cellular regulation including cellular senescence.

In support of these findings, Chang and Huang[92] have demonstrated a significant reduction in PKC activation in old IMR fibroblasts, as determined by the degree of TPA-induced phosphorylation of myristolated alanine-rich C kinase substrate (MARCKS). In parallel to the reduction of PKC activity, the generation of diacylglycerol in response to growth factor stimulation also declined in aged fibroblasts. Selective alterations of the PKC isoforms in T cells from old adults have also been reported.[93] Therefore, regulation in PKC protein expression may contribute to the abnormal function in signal transduction in the elderly.

In an attempt to understand the mechanisms underlying the ceramide effect on Rb dephosphorylation, Lee and colleagues[94] demonstrated that ceramide indirectly inactivates PKCα activity. Activation of PKC resulted in Rb phosphorylation, suggesting one component of induction of Rb dephosphorylation by ceramide may be through inhibition of PKC-dependent pathway. The mechanisms responsible for such inhibition are currently unknown. Dobrowsky and Hannun[95] have identified a ceramide-activated protein phosphatase (CAPP) in vitro. Incubation of ceramide with crude cytosol from T9 glioma cells hydrolyzed [^{32}P] phosphohistone in a concentration and time dependent manner. The activation was specific to ceramide where dihydroceramide was inert. CAPP belongs to the PP2A heterotrimeric subfamily of the serine/threonine protein phosphatases and is potently inhibited by okadaic acid with an IC_{50} of 1-10 nM. Since okadaic acid reversed the action of ceramide on PKC activity inhibition,[94] this suggests that the activation of PP2A by ceramide may represent a global mechanism for the subsequent intracellular events. It is now evident that ceramide also inhibits CDK2 kinase activity (Lee and Obeid, unpublished observations). It is conceivable that inhibition of CDK2 kinase activity and ultimate growth suppression by ceramide reflects the existence of a complex array of PKC-CDK signaling pathways as seen in human vascular endothelial cells.[96]

Role of Ceramide on Phospholipase D and Cellular Senescence

Senescent cells have higher basal levels of DAG and are less responsive to serum stimulation as compared to young cells. More interestingly, phospholipase D (PLD) activity was inhibited in senescent cells in response to serum as determined by the incorporation of exogenous ethanol into phosphatidylethanol, a measure of the transphosphatidylation reaction of PLD. Addition of exogenous ceramide to young cells resulted in almost complete inhibition of DAG production in response to serum. Cells treated with C_6-ceramide failed to activate PLD in response to serum or PMA, mimicking senescent cells. These results suggest that ceramide may induce a senescent phenotype by inhibiting the production of DAG through PLD activation. In a cell-free system, it was found that PLD is activated by GTPγS. Addition of PKC activators synergistically enhanced activation of PLD by GTPγS while ceramide was able to inhibit this effect. Moreover, recombinant ARF and PKC were able to reconstitute membrane PLD activity. Ceramide completely abolished the synergistic activation by PKC, although ceramide was ineffective on ARF activable PLD. These results supported the hypothesis that ceramide interferes with PKC-mediated activation of PLD.[97]

CONCLUSIONS AND FUTURE DIRECTIONS

Clearly, cellular senescence represents a distinct and yet complicated biological phenomenon. The different models of cellular senescence could overlap and reinforce each other. For example, DNA damage accumulated with age may change the genetic program controlling cellular senescence. On the other hand, alterations of genetic program could lead to deficiency of DNA repair function, accumulate DNA damage and cause aging. Recent identification of Werner's syndrome gene as a DNA helicase has provided new evidence on the connection between DNA damage or mutations and the phenotype of premature aging. Recessive mutations in Werner's gene could lead to cellular defects. Certainly, identification of specific genes that are differentially expressed in senescent cells could shed some light on the regulation of senescence.

Cellular senescence is involved with inhibition of growth stimulatory genes and activation of growth inhibitory genes. Other factors regulating CDK activity such as CDK activating kinase and CDK phosphatase(s) may also play important roles in the regulation of cellular senescence. The challenge ahead is to understand what activates growth inhibitory genesand how, and the ultimate effect of gene regulation on cellular senescence in humans.

Oxidative damage appears to be associated with cellular senescence. Future studies addressing the changes in gene expression and activity in cells cultured under a low O_2 concentration in comparison with those seen in senescent cells would provide useful information on the basic biology, including senescence. Studies using model systems such as yeast, *Drosophila*, and *C. elegans* may also add to our knowledge

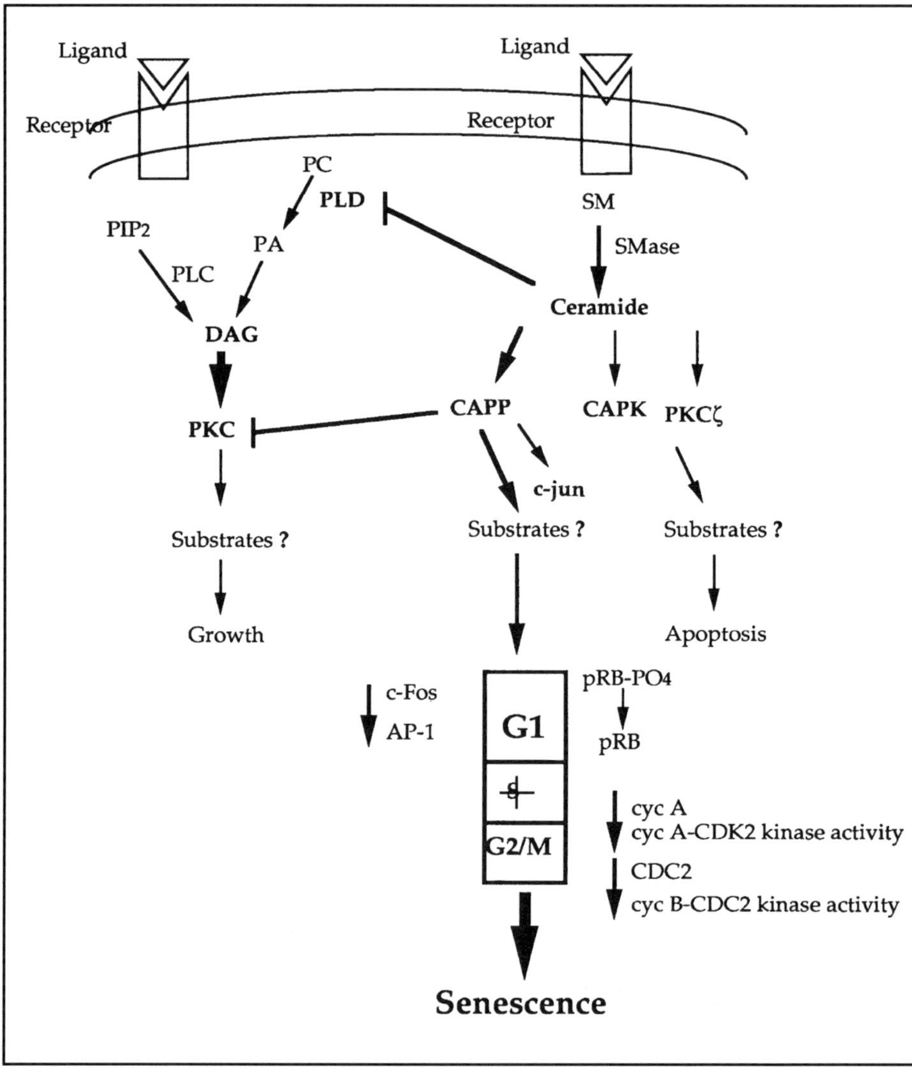

Fig. 5.1. Proposed diagramatic scheme for ceramide signaling. Exposure to extracellular ligands or stress signals leads to hydrolysis of sphingomyelin (SM) and formation of ceramide by sphingomyelinase (SMase). Once generated, ceramide mediates growth arrest, leads to cell death and induces senescence through activation of a protein phosphatase (CAPP), a protein kinase (CAPK) and Protein Kinase C ζ (PKCζ). In addition, ceramide induces Rb dephosphorylation, inhibits transcription factors AP-1 and c-Fos, arrests cells at G0/G1 and G2/M phases of the cell cycle, down regulates the expression of cyclin A (cyc A) and CDC2 and inhibits cyclin-dependent kinase (CDK2/CDC2) activities, mimicking cellular senescence. A number of extracellular ligands lead to the activation of phospholipase C (PLC) and phospholipase D (PLD) to generate diacylglycerol (DAG) which then activates PKC and regulates cell growth. Ceramide inhibits the PLD and PKC activities.

of the poorly understood mechanisms of senescence.

Sphingolipids and metabolites have emerged as signaling molecules regulating cell growth, differentiation, apoptosis and senescence. Owing to the complexity of sphingolipid species and different enzymatic pathways in metabolism, it is clear that the interactions between lipids and different signaling pathways could influence cellular functions including senescence. Understanding the interactions of different signaling pathways would disclose novel components involved in the regulation of cell growth.

The lipid second messenger, ceramide, is believed to be a key mediator in transducing various messages for cell function. Figure 5.1 represents a hypothetical scheme for the role of ceramide in modulating cell growth and cellular senescence. Little is known about the regulation of ceramide production and molecular mechanisms of ceramide action, particularly in cellular senescence. Further studies comparing the changes in gene expression and function in senescent cells and cells treated with ceramide would help to define the potential role of ceramide in cellular senescence and to gain insight on the fundamentals of the senescence process.

Despite the growing number of potential targets of ceramide being found, the lack of bona fide targets has limited our knowledge of ceramide function in cellular senescence. Once the true targets have been identified, the regulation of ceramide function and the subsequent intracellular signaling events can be evaluated.

Finally, it is highly likely the cross-talk between the proliferation and antiproliferation transduction pathways may regulate cell function. Tilting the balance between the DAG/PKC pathway and ceramide/CAPP pathway would alter cellular function. Further development of our understanding on the balance between these two pathways could provide new insight on senescene process and specific treatment for age-related diseases.

Acknowledgment

We thank Dr. Yusuf Hannun for helpful discussions and careful review of the manuscript, and Lisa Leonhardt for technical assistance.

References

1. Hannun YA, Loomis CR, Merrill AH Jr, Bell RM. Sphingosine inhibition of protein kinase C activity and of phorbol dibutyrate binding in vitro and human platelets. J Biol Chem 1986; 261: 12604-12609.
2. Okazaki T, Bell RM, Hannun YA. Sphingomyelin turnover induced by vitamin D_3 in HL-60 cells. Role in cell differentiation. J Biol Chem 1989; 264:19076-19080.
3. Obeid LM, Linardic CM, Karolak LA, Hannun YA. Programmed cell death induced by ceramide. Science 1993; 259: 1769-1771.
4. Jayadev S, Liu B, Bielawska AE, Lee JY, Nazaire F, Pushkareva MY, Obeid LM, Hannun YA. Role for ceramide in cell cycle arrest. J Biol Chem 1995; 270:2047-2052.
5. Venable ME, Lee JY, Smyth MJ, Bielawska A, Obeid LM. Role of ceramide in cellular senescence. J Biol Chem 1995; 270:30701-30708.
6. Hayflick, L, Moorhead PS. The serial cultivation of human diploid strains. Exp Cell Res 1961; 25:585-621.
7. Hayflick L. The limited in vitro lifetime of human diploid cell strains. Exp Cell Res 1965; 37:614-636.
8. Campisi J, Dimri G, Hara E. Control of replicative senescence. In: Schneider E, Rowe J, eds. Handbook of the Biology of Aging, Fourth Edition. New York: Academic Press, 1996:121-149.
9. Cristofalo VJ, Pignolo RJ. Replicative senescence of human fibroblast-like cells in culture. Physiological Reviews 1996; 73:617-638.
10. Stanulis-Praeger B. Cellular senescence revisited: A review. Mechanisms of Aging and Development 1987; 38:1-48.
11. Martin GM, Sprague CA, Epstein CJ. Replicative life-span of cultivated human cells. Effects of donor's age, tissue and genotype. Laboratory Investigation 1970; 23:86-92.

12. Goldstein S. Aging in vitro: growth of cultivated cells from the Galapagos tortoise. Exp Cell Res 1974; 83:297-302.
13. Goldstein S. Human genetic disorders that feature premature onset and accelerated progression of biological aging. In: Schneider EL, ed. The Genetics of Aging. New York: Plenum Press, 1978: 171-224.
14. Martin GM. Genetic syndromes in man with potential relevance to the pathology of aging. In: Bergsma D, Harrison DE, eds. Genetic Effects on Aging, Birth Defects: Original Article Series. (New York: Alan Liss, 1978:5-39.
15. Dimri GP, Lee X, Basile G, Acosta M, Scott G, Roskelley C, Medrano EE, Linskens M, Rubely I, Pereira-Smith O, Peacocke M, Campisi J. A Bio-marker that identifies senescent human cells in culture and in aging skin in vivo. Proc Natl Acad Sci 1995; 92:9363-9367.
16. Norwood TH, Pendergrass WR, Sprague CA, Martin GM. Dominance of the senescent phenotype in heterkaryons between replicative and post-replicative human fibroblast-like cells. Proc Natl Acad Sci USA 1974; 71:2231-2235.
17. Pereira-Smith OM, Smith JR. Genetic analysis of indefinite division in human cells: identification of four complementation groups. Proc Natl Acad Sci 1988; 85:6042-6046.
18. Dimri GP, Campisi J. Molecular and cell biology of replicative senescence. Cold Spring Harbor Symp Quant Biol 1994; 59:67-73.
19. Afshari CA, Vojta PJ, Annab LA, Futreal PA, Willard TB, Barrett JC. Investigation of the role of G1/S cell cycle mediators In cellular senescence. Exp Cell Res 1993; 209:231-237.
20. Seshadri T, Campisi J. c-*fos* repression and an altered genetic program in senescent human fibroblasts. Science 1990; 247:205-209.
21. Riabowol K, Schiff J, Gilman MZ. Transcription factor AP-1 activity is required for initiation of DNA synthesis and is lost during cellular aging. Proc Natl Acad Sci USA 1992; 89:157-161.
22. Benezra R, Davis RL, Lockshon D, Turner DL, Weintraub H. The Protein Id: a negative regulator of helix-loop-helix DNA binding proteins. Cell 1990; 61:49-59.
23. Sun XH, Copeland NG, Jenkins NA, Baltimore D. Id proteins Id1 and Id2 selectively inhibit DNA binding by one class of helix-loop-helix proteins. Molec Cell Biol 1991; 13:7874-7880.
24. Smith JR, Pireira-Smith OM. Replicative senescence: implications for in vivo aging and tumor suppression. Science 1996; 273:63-67.
25. Gorman SD, Cristofalo VJ. Reinitiation of cellular DNA synthesis in BrdU-selected nondividing senescent WI-38 cells by simian virus 40 infection. J Cell Physiol 1985; 125:122-126.
26. Ide T, Tsuyi Y, Ishibashi S, Mitsui Y. Reinitiation of host DNA synthesis in senescent human diploid cells by interaction with simian virus 40. Exp Cell Res 1983; 143:343-349.
27. Buckkovich K, Duffy LA, Harlow ED. The retinoblastoma protein is phosphorylated during specific phases of the cell cycle. Cell 1989; 58:1097-1105.
28. Chen P-L, Scully P, Shew J-Y, Wang JYJ, Lee W-H. Phosphorylation of the retinoblastoma gene product is modulated during the cell cycle and cellular differentiation. Cell 1989; 58:1193-1198.
29. DeCaprio JA, Ludlow JW, Lynch D, Furukawa Y, Griffin J, Piwnica-Worms H, Huang, C-M, Livingston DM. The product of the retinoblastoma susceptibility gene has properties of a cell cycle regulatory element. Cell 1989; 58:1085-1095.
30. Hinds PW, Mittnacht S, Dulic V, Arnold A, Reed SI, Weinberg RA. Regulation of retinoblastoma protein functions by ectopic expression of human cyclins. Cell 1992; 70:993-1006.
31. Dowdy SF, Hinds PW, Louie K, Reed SI, Arnold A, Weinberg RA. Physical interaction of the retinoblastoma protein with human D cyclins. Cell 1993; 73:499-511.
32. Ewen ME, Sluss HK, Sherr CJ, Matsushime H, Kato J-Y, Livingston DM. Functional interactions of the retinoblastoma protein with mammalian D-type cyclins. Cell 1993; 73:489-497.
33. Stein GH, Drullinger LF, Robetorye RS, Pereira-Smith OM, Smith JR. Senescent cells fail to express cdc2, cycA, and sycB

in response to mitogen stimulation. Proc Natl Acad Sci USA 1991; 88:11012-11016.
34. Dulic V, Drullinger LF, Lees E, Reed SI, Stein GH. Altered regulation of G_1 cyclins in senescent human diploid fibroblasts: accumulation of inactive cyclin E-Cdk2 and cyclin D1-Cdk2 complexes. Proc Natl Acad Sci USA 1993; 90:11034-11038.
35. Yamamoto, T, Nikaido T. Effect of tumor suppressors on cell cycle regulatory genes: RB suppresses p34/cdc2 expression and normal p53 suppresses cyclin A expression. Exp Cell Res 1994; 210:94-101.
36. Sherr CJ. Mammalian G1 cyclins. Cell 1993; 73:1059-1065.
37. Pines J. Cyclins and their associated cyclin-dependent protein kinases in the human cell cycle. Biochem Soc Trans 1993; 21:921-925.
38. Guadagno TM, Newport JW. CDK2 kinase is requied for entry into mitosis as a positive regulator of CDC2-cyclin B kinase activity. Cell 1996; 84:73-82.
39. Noda A, Ning Y, Venable SF, Pereira-Smith OM, Smith JR. Cloning of senescent cell-derived inhibitors of DNA synthesis using an expression screen. Exp Cell Res 1994; 211:90-98.
40. Harley CB, Futcher AB, and Greider CW. Telomeres shorten during ageing of human fibroblasts. Nature 1990; 345: 458-460.
41. Harley CB, Villeponteau B. Telomeres and telomerase in aging and cancer. Current Opinion in Genetics and Development 1995; 5:249-255.
42. Counter CM, Hirte HW, Bacchetti S, Harley CB. Telomerase activity in human ovarian carcinoma. Proc Natl Acad Sci 1994; 91:2900-2904.
43. Harley CB, Vaziri H, Counter CM, Allsopp RC. The telomere hypothesis of cellular aging. Exp Gerontol 1992; 27: 375-382.
44. Murnane JP, Sabatier L, Marder BA, Morgan WF. Telomere dynamics in an immortal human cell line. EMBO J 1994; 13:4953-4962.
45. Bryan TM, Englezou A, Gupta J, Bacchetti S, Reddel RR. Telomere elongation in immortal human cells without detectable telomerase activity. EMBO J 1995; 14:4240-4248.
46. Broccoli D, Young JW, DeLange T. Telomerase activity in normal and malignant hematopoietic cells. Proc Natl Acad Sci 1995; 92:9082-9086.
47. Counter CM, Gupta J, Harley CB, Lebe B, Bacchetti S. Telomerase activity in normal leukocytes and in hematologic malignancies. Blood 1995; 85:2315-2320.
48. D'Mello NP, Jazwinski SM. Telomere length constancy during aging of Saccharomyces cerevisiae. J Bacteriol 1991; 173:6709-6713.
49. Wang E, Lee, M-J, Pandey S. Control of fibroblast senescence and activation of programmed cell death. J Cell Biochem 1994; 54:432-439.
50. Hockenbery D, Nunez G, Milliman C, Schreiber RD, Korsmeyer SJ. Bcl-2 is an inner mitochondrial membrane protein that blocks programmed cell death. Nature 1990; 348:334-336.
51. Reed JC. Bcl-2 and the regulation of programmed cell death. J Cell Biol 1994; 124:1-6.
52. Meikrantz W, Gisselbrecht S, Tam SW, Schlegel R. Activation of cyclin A-dependent protein kinases during apoptosis. Proc Natl Acad Sci 1994; 91:3754-3758.
53. Hoang AT, Cohen KJ, Barrett JF, Bergstrom DA. Participation of cyclin A in Myc-induced apoptosis. Proc Natl Acad Sci 1994; 91:6875-6879.
54. Li CJ, Friedman DJ, Wang C, Metelev V, Pardee AB. Induction of apoptosis in uninfected lympocytes by HIV-1 Tat protein. Science 1995; 268:429-431.
55. Shi L, Nishioka WK, Th'ng J, Morton Bradbury E, Litchfield DW, Greenberg AH. Premature $p34^{cdc2}$ activation required for apoptosis. Science 1994; 263:1143-1145.
56. Haupt Y, Rowan S, Oren M. p53-mediated apoptosis in Hela cells can be overcome by excess pRB. Oncogene 1995; 10:1563-1571.
57. Haas-Kogan DA, Kogan SC, Levi D, Dazin P, T'Ang A, Fung Y-KT, Israel MA. Inhibition of apoptosis by the retinoblastoma gene product. EMBO J 1995; 14:461-472.
58. Wang J, Walsh K. Resistance to apoptosis conferred by CDK inhibitors during myocyte differentiation. Science 1996; 273:359-361.

59. Bernstein H, Gensler HL. DNA damage and aging. In: Yu BP, ed. Free Radicals in Aging. Boca Raton, FL: CRC Press 1993:89-122.
60. Randerath K, Randerath E, Filburn C. Genomic and mitochodrial DNA alterations with aging. In: Schneider E, Rowe J, eds. Handbook of the Biology of Aging, Fourth Edition. New York: Academic Press, 1996:121-149.
61. Chen Q, Fischer A, Reagan JD, Yan LJ, Ames BN. Oxidative DNA damage and senescence of human diploid fibroblast cells. Proc Natl Acad Sci 1995; 92:4337-4341.
62. Guarente L. Do changes in chromosomes cause aging? Cell 1996; 86:9-12.
63. Jazwinski SM. Longevity, genes, and aging. Science 1996; 273:54-59.
64. Lithgow GJ. Molecular genetics of Caenorhabditis elegans aging. In: Schneider E, Rowe J, eds. Handbook of the Biology of Aging, Fourth Edition. New York: Academic Press, 1996:55-73.
65. Larsen PL. Aging resistance to oxidative damage in Caenorhabditis elegans. Proc Natl Acad Sci 1993; 90:8905-8909.
66. Vanfleteren JR. Oxidative stress and ageing in Caenorhabditis elegans. Biochem J 1993; 292:605-608.
67. Melov S, Lithgow GJ, Fisher DR, Tedesco PM, Johnson TE. Increase frequency of deletions in the mitochondrial genome with age of Caenorhabditis elegans. Nucleic Acids Res 1995; 23:1419-1425.
68. Fleming JE, Rose MR. Genetics of aging in Drosophila. In: Schneider E, Rowe J, eds. Handbook of the Biology of Aging. New York: Academic Press, 1996:74-93.
69. Orr WC, Sohal RS. Extension of the lifespan by overexpression of superoxide dismutase and catalase in *Drosophila* melanogaster. Science 1994; 263: 1128-1130.
70. Miquel J, Economos AC, Bensch KG. Insect versus mammalian aging. Aging and Cell Structure 1981; 1:347-379.
71. Lewin MB, Timiras PS. Lipid changes with aging in cardiac mitochondrial membranes. Mechanisms of Ageing and Development 1984; 24:343-351.
72. Yechiel, E, Barenholz Y. Relationships between membrane lipid composition and biological properties of rat myocytes. J Biol Chem 1985; 260:9123-9131.
73. Rouser G, Kritchevsky G, Yamamoto A, Baxter DV. Lipids in the nervous system of different speicies as a function of age. Adv Lipid Res 1972; 10:261-360.
74. Alberghina, M, Viola M. Region-selective decline of in vivo lipid synthesis in the aged rat visual system. Molecular and Chemical Neuropathology 1989; 11: 1102-1109.
75. Giusto NM, Roque ME, Lincheta de Boschero MG. Effects of aging on the content, composition and synthesis of sphingomyelin in the central nervous system. Lipids 1992; 27:835-839.
76. Prisco D, Rogasi PG, Matucci M, Paniccia R, Abbate R, Gensini GF, Neri Serneri GG. Age related changes in platelet lipid composition. Thromb Res 1986; 44:427-437.
77. Denda M, Koyama J, Hori J, Horii I, Takahashi M, Hara M, Tagami H. Age- and sex-dependent change in stratum corneum sphingolipids. Archives of Dermatological Research 1993; 285:415-417.
78. Ghadially R, Brown BE, Sequeira-Martin SM, Feingold KR, Elias PM. The Aged epidermal permeability barrier. J Clin Invest 1995; 95:2281-2290.
79. Petkova DH, Momchilova-Pankova AB, Markovska TT, Koumanov KS. Age-related changes in rat liver plasma membrane sphingomyelinase activity. Exp Gerontol 1988; 23:19-24.
80. Petkova DH, Momchiliva AB, Koumanov KS. Age-related changes in rat liver plasma membrane phospholipase A2 activity. Exp Gerontol 1986; 21:187-193.
81. Jayadev S, Linardic CM, Hannun YA. Identification of arachidonic acid as a mediator of sphingomyelin hydrolysis in response to tumor necrosis factor α. J Biol Chem 1994; 269:5757-5763
82. Tamiya-Koizumi, K, Kojima K. Activation of magnesium-dependent neutral sphingomyelinase by phosphatidylserine. J Biochem (Tokyo) 1986; 99:1803-1806
83. Dbaibo G, Obeid LM, Hannun YA. TNFα signal transduction through ceramide: dissociation of growth inhibitory effects of TNFα from activation of NF-κB. J Biol Chem 1993; 268:17762-17766.

84. Dobrowsky RT, Werner MH, Castellino AM, Chao MV, Hannun YA. Activation of the sphingomyelin cycle through the low-affinity neurotrophin receptor. Science 1994; 265:1596-1599.
85. Dbaibo GS, Pushkareva MY, Jayadev S, Schwarz JK, Horowitz JM, Obeid LM, Hannun YA. Retinoblastoma gene product as a downstream target for a ceramide-dependent pathway of growth arrest. Proc Natl Acad Sci 1995; 92:1347-1351.
86. Rani CSS, Abe A, Chang Y, Rosenzweig N, Saltiel AR, Radin NS, Shayman JA. Cell cycle arrest induced by an inhibitor of glucosylceramide synthesis: correlation with cyclin-dependent kinases. J Biol Chem 1995; 270:2859-2867.
87. Carroll MP, May WS. Protein kinase C-mediated serine phosphorylation directly activates Raf-1 in murine hematopoietic cells. J Biol Chem 1994; 269:1249-1256.
88. Johnson MD, Wang HY, Friedman E. Protein kinase C activity and contractile responsiveness in senescent blood vessels. Eur J Pharmacol 1990; 189:405-410.
89. Venable ME, Blobe GC, Obeid LM. Identification of a defect in the phospholipase D/ diacylglycerol pathway in cellular senescence. J Biol Chem 1994; 269: 26040-26044.
90. De Tata V, Ptasznik A, Cristafalo VJ. Effect of the tumor promoter phorbol 12-myristate 13-acetate (PMA) on proliferation of young and senescent Wi-38 human diploid fibroblasts. Exp Cell Res 1993; 205:261-269.
91. Lipschitz DA, Udupa KB, Indelicato SR, Das M. Effects of age on second messenger generation in neutrophils. Blood 1991; 78:1347-1354.
92. Chang ZF, Huang DY. Decline of protein kinase C activation in response to growth stimulation during senescence of IMR-90 human diploid fibroblasts. Biochem Biophys Res Commun 1994; 200: 16-27.
93. Whisler RL, Newhouse YG, Grants IS, Hackshaw KV. Differential expression of the α- and β- isoforms of protein kinase C in peripheral blood T and B cells from young and elderly adults. Mechanisms of Aging and Development 1995; 77:197-211.
94. Lee JY, Hannun YA, Obeid LM. Ceramide inactivates cellular protein kinase Cα. J Biol Chem 1996; 271:13169-13174.
95. Dobrowsky RT, Hannun YA. Ceramide stimulates a cytosolic protein phosphatase. J Biol Chem 1992; 267:5048-5051.
96. Zhou W, Takuwa N, Kumada M, Takuwa Y. Protein kinase C-mediated bidirectional regulation of DNA synthesis, RB protein phosphoryaltion, and cyclin-dependent kinases in human vascular endothelial cells. J Biol Chem 1993; 268:23041-23048.
97. Venable ME, Bielawska A, Obeid LM. Ceramide inhibits phospholipase D in a cell-free system. J Biol Chem 1996; 271:24800-24805.

CHAPTER 6

Ceramide, a Mediator of Cytosine Arabinoside Induced Apoptosis

Susan P. Whitman and Larry W. Daniel

Many diseases including cancer arise from alterations in signaling pathways. The cellular machinery controlling whether a cell is quiescent, proliferates, differentiates, or dies is no longer "correctly" regulated in transformed cells.[1] However, some cancer cells retain the ability to undergo differentiation and cell death (primarily those cells of the hematopoietic and immune systems). The signaling mechanisms by which current anticancer agents induce differentiation and cell death are not completely understood. However, because of their central role in controlling cell growth and differentiation these signaling pathways are attractive targets for the improvement of therapy regimens and the development of new drugs.[2,3] It is our aim to determine the signaling pathways in response to chemotherapeutic agents with the goal of developing more effective drugs and new combinations of existing drugs.

APOPTOSIS

Conventional cytotoxic chemotherapy has traditionally targeted DNA replication and mitotic processes. It is now widely recognized that cancer cells may respond to cytotoxic concentrations of these inhibitors by inducing a program of cell death termed apoptosis.[4] This physiological process of cell death has a central role in the control of cell populations during carcinogenesis, embryonic development and normal tissue homeostasis.[5] During apoptosis, Ca^{2+}-dependent endonucleases cut genomic DNA into distinct fragments.[6] Additionally, chromatin is condensed, the cell volume is reduced, and cytoplasmic blebbing occurs. Importantly, internal and external membranes are not disrupted and the cellular contents remain sealed inside. The dying cells fragment into small apoptotic bodies, which are engulfed by phagocytic cells. As a result, apoptosis and phagocytosis limit the acute inflammatory response to injury by preventing the release of toxic enzymes and inflammatory mediators (for review, see ref. 7). In contrast, necrosis is a passive mode of cell death.[8-10] In this process, there is general lysis of the cell membrane and widespread cleavage of DNA, resulting in a release of intracellular components. This induces an inflammatory response in the tissue. Our understanding of the underlying mechanisms that are induced by a variety of apoptotic stimuli is incomplete.[11]

Sphingolipid-Mediated Signal Transduction, edited by Yusuf A. Hannun.

Apoptotic cell death can be induced in cultured cells by several mechanisms including elevation of intracellular calcium ions, elevations in cAMP, irradiation, inhibition of DNA synthesis, oxidative stress,[12] growth factor deprivation,[13] cytokines,[14-17] and glucocorticoids.[4,18] It has been reported that some unknown cytoplasmic factor(s) can regulate apoptosis, since cytoplasmic extracts from cells exposed to topoisomerase inhibitors were found to trigger DNA fragmentation in isolated nuclei from HL-60 cells.[19]

Several gene products appear to have a role in the regulation of apoptosis. These include interleukin-1β converting enzyme (ICE) and related proteases, Bcl-2,[20] c-myc, and p53. The cysteine protease, ICE, cleaves the inactive precursor, interleukin-1β to generate the active cytokine resulting in programmed cell death.[21-23] Bcl-2α is a potent suppressor of apoptosis and its overexpression can prevent apoptosis in response to diverse stimuli.[20] Other members of the Bcl-2 family such as Bax have been identified. Bax, a 21 kDa protein, exists as a homodimer but can form heterodimers with Bcl-2α. Bax functions to counteract apoptosis suppression by Bcl-2. The *c-myc* gene product appears to have a dual role in promoting cell proliferation and death.[24-26] In some cell types, downregulation of *c-myc* expression, rather than induction, correlates with induction of apoptosis. The most commonly disrupted gene in human malignancies[27] is p53, which functions as a tumor suppressor. The normal function of p53 is to arrest cells at the G1 or G2 checkpoints of the cell cycle in response to DNA damage.[28-30] p53 can directly affect apoptosis by inducing downregulation of Bcl-2 and upregulation of Bax. However, p53 is not essential for viability or apoptosis in HL-60 cells, which do not express p53.[31,32]

The wide variety of signals and downstream events which induce apoptosis, implies that there are multiple signaling pathways that converge upstream of a common sequence of events which lead to death of the cells. Studies aimed at identifying common components in the apoptosis signaling mechanisms will aid in furthering our understanding of this cellular process. Some of the common components of these signaling pathways involve protein kinases including the c-Jun kinase (JNK)[33] also called the stress-activated protein kinase (SAPK). JNK can phosphorylate and activate c-Jun. The c-Jun transcription factor is required for apoptosis in a variety of cell types.[33-35] This was further demonstrated by the inhibition of apoptosis by dominant-negative c-Jun.

DIFFERENTIATION

Within the last 15 years, it has been recognized that inducing leukemic cell maturation to a mature, slower growing phenotype may be an alternative to conventional cytotoxic chemotherapy.[36-38] Myeloid cells can be induced to differentiate to phenotypes with normal characteristics of macrophages, granulocytes, or erythrocytes. Acute myeloid leukemia (AML) has been described as a neoplastic transformation at a hematopoetic stem cell stage which results in the inability to differentiate to maturity. This leads to proliferation and accumulation of leukemic cells. However, AML cells have been reported to differentiate in response to diverse agents.[37] The HL-60 cell line, derived from a patient with AML, has been a useful system for studying signal transduction mechanisms and the regulation of gene expression which lead to cellular differentiation.[39] In response to a variety of soluble agents, these cells are capable of differentiation to either monocyte/macrophage-like cells or granulocyte-like cells. For example, while not mediators of normal hematopoiesis, dimethylsulfoxide and pharmacological concentrations of retinoic acid induce granulocytic differentiation of HL-60 cells.

One of the most characterized signal transduction pathways leading to terminal differentiation of HL-60 cells to adherent macrophages involves a sequential activation of kinases in response to the tumor promoting phorbol ester, 12-*O*-tetradecanoylphorbol-13-acetate (TPA).[40-42] This cascade is initiated by the binding of TPA to protein kinase C (PKC), (PKCβII in HL-60 cells), and translocation of the enzyme from the cytosol to the membranes. PKCβII, directly or indirectly, stimulates the phosphorylation and activation of Raf-1 kinase which leads to the activation of the mitogen-activated protein kinase kinase(s) (MEKK1 and MEKK2). This is followed by activation of members of the mitogen-activated protein kinase family, termed the extracellular regulated protein kinases (ERK1 and ERK2).[43,44] The ERKs can phosphorylate and activate c-Jun and c-Fos to stimulate new gene expression. In response to TPA, the activation of the ERK pathway leads to the formation of monocytes and adherent macrophages from parental HL-60 cells. Also, the RAF-MEK-ERK pathway has been demonstrated to be activated in response to a variety of growth factors[45] and serum and results in a proliferative, cell survival response.[46]

DIACYLGLYCEROL AND PROTEIN KINASE C

Protein kinases are an integral part of cellular responses and their activation by second messengers is an early component of signal transduction pathways. The best characterized second messengers include Ca^{2+}, cAMP, free fatty acids, and diacylglycerol (DAG). The formation of DAG from the hydrolysis of membrane phospholipids can occur in response to growth factors. DAG acts as a second messenger by binding to and activating members of the PKC family (PKC βI, βII, α, δ, ε and γ).[47-50] Binding of DAG to PKC results in a decrease in the Ca^{2+} requirement of the Ca^{2+}-dependent PKC isoforms to physiological levels.[51,52] PKC activation appears to have a central role in signal transduction pathways involved in mitogenic and proliferative responses, tumor progression, and embryogenesis. In the case of phorbol ester activated PKC, myeloid leukemia cells respond by inducing a specific program of differentiation.[53]

It is now evident that stimulation of a cell surface receptor initiates a hydrolytic cascade of various membrane lipid components.[50] The rapid and transient formation of DAG by hydrolysis of inositol phosphoglycerides by PI-specific phospholipase C, and activation of PKC can be followed by a second wave of sustained increase in DAG by hydrolysis of choline-containing phosphoglycerides by the sequential activities of PLD and phosphatidic acid phosphohydrolase. This second wave is observed often in response to signals from growth factors, cytokines and phorbol esters.[54] Thus, this sustained increase in DAG levels has been associated with cell differentiation and cell growth. However, it is not yet clear whether sustained PKC activity or perhaps repetitive PKC activation is necessary for these long-term cellular responses.

SECOND MESSENGER ROLE OF CERAMIDE

Signaling roles for sphingolipids were initially proposed when sphingosine was demonstrated to be a potent inhibitor of PKC activity. Studies which followed demonstrated the importance of sphingolipids in signal transduction. Hydrolysis of sphingomyelin to ceramide and phosphorylcholine by the action of either a neutral sphingomyelinase (activated in HL-60 cells, for example) or an acidic sphingomyelinase (observed in Jurkat T cells) was demonstrated in response to differentiation inducing concentrations of vitamin D_3 and tumor necrosis factor α (TNFα)[55-57] respectively. The transient and reversible hydrolysis of membrane sphingomyelin with concomitant ceramide generation is termed the "sphingomyelin cycle".[58] Other inducers of this signaling pathway have been identified and include γ-interferon,[59] dexamethasone, interleukin-1,[60] and nerve growth factor. Similar to the discussion with sustained generation of DAG, recent work in MOLT-4 leukemia cells suggests that a prolonged increase in ceramide may have a role in long term cellular responses to environmental stress.[61]

The demonstration that the cell permeable ceramide analog D-*erythro*-N-acetylsphingosine (D-*erythro*-C_2-ceramide) could mimic 1,25-dihydroxy-vitamin D_3 and TNFα induction of monocytic differentiation of HL-60 cells indicated that ceramide was a second messenger. In addition,

D-*erythro*-dihydroceramide, lacking the 4,5-trans double bond in the sphingosine backbone, is not active, which demonstrates the specificity of the D-*erythro*-ceramide structure.[56,62] Thus, ceramide appears to participate in a selective program of monocytic differentiation because granulocytic inducers do not induce ceramide formation by the sphingomyelin cycle. Since these initial studies, stimulated ceramide formation has been shown to have a role in leukemic cell growth suppression,[61] tumor progression, differentiation,[63] and apoptosis.[58,64-69]

While PKC isoforms are the target for DAG, the direct intracellular target(s) for ceramide has not yet been conclusively determined. Hannun and co-workers[70] demonstrated that C_2-ceramide treatment induced the activity of a phosphoprotein phosphatase (PP2A) in HL-60 cells, and this was associated with *c-myc* downregulation and dephosphorylation of the retinoblastoma protein, a regulator of cell cycling. C_2-ceramide also stimulates a serine/threonine protein kinase in intact cells and in cell free systems.[71,72] However, direct activation of this proline-directed kinase did not occur in an in vitro assay using partially purified kinase.[73] Recent studies by Yao and co-workers[74] suggest that the ceramide-activated protein kinase activates Raf by phosphorylation on Thr 269. Thus, the ceramide stimulated protein kinase can be termed a Raf- 1 kinase kinase. This may activate the "ERK" pathway and mediate several of the inflammatory responses induced by TNFα.[75] The p21ras protein, which docks Raf-1 to membranes during activation by tyrosine kinases, may be necessary for ceramide induced apoptosis signaling, since inhibition of ras using dominant-negative ras mutants inhibits ceramide induced apoptosis (see ref. 65 and references therein). R-Ras, a member of the ras family, may also function in apoptotic pathways.[13] However, p21ras appears to have a role in both cell survival and apoptosis.[76]

Ceramide has been shown to activate the PKCζ isoform which may be involved in activation of NF-κB, a transcription factor associated with differentiation.[77,78] However, in other reports NF-κB activation did not occur in intact cells in response to cell permeable ceramide treatment.[79,80] The conflicting evidence for ceramide-induced NF-κB activation may be a result of experimental conditions used or separate pools for ceramide generation and subsequent function within a cell.[80] Therefore, the role of NF-κB and PKCζ in the cellular responses to ceramide remain to be determined.

Cell permeable ceramides also mediate interleukin-1 induced prostaglandin E_2 production[81] and inhibit phospholipase D.[82] These studies and others suggest a relationship between ceramide and DAG generation. In this regard, sphingomyelin synthesis is thought to occur primarily by the action of sphingomyelin synthase. This enzyme catalyzes the transfer of the phosphocholine headgroup of PC to ceramide, generating sphingomyelin and diglyceride. Therefore, this represents one mechanism interrelating the levels of DAG and ceramide synthesis within cells. Cellular outcomes such as apoptosis and differentiation may be determined by the relative levels of ceramide and DAG, and not the absolute amounts. These studies collectively suggest that ceramide has a signaling role in antiproliferative cellular responses to stress signals, while DAG, by activating PKC, is generally involved in proliferative or cell survival responses as well as myeloid leukemia cell differentiation.

CELLULAR RESPONSES TO CYTOSINE ARABINOSIDE

Cytosine arabinoside (ara-C) is a structural analog of 2'-deoxycytidine.[80,83,84] It differs from its parent compound by the presence of a β-OH group in the 2' position of the ribose sugar moiety. Although ara-C is highly effective as a single agent in the treatment of leukemia, including acute myelogenous leukemias (AML),[85] the mechanisms which lead to ara-C induced responses are not well defined.

Ara-C is transported across the cell membrane by a nucleotide transporter and is then converted to ara-CTP by mono- and diphosphate kinases and deoxycytidine kinase. The best understood process leading to ara-C induced lethality is the incorporation of its active metabolite, ara-CTP, into DNA, which inhibits DNA replication.[86-90] Ara-C lethality may arise from a combination of events including inhibition of DNA polymerase activity, interference with DNA chain initiation and elongation, and aberrant chromosomal reduplication.

Relatively little is known about the signal transduction mechanisms which ultimately lead to ara-C induced growth arrest, cellular differentiation,[86,87,89] and cell death by apoptosis.[91] Low concentrations of ara-C (0.05-1.0 μM) induce monocytic differentiation of HL-60 cells[92] which have been useful in determining the signal transduction mechanisms involved in differentiation and programmed cell death in response to diverse agonists.[39] Kharbanda and co-workers[93] showed that ara-C stimulates a serine/threonine kinase activity in HL-60 cells and also in a TPA unresponsive, HL-60 cell variant (HL-525).[94] In both cell lines, ara-C was demonstrated to transcriptionally activate the *c-jun* oncogene.[93-95] However, the protein kinase activated in HL-60 and HL-525 cells in response to ara-C was not identified and characterized. Recent studies suggest that ara-C induces proteolysis of PKCδ in the myeloid leukemia cell line, U937, resulting in the release of an active catalytic fragment.[96] However, under our experimental conditions, ara-C stimulates PKCβII and not PKCδ, in the HL-60 cell line.[97]

Ara-C Induces Diglyceride and Ceramide Formation in HL-60 cells

We have previously shown that ara-C stimulates both ceramide and DAG formation in HL-60 cells.[98] This was the first demonstration that a chemotherapeutic drug could stimulate ceramide formation and indicated that sphingomyelin turnover may be important in the response of cells to ara-C and other chemotherapeutic agents. Thus ara-C causes an increase in two lipid second messengers. Therefore, to further our understanding of the regulation of leukemic cell growth, differentiation, and cell death, it is of interest to determine the relative contributions of DAG and ceramide in HL-60 cells treated with ara-C.

The mechanism by which ara-C induces hydrolysis of membrane phospholipids is unclear. Inhibition of DNA synthesis may be the initial signal induced by ara-C, which is then propagated to the cytoplasm by induction of diglyceride and ceramide formation. Strum et al[98] demonstrated that ara-C stimulates a neutral sphingomyelinase activity which increases concurrently with ceramide generation and sphingomyelin hydrolysis. However, the origin of the ara-C induced DG synthesis is less clear. Ara-C can be used in the reverse reaction catalyzed by ara-CDPcholine-DAG phosphocholine transferase.[99,100] This results in the formation of ara-CDPcholine and DAG from phosphatidylcholine and ara-CMP (Fig. 6.1). The DAG generated by this mechanism could then activate PKC. These data suggest

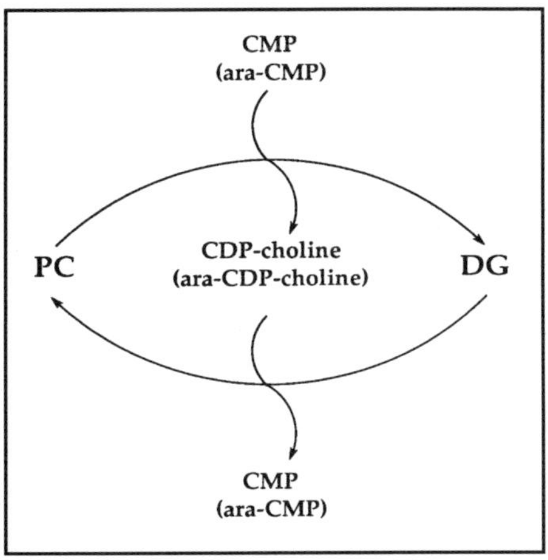

Fig. 6.1. Synthesis of phosphatidylcholine (PC) by CDP-choline:1,2-diacylglycerol phosphocholinetransferase. Cells treated with ara-C accumulate ara-CDP-choline which was originally suggested to be a product of CTP-phosphocholine cytidylyltransferase. However, Kucera and Capizzi[99] found that ara-CMP was a substrate for the reverse reaction of choline phosphotransferase. This provides a potential mechanism for the ara-C stimulated synthesis of diacylglycerol (DG). However, the quantitative significance of this reaction in ara-C treated cells is not known.

a possible mechanism for DAG formation in response to ara-C in HL-60 cells. However, the relatively high Km for this reaction in the murine leukemia cell line used in the study suggests that this mechanism may be minor in ara-C induced cellular responses since 0.1-10 μM ara-C is the range typically used to demonstrate differentiation and apoptosis in the majority of leukemic cells studied (HL-60, U937, KG-1).

The relative contributions of DG and ceramide in ara-C-induced signaling pathway(s) which lead to apoptosis and differentiation of HL-60 cells are unclear. It has been shown that PKC activation is antagonistic to ceramide-induced apoptosis.[101] In HL-60 cells, exogenous ceramide induces apoptosis in a few hours.[66] However, DAG (DiC_8) or TPA inhibits ceramide-induced apoptosis.[66,67,101] Therefore, ara-C induces two lipid mediators that are potentially involved in opposing pathways leading to the induction of cell death by apoptosis or to monocytic differentiation.

Inhibition of PKC Enhances Apoptosis in Response to ara-C

Since ara-C apparently stimulates competing pathways we sought to determine if inhibition of the PKC-dependent pathway of cell differentiation would enhance ara-C-induced apoptosis. For these studies, we used 1-*O*-octadecyl-2-*O*-methyl-rac-glycero-3-phosphocholine (ET-18-OCH_3) which is representative of the ether lipid class of chemotherapeutic agents (Fig. 6.2) (reviewed in ref. 102). All compounds in this class are selective for growth inhibition of neoplastic cells and have been used in phase 1 and 2 clinical trials with little problems from limiting toxicity. The active ether lipids are inhibitors of PKC and are competitive inhibitors of phosphatidylserine binding.[103] As expected, internucleosomal DNA fragmentation induced by ara-C (0.1-10 μM) was markedly enhanced by the PKC inhibitor ET-18-OCH_3.[97] ET-18-OCH_3 as a single agent has been shown to induce apoptosis in leukemic cells at 24-48 hours with concentrations above 10 μM.[104-107] Recently, it has been demonstrated the PKC inhibitors, chelerythrine and calphostin C, while inhibiting PKC activity, also induce ceramide formation and this is associated with induction of apoptosis.[108] However, at the concentrations and times of exposure used in our experiments, ET-18-OCH_3 did not induce DNA fragmentation or cause cell death (0.1-2 μM for up to 24 hours).[97] The increased apoptosis in ara-C plus ET-18-OCH_3 treated cells was associated with a decrease in ara-C-stimulated membrane-bound PKCβII in the presence of ET-18-OCH_3.[97]

We also studied the effect of ET-18-OCH_3 on ara-C induced downregulation of Bcl-2α. The mechanism by which Bcl-2α suppresses apoptosis is not known, however, increased levels are negatively correlated with sensitivity of leukemic cells to chemotherapeutic drugs. One possible mechanism involves phosphorylation of Bcl-2α by a PKC-dependent mechanism and this has been associated with suppression of apoptosis.[109] However, some controversy exists, since the phosphatase inhibitor, okadaic acid, and the chemotherapeutic agents, Taxol and 5'-fluorouracil were demonstrated to induce phosphorylation of Bcl-2α while inducing apoptosis in lymphoid cells.[110] In HL-60 cells, we demonstrated that ara-C-induced downregulation of Bcl-2α is mediated by ceramide, since we observe decreased Bcl-2α levels with ceramide treatment but not with PKC activators, TPA and DiC_8. Furthermore, the mechanism of Bcl-2α downregulation is at the level of transcription for both ara-C and C_2-ceramide.[111,112] The data demonstrate that ceramide in response to ara-C has a positive signaling role in ara-C induced apoptosis, mediating downregulation of Bcl-2α and induction of DNA fragmentation. Downregulation of Bcl-2 in response to ara-C was enhanced by simultaneous treatment with ET-18-OCH_3.[97] In contrast, activation of PKCβ by phorbol ester resulted in an increase in Bcl-2, and this effect was abolished by ET-18-OCH_3. However, PKCβ was not affected by exogenously added C_2-ceramide. Also, ET-18-OCH_3 did not affect levels of DNA fragmentation nor Bcl-2 down-

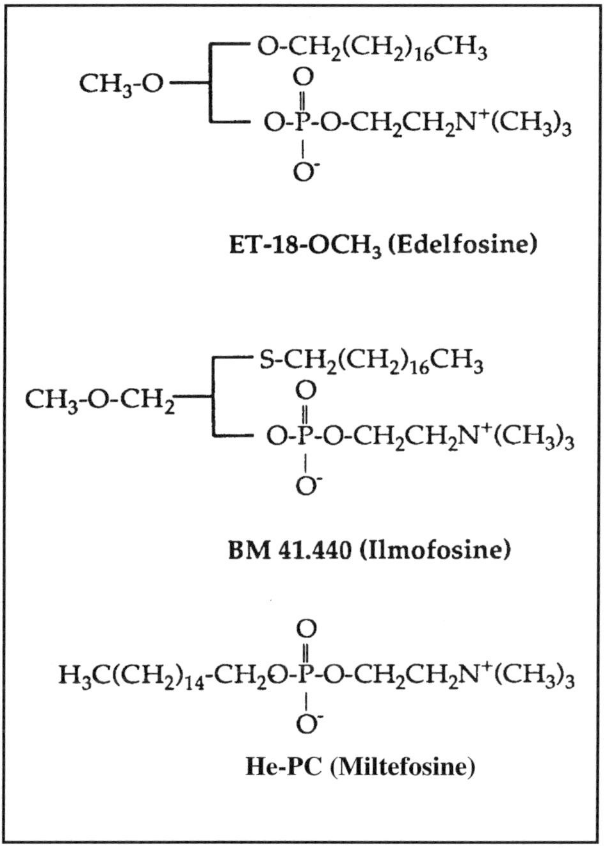

Fig. 6.2. Structures of representative ether linked lipids. The ether lipids shown have all been used in phase-1 and phase-2 clinical trials. In general, these compounds are well tolerated and are without unmanageable side effects. However, despite their initial promise in in vitro studies they have not been highly effective as single agents. ET-18-OCH3 is the most frequently used representative of this class of compounds. There is general agreement that all the ether lipids have similar mechanisms of action.

regulation induced by C_2-ceramide. In contrast, cell permeable DAG inhibited downregulation of Bcl-2 by C_2-ceramide.[97] Taken together, our data indicate that ara-C-induced ceramide formation stimulates the induction of DNA fragmentation and Bcl-2 downregulation, while DAG-activated PKCβ has an opposing role in ara-C and ceramide induced apoptosis. Other PKC inhibitors (staurosporine, H7, and GF109203X) have also been shown to enhance ara-C induced apoptosis,[113] which supports the role of PKC in limiting the apoptotic response to ara-C in myeloid leukemia.

Since the ether lipids have other cellular effects in addition to PKC inhibition[2,3,114-117] we have also used antisense oligodeoxynucleotides directed toward specific PKC isoforms as a second approach to inhibiting PKC signaling pathways. Antisense oligodeoxyncleotides inhibit the synthesis of the protein of interest by binding to complementary mRNA and inhibiting translation.[118-120] This method was recently applied to show that the PKCβII isoform is important in the monocytic differentiation of HL-60 cells induced by vitamin D_3.[121] The antisense but not the sense oligodeoxynucleotides, directed toward PKCβII stimulated ara-C-induced DNA fragmentation.[97] This supports the hypothesis that PKCβII activation is antagonistic to ara-C induced apoptosis and indicates that PKC inhibition by either ET-18-OCH_3 or antisense oligodeoxynucleotides result in enhancing ceramide-mediated apoptosis in ara-C treated cells.

A possible consequence of two opposing signaling pathways in leukemic cells in response to ara-C is the potential for acquired resistance to ara-C in relapsed leukemia. In addition to decreased deoxycytidine kinase activity, increased deaminase activity and decreased transport, another possible mechanism for the occurrence of a resistant phenotype could be a result of the antagonistic effects of PKC-dependent pathways on anti-proliferative ceramide pathways, allowing a fraction of cells to survive the initial treatment. In MOLT-4 leukemic cells, serum deprivation leads to the generation of ceramide and apoptosis. DAG, which is generated at a later time and to a lesser extent than ceramide, could reduce cell death.[61] Notably, resistance to adriamycin has been associated with increases in PKC.[122,123] Therefore, the relative intracellular levels of DG and ceramide, and not the absolute amounts, may determine the cellular response.

Since ceramide also induces differentiation, and experiments have shown the combination of ceramide and ara-C is additive for induction of monocytes (unpublished data), a role for ceramide signaling in ara-C stimulated HL-60 cell differentiation is therefore possible. Ceramide has been demonstrated to induce a Mg^{2+}-dependent serine/threonine kinase activity in HL-60 cells[72,124] and tyrosine phosphorylation of p42/44 MAP kinase[75] in HL-60 cells, demonstrating convergence in some signaling pathways. Therefore, it is probable that two pathways, PKC-dependent and ceramide-dependent, are activated in response to ara-C and lead to the ara-C induced differentiated phenotype (Fig. 6.3). In this regard, ceramide anabolites have been demonstrated to inhibit TPA induced adherence of differentiated cells, and not other common markers of monocytic differentiation.[125] Therefore, the appearance of monocytes and not terminally differentiated adherent macrophages may be due to the activation of both PKCβ and ceramide-dependent pathways in response to low concentrations of ara-C.

In the phorbol ester-resistant HL-525 cells, activation of a differentiation and apoptosis-associated transcription factor, c-Jun, occurs in response to ara-C and not TPA,[94,126] and this may be due to the increase in PKCβ we observe in this cell line. However, Grant and co-workers[113] have shown that coexposure of ara-C with staurosporine does not inhibit *c-jun* upregulation in HL-60 cells. Recently, ceramide has been demonstrated to transcriptionally activate *c-jun*. These data collectively support the suggestion that there may be convergence in

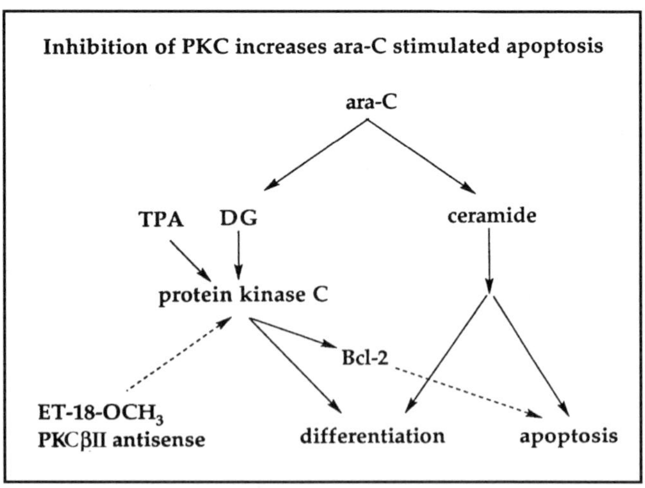

Fig. 6.3. Ara-C stimulates both ceramide-dependent and DG-dependent signaling in HL-60 cells. Treatment of HL-60 cells with ara-C results in increases in DG and ceramide.[98] Stimulation of PKCβII with TPA or exogenous DG causes an increase in Bcl-2 which inhibits apoptosis (dotted line).[97] In contrast, exogenous ceramide causes a decrease in Bcl-2 and stimulates apoptosis. Thus, ara-C stimulates synthesis of two second-messengers which have opposing effects on Bcl-2 and apoptosis. Inhibition of PKCβII with ET-18-OCH_3, or with PKCβII-directed antisense oligonucleotides decreases Bcl-2 and increases apoptosis in response to ara-C.[97]

pathways which lead to some ara-C dose-dependent responses.

In summary, these in vitro studies demonstrate enhancement of leukemic cell death and further define the signaling roles of diacylglycerol and ceramide in HL-60 cells in response to ara-C (Fig. 6.3). The recent observation that vincristine, another chemotherapeutic drug, induces ceramide formation and apoptosis[108] suggests that ceramide generation may be a common step leading to apoptosis in response to a variety of chemotherapeutic agents. Our results also suggest an additional use of ether lipids in achieving greater chemotherapeutic results by PKC inhibition. With myelosuppression and neurocytotoxicity as the major obstacles to high dose ara-C treatment, a combination of low dose ara-C and ET-18-OCH_3 may prove clinically beneficial.

References

1. Pawson T, Hunter T. Signal transduction and growth control in normal and cancer cells. Curr Opinion in Gen & Dev 1994; 4:1-4.
2. Powis G, Workman P. Signalling targets for the development of cancer drugs. Anti-Cancer Drug Design 1994; 9: 263-277.
3. Powis G. Signalling pathways as targets for anticancer drug design. Pharmac & Thera 1994; 62:57-95.
4. Wyllie AH. Apoptosis: cell death in tissue regulation. J Pathol 1987; 153:313.
5. Solary E, Bertrand R, Pommier Y. Apoptosis of human leukemic HL-60 cells induced to differentiate by phorbol ester treatment. Leukemia 1994; 8:792-797.
6. Compton MM. A biochemical hallmark of apoptosis: internucleosomal degradation of the genome. Cancer Metastasis Rev 1992; 11:105-119.
7. Darzynkiewicz Z. Apoptosis in antitumor strategies: Modulation of cell cycle or differentiation. J Cell Biochem 1995; 58:151-159.
8. Wyllie AH. Cell death: a new classification separating apoptosis from necrosis. In: Bowen ID, ed. Cell Death in Biology and Pathology. London, New York: Chapman and Hall, 1981: 9-34.
9. Gerschenson LE, Rotello RJ. Apoptosis, a different type of cell death. FASEB J 1992; 6:2450-2455.
10. Kerr J, Searle S. Shrinkage necrosis: a distinct mode of cell death. J Pathol 1972; 105:13-18.
11. McKonkey D, Orrenius S. Signal transduction pathways to apoptosis. Trends in Cell Biology 1994; 4:370-374.
12. Buttke TM, Sandstrom PA. Oxidative stress as a mediator of apoptosis. Immunology Today 1994; 15:7-10.
13. Wang H-G, Millan JA, Cox AD et al. R-Ras promotes apoptosis caused by growth factor deprivation via a Bcl-2 suppressible mechanism. J Cell Biol 1995; 129:1103-1114.
14. Tartaglia LA, Ayres TM, Wong GH et al. A novel domain within the 55 kd TNF receptor signals cell death. Cell 1993; 74:845-853.
15. Yonehara S, Ishii A, Yonehara M. A cell-killing monoclonal antibody (anti-Fas) to a cell surface antigen co-downregualted with the receptor of tumor necrosis factor. J Exp Med 1989; 169:1747-1746.
16. Trauth BC, Klas C, Peters AM et al. Monoclonal antibody-mediated tumor regression by induction of apoptosis. Science 1989; 245:301-305.
17. Itoh N, Nagata S. A novel protein domain required for apoptosis. Mutational analysis of human Fas antigen. J Biol Chem 1993; 268:10932-10937.
18. Wyllie AH. Glucocorticoid-induced thymocyte apoptosis is associated with endonuclease activation. Nature 1980; 284:554-556.
19. Schulze-Osthoff K, Walczak H, Droge W et al. Cell nucleus and DNA fragmentation are not required for apoptosis. J Cell Biol 1994; 127:15-20.
20. Nunez G, Clarke MF. The Bcl-2 family of proteins: Regulators of cell death and survival. Trends in Cell Biology 1994; 4:399-403.
21. Thornberry NA, Bull HG, Calaycay JR et al. A novel heterodimeric cysteine protease is required for interleukin-1 beta processing in monocytes. Nature 1992; 356:768-774.
22. Miura M, Zhu H, Rotello R et al. Induction of apoptosis in fibroblasts by IL-1

beta-converting enzyme, a mammalian homolog of the *C. elegans* cell death gene ced-3. Cell 1993; 78:653-660.
23. Gagliardini V. Prevention of neuronal cell death by the crmA gene. J Biol Chem 1994; 268:826-828.
24. Evan GI, Wyllie AH, Gilbert CS et al. Induction of apoptosis in fibroblasts by c-myc protein. Cell 1992; 69:119-128.
25. Evan G, Harrington E, Fanidi A et al. Integrated control of cell proliferation and cell death by the c-myc oncogene. Philos Trans R Soc Lond B Biol Sci 1994; 345:269-275.
26. Bissonnette RP, McGahon A, Mahboubi A et al. Functional Myc-Max heterodimer is required for activation-induced apoptosis in T cell hybridomas. J Exp Med 1994; 180:2413-2418.
27. Greenblatt MS, Bennett WP, Hollstein M et al. Mutations in the p53 tumor suppressor gene: clues to cancer etiology and molecular pathogenesis. Cancer Res 1994; 54:4855-4878.
28. Kastan MB, Onyekwere O, Sidransky D et al. Participation of p53 protein in the cellular response to DNA damage. Cancer Res 1991; 51:6304-6311.
29. Kuerbitz SJ, Plunkett BS, Walsh WV et al. Wild-type p53 is a cell cycle checkpoint determinant following irradiation. Proc Natl Acad Sci USA 1992; 89:7491-7495.
30. Kastan MB, Zhan Q, el-Deiry WS et al. A mammalian cell cycle check point pathway utilizing p53 and GADD45 is defective in ataxia-telangiectasia. Cell 1992; 71:587-597.
31. Donehower LA, Harvey M, Slagle BL et al. Mice deficient for p53 are developmentally normal but susceptible to spontaneous tumours. Nature 1992; 356: 215-221.
32. Clarke AR, Gledhill S, Hooper ML et al. p53 dependence of early apoptotic and proliferative responses within the mouse intestinal epithelium following gamma-irradiation. Oncogene 1994; 9:1767-1773.
33. Chen Y-R, Meyer CF, Tan T-H. Persistent activation of c-Jun N-terminal kinase (JNK 1) by γ radiation-induced apoptosis. J Biol Chem 1996; 271: 631-634.
34. Sawai H, Okazaki T, Yamamoto H et al. Requirement of AP-1 for ceramide-induced apoptosis in human leukemia HL-60 cells. J Biol Chem 1995; 270: 27326-27331.
35. Colotta F, Polentarutti N, Sironi M et al. Expression and involvement of c-fos and c-jun protooncogenes in programmed cell death induced by growth factor depreivation in lymphoid cell lines. J Biol Chem 1992; 267:18278-18283.
36. Hozumi M. Fundamentals of chemotherapy of myeloid leukemia by induction of leukemia cell differentiation. Adv Cancer Res 1983; 38:121-169.
37. Hassan HT. Differentiation induction therapy of acute myelogenous leukemia. Haematologia 1988; 21:141-150.
38. Koeffler HP. Induction of differentiation of human acute myelogenous leukemia cells: therapeutic implications. Blood 1983; 62:709-721.
39. Collins SJ. The HL-60 promyelocytic leukemia cell line: proliferation, differentiation, and cellular oncogene expression. Blood 1987; 70:1233-1244.
40. Tonetti DA, Horio M, Collart FR et al. Protein kinase C beta gene expression is associated with susceptibility of human promyelocytic leukemia cells to phorbol ester-induced differentiation. Cell Growth & Diff 1992; 3:739-745.
41. Tonetti DA, Henning-Chubb C, Yamanishi DT et al. Protein kinase C-beta is required for macrophage differentiation of human HL-60 leukemia cells. J Biol Chem 1994; 269:23230-23235.
42. Macfarlane DE, Manzel L. Activation of β-isozyme of protein kinase C (PKCβ) is necessary and sufficient for phorbol-ester-induced differentiation of HL-60 promyelocytes. J Biol Chem 1994; 269: 4327-4331.
43. Kharbanda S, Saleem A, Emoto Y et al. Activation of Raf-1 and mitogen-activated protein kinase during monocytic differentiation of human myeloid leukemic cells. J Biol Chem 1994; 269:872-878.
44. Adams PD, Parker PJ. TPA-induced activation of MAP kinase. FEBS Letters 1991; 290:77-82.
45. Rapp UR. Role of Raf-1 serine/threonine

protein kinase in growth factor signal transduction. Oncogene 1991; 6:495-500.
46. Haimovitz-Friedman A, Balaban M, McLoughlin M et al. Protein kinase C mediates basic fibroblastic growth factor protection of endothelial cells against radiation-induced apoptosis. Cancer Res 1994; 54:2591-2597.
47. Ganong BR, Loomis CR, Hannun YA et al. Specificity and mechanism of protein kinase C activation by *sn*-1,2-diacylglycerols. Proc Natl Acad Sci USA 1986; 83:1184-1188.
48. Blumberg PM, Acs G, Areces LB et al. Protein kinase C in signal transduction and carcinogenesis. In: Spitzer HL, Ed. Receptor-mediated biological processes: Implications for evaluating carcinogenesis. New York: Wiley-Liss, Inc, 1994:3-19.
49. Blobe GC, Obeid LM, Hannun YA. Regulation of protein kinase C and role in cancer biology. Cancer Metastasis Rev 1994; 13:411-431.
50. Asaoka Y, Nakamura S, Yoshida K et al. Protein kinase C, calcium, and phospholipid degradation. TIBS 1992; 17:414-417.
51. Nishizuka Y. The role of protein kinase C in cell surface signal transduction and tumour promotion. Nature 1984; 261: 693-698.
52. Nishizuka Y. Intracellular signalling by hydrolysis of phospholipids and activation of protein kinase C. Science 1992; 258:607-614.
53. Hubermann E, Callaham MF. Induction of terminal differentiation in human promyelocytic leukemia cells by tumor-promoting agents. Proc Natl Acad Sci USA 1979; 76:1293-1297.
54. Nishizuka Y. Protein kinase C and lipid signaling for sustained cellular responses. FASEB J 1995; 9:484-496.
55. Okazaki T, Bell RM, Hannun YA. Sphingomyelin turnover by vitamin D_3 in HL-60 cells: role in cell differentiation. J Biol Chem 1989; 264:19076-19080.
56. Okazaki T, Bielawska A, Bell RM et al. Role of ceramide as a lipid mediator of 1α-25-dihydroxyvitamin D_3-induced HL-60 cell differentiation. J Biol Chem 1990; 265:15823-15831.
57. Dbaibo GS, Obeid LM, Hannun YA. Tumor necrosis factor alpha (TNFα) signal transduction through ceramide. J Biol Chem 1993; 268:17762-17766.
58. Hannun YA. The sphingomyelin cycle and the second messenger function of ceramide. J Biol Chem 1994; 269:3125-3128.
59. Kim M-Y, Linardic C, Hannun Y. Identification of sphingomyelin turnover as an effector mechanism for the action of tumor necrosis factor alpha and gamma interferon. J Biol Chem 1991; 266: 484-489.
60. Mathias S, Younes A, Kan C-C et al. Activation of the sphingomyelin signalling pathway in intact EL4 cells and in a cell free system by IL-1β. Science 1993; 259:519-522.
61. Jayadev S, Liu B, Bielawaska AE et al. Role for ceramide in cell cycle arrest. J Biol Chem 1995; 270:2047-2052.
62. Bielawska A, Crane HM, Liotta D et al. Selectivity of ceramide mediated biology. Lack of activity of erthryo-dihydroceramide. J Biol Chem 1993; 268:26226-26232.
63. Bielawaska A, Linardic CM, Hannun YA. Modulation of cell growth and differentiation by ceramide. FEBS Letters 1992; 307:211-214.
64. Obeid LM, Hannun YA. Ceramide: a stress signal and mediator of growth suppression and apoptosis. J Cell Biochem 1995; 58:191-198.
65. Jarvis WD, Grant S, Kolesnick RN. Ceramide and the induction of apoptosis. Clinical Cancer Res 1996; 2:1-6.
66. Obeid LM, Linardic CM, Karolak LA et al. Programmed cell death induced by ceramide. Science 1993; 259:1769-1771.
67. Jarvis WD, Kolesnick RN, Fornari FA et al. Induction of apoptotic DNA damage and cell death by activation of the sphingomyelin pathway. Proc Natl Acad Sci USA 1994; 91:73-77.
68. Kolesnick RN, Friedman-Haimovitz A, Fuks Z. The sphingomyelin signal transduction pathway mediates apoptosis for tumor necrosis factor, Fas, and ionizing radiation. Biochem Cell Biol 1994; 72: 471-474.
69. Kolesnick RN, Fuks Z. Ceramide: a signal for apoptosis or mitogenesis. J Expt Med 1995; 181:1949-1952.

70. Wolff RA, Dobrowsky RT, Bielawska A et al. Role for ceramide-activated protein phosphatase in ceramide-mediated signal transduction. J Biol Chem 1994; 269: 19605-19609.
71. Dressler KA, Mathias S, Kolesnick RN. Tumor necrosis factor α activates the sphingomyelin signal transduction pathway in a cell free system. J Biol Chem 1993; 268:20520-20523.
72. Mathias S, Dressler KA, Kolesnick RN. Characterization of a ceramide-activated protein kinase. Proc Natl Acad Sci USA 1991; 88:10009-10013.
73. Liu J, Mathias S, Yang Z et al. Renaturation and tumor necrosis factor-α stimulation of a 97-kDa ceramide activated protein kinase. J Biol Chem 1994; 269: 3047-3052.
74. Yao B, Zhang Y, Delikat S et al. Ceramide activated protein kinase is a Raf kinase. Nature 1995; 378:307-310.
75. Raines MA, Kolesnick RN, Golde DW. Sphingomyelinase and ceramide activate mitogen-activated protein kinase in myeloid HL-60 cells. J Biol Chem 1993; 268:14572-14575.
76. Chen CY, Faller DV. Direction of p21ras-generated signals towards cell growth or apoptosis is determined by protein kinase C and Bcl-2. Oncogene 1995; 11: 1487-1498.
77. Lozano JB, Municio E MM, Diaz-Meco MT et al. Protein kinase C ζ isoform is critical for κB-dependent promoter activation by sphingomyelinase. J Biol Chem 1994; 269:19200-19202.
78. Diaz-Meco MT, Berra F, Munico MM et al. A dominant negative protein kinase C zeta subspecies blocks NF-κB activation. Mol Cell Bio 1993; 13:4770-4775.
79. Betts JC, Agranoff AB, Nabel GJ et al. Dissociation of endogenous cellular ceramide from NF-κB activation. J Biol Chem 1994; 269:8455-8458.
80. Westwick JK, Bielawaska AE, Dbaibo G et al. Ceramide activates the stress-activated protein kinases. J Biol Chem 1995; 270:22689-22692.
81. Ballou LR, Chao CP, Holness MA et al. Interleukin-1-mediated PGE2 production and sphingomyelin metabolism. Evidence for the regulation of cyclooxygenase gene expression by sphingosine and ceramide. J Biol Chem 1992; 267:20044-20050.
82. Venable ME, Bielawska A, Obeid LM. Ceramide inhibits phospholipase D in a cell-free system. J Biol Chem 1996; 271:24794-24799.
83. Grant S. Biochemical modulation of cytosine arabinoside. Pharmac Ther 1990; 48:29-44.
84. Bergmann W, Feeney R. Contributions to the study of marine products XXXII. The arabinosides of sponges. J Org Chem 1951; 16:981-987.
85. Mayer RJ, Davis RB, Schiffer CA et al. Intensive postremission chemotherapy in adults with acute myeloid leukemia. New Eng J Med 1994; 331:895-903.
86. Kufe DW, Major PP, Egan EM et al. Correlation of cytotoxicity with incorporation of ara-C into DNA. J Biol Chem 1980; 255:8997-9000.
87. Kufe DD, Spriggs EM, Munroe D. Relationships among ara-CTP pools, formation of (ara-C)DNA, and cytotoxicity of human leukemia cells. Blood 1984; 64: 54-58.
88. Furth JJ, Cohen SS. Inhibition of mammalian DNA polymerase by the 5'-triphosphate of 1-β-D-arabinofuranosylcytosine and the 5'-triphosphate of 9-β-D-arabinofuranosyladenine. Cancer Res 1968; 28:2061-2067.
89. Gale RP. Advances in the treatment of acute myelogenous leukemia. New Engl J Med 1979; 300:1189-1199.
90. Fram RJ, Kufe DW. DNA strand breaks caused by inhibitors of DNA synthesis: 1-β-D-arabinofuranosylcytosine and aphidicolin. Cancer Res 1982; 42:4050-4053.
91. Gunji H, Kharbanda S, Kufe D. Induction of internucleosomal DNA fragmentation in human myeloid leukemia cells by 1-B-D-arabinofuranosylcytosine. Cancer Res 1991; 51:741-743.
92. Collins SJ, Bodner A, Ting R et al. Induction of morphological and functional differentiation of human promyelocytic leukemia cells (HL-60) by compounds which induce differentiation of murine leukemia cells. Int J Cancer 1980; 25: 213-218.
93. Kharbanda S, Datta R, Kufe D. Regula-

tion of c-jun gene expression in HL-60 leukemia cells by 1-β-D-arabinofuranosylcytosine. Potential involvement of a protein kinase C-dependent mechanism. Biochemistry 1991; 30:7947-7951.
94. Kharbanda S, Emoto Y, Kisaki H et al. 1-β-D-arabinofuranosylcytosine activates serine/threonine protein kinase and c-jun gene expression in phorbol-ester-resistant myeloid leukemia cells. Mole Pharm 1994; 46:67-72.
95. Kharbanda SM, Sherman ML, Kufe DW. Transcriptional regulation of c-jun gene expression by arabinofuranosylcytosine in human myeloid leukemia cells. J Clin Invest 1990; 11:1517-1523.
96. Emoto Y, Kisaki H, Manome Y et al. Activation of protein kinase C delta in human myeloid leukemia cells treated with 1-beta-D-arabinofuranosylcytosine. Blood 1996; 87:1990-1996.
97. Whitman SP, Civoli F, Daniel LW. 1-β-D-arabinofuranosylcytosine stimulates antagonistic signalling pathways in HL-60 cells. Manuscript submitted
98. Strum JC, Small GW, Pauig SB et al. 1-β-D-arabinofuranosylcytosine stimulates ceramide and diglyceride formation in HL-60 cells. J Biol Chem 1994; 269: 15493-15497.
99. Kucera GL, Capizzi RL. 1-β-D-arabinofuranosylcytosine-diphosphate choline is formed by the reversal of cholinephosphotransferase and not via cytidylyltransferase. Cancer Res 1992; 52:3886-3891.
100. Sasvari-Szekely M, Spasokukotskaja T, Sooki-Toth A et al. Deoxycytidine is salvaged not only into DNA but also into phospholipid precursors. II. Ara-C does not inhibit the later process in lymphoid cells. Biochem Biophys Res Comm 1989; 163:1158-1167.
101. Jarvis WD, Fornari FA, Browning JL et al. Attenuation of ceramide-induced apoptosis by diglyceride in human myeloid leukemia cells. J Biol Chem 1994; 269:31685-31692.
102. Daniel LW. Ether lipids in experimental cancer chemotherapy. In: Hickman JA, Tritton TR, eds. Cancer Chemotherapy. Oxford: Blackwell, 1991:146-178.
103. Marasco CJ, Piantadosi C, Mayer KL et al. Synthesis and biological activity of novel quaternary ammonium derivatives of alkylglycerol derivatives of alkylglycerols as potent inhibitors of protein kiase C. J Med Chem 1990; 33:985-992.
104. Araki S, Tsuna I, Kaji K et al. Programmed cell death in response to alkyllysophospholipids in endothelial cells. J Biochem 1994; 115:245-247.
105. Diomede L, Colotta F, Piovani B et al. Induction of apoptosis in human leukemic cells by the ether lipid 1-octadecyl-2-methyl-*rac*-glycero-3-phosphocholine. A possible mechanism for its selective action. J Cancer 1993; 53:124-130.
106. Diomede L, Piovani B, Re F et al. The induction of apoptosis is a common feature of the cytotoxic action of ether-linked glycerophospholipids in human leukemic cells. Int J Cancer 1994; 57: 645-649.
107. Mollinedo F, Martinez-Dalmau R, Modolell M. Early and selective induction of apoptosis in human leukemic cells by the alkyl-lysophospholipid ET-18-OCH3. Biochem Biophys Res Comm 1993; 192:603-609.
108. Zhang J, Alter N, Reed JC et al. Bcl-2 interrupts the ceramide-mediated pathway of cell death. PNAS, USA 1996; 93:5325-5328.
109. May WS, Tyler PG, Armstrong DK et al. Role for serine phosphorylation of Bcl-2 in an anti-apoptotic signaling pathway triggered by IL-3, erythropoietin, and bryostatin 1. Blood 1993; 82:1738.
110. Blagosklonny MV, Schulte T, Nguyen P et al. Taxol-induced apoptosis and phosphorylation of Bcl-2 protein involves c-Raf-1 and represents a novel c-Raf-1 signal transduction pathway. Cancer Res 1996; 56:1851-1854.
111. Delia D, Aiello A, Soligo D et al. Bcl-2 proto-oncogene expression in normal and neoplastic human myeloid cells. Blood 1992; 79:1291-1298.
112. Chen M, Quintans J, Fuks Z et al. Suppression of Bcl-2 messenger RNA production may mediate apoptosis after ionizing radiation, tumor necrosis factor α, and ceramide. Cancer Res 1995; 55: 991-994.
113. Grant S, Turner AJ, Bartimole TM et al.

1-β-D-arabinofuranosyl cytosine-induced apoptosis in human myeloid leukemia cells by staurosporine and other pharmacological inhibitors of protein kinase C. Oncol Res 1994; 6:87-99.

114. Surette ME, Winkler JD, Fonteh AN et al. Relationship between arachidonate-phospholipid remodeling and apoptosis. Biochemistry 1996; 35:9187-9196.

115. Seewald MJ, Olsen RA, Sehgal I et al. Inhibition of growth-factor dependent inositol phosphate Ca^{2+} signaling by antitumor ether lipid analogues. Cancer Res 1990; 50:4458-4463.

116. Boggs KP, Rock CO, Jackowski S. Lysophosphatidylcholine attenuates the cytotoxic effects of the antineoplastic phospholipid1-O-octadecyl-2-O-methyl-rac-glycero-3-phosphocholine. J Biol Chem 1995; 270:11612-11618.

117. Boggs KP, Rock CO, Jackowski S. Lysophosphatidylcholine and 1-O-octadecyl-2-O-methyl-rac-glycero-3-phosphocholine inhibit the CDP-choline pathway of phosphatidylcholine synthesis at the CTP:phosphocholine cytidylyltransferase step. J Biol Chem 1995; 270: 7757-7764.

118. Baserga R, Reiss K, Alder H et al. Inhibition of cell cycle progression by antisense oligodeoxynucleotides. Ann NY Acad Sci 1992; 660:64-69.

119. Ferrari S, Manfredini R, Grande A et al. Antisense strategies to characterize the role of genes and oncogenes involved in myeloid differentiation. Annals New York Academy of Sciences 1992; 11-25.

120. Neckers L, Rosolen A, Fahmy B et al. Specific inhibition of oncogene expression in vitro and in vivo by antisense oligonucleotides. Ann of NY Acad Sci 1992; 660:37-44.

121. Gamard CJ, Blobe GC, Hannun YA et al. Specific role for protein kinase C β in cell differentiation. Cell Growth and Diff 1994; 5:405-409.

122. Aquino A, Hartman KD, Knode MC et al. Role of protein kinase C in phosphorylation of vinculin in adriamycin resistant HL-60 leukemia cells. Cancer Res 1988; 48:3324-3329.

123. Psada J, Tritton TR. Protein kinase C in adriamycin action and resistance in mouse sarcoma 180 cells. Cancer Res 1989; 49:6634-6639.

124. Joseph CK, Byun H-S, Bittman R et al. Substrate recognition by ceramide-activated protein kinase. J Biol Chem 1993; 268: 20002-20006.

125. Kan C-C, Kolesnick RN. A synthetic ceramide analog, d-threo-1-phenyl-2-decanoylamino-3-morpholino-1-propanol, selectively inhibits adherence during macrophage differentiation of human leukemia cells. J Biol Chem 1992; 267: 9663-9667.

126. Kharbanda S, Huberman E, Kufe D. Activation of the jun-D gene during treatment of human myeloid leukemia cells with 1-β-D-arabinofuranosylcytosine. Bioch Pharm 1993; 45:2055-2061.

CHAPTER 7

The Use of Cerebroside Synthase Inhibitors as Probes for Assessing the Metabolism and Function of Sphingolipids

James A. Shayman and Norman S. Radin

INTRODUCTION

Although sphingolipids have been the object of active investigation for over 50 years, the intensity of interest in these ubiquitous lipids has increased recently. This interest is the result of two significant developments. First, several agonists have been identified as effectors of sphingolipid metabolism. Second, several critical cellular functions have been associated with sphingolipid products, consistent with the view that sphingolipids have distinct second messenger functions.

Sphingomyelin in particular has been identified as a target for agonist-stimulated sphingomyelinase activation. Tumor necrosis factor α[1] and agonists with structurally related receptors, including IL-1,[2] Fas,[3] and NGF,[4] release ceramide following binding to cellular receptors. Other agonists, including dihydroxyvitamin D3,[5] formylated peptides,[6] and complement,[7] are also reported to stimulate sphingomyelin hydrolysis.

In addition, long chain bases such as sphingosine and related metabolites such as sphingosine-1-phosphate[8] and N,N-dimethylsphingosine[9] may also be formed following agonist stimulation.

Demonstrating a particular functional role for a candidate sphingolipid has proven challenging. Investigators have often relied on the exogenous addition of sphingolipids to cultured cells as a demonstration of biological function. These exogenously added sphingolipids are usually truncated by chemical synthesis in order to reduce the problem of poor solubility of normal chain length sphingolipids in aqueous solutions. But even these truncated lipids are readily metabolized to other sphingolipids some of which have independent potential activities. Whether these added sphingolipids localize to functionally relevant sites

Sphingolipid-Mediated Signal Transduction, edited by Yusuf A. Hannun.

within cells is not well established. In addition, few sphingolipid, dependent enzymes have been cloned and thus the critical necessity of a given proposed signaling pathway is often difficult to demonstrate as has been done in other lipid signaling systems.

An alternative means for assessing sphingolipid function is through the use of inhibitors of sphingolipid metabolism. Several natural and synthetic inhibitors of simple sphingolipid metabolism have now been described. These include the fungal toxin fumonisin B1[10] (an inhibitor of sphingoid base acylation), conduritol B epoxide[11] (an inhibitor of β-glucosidase), and β-halogenated alanines[12] (inhibitors of long chain base synthesis).

Another inhibitor, 1-phenyl-2-decanoylamino-3-morpholino-1-propanol blocks glucosylceramide synthesis. This compound was designed and synthesized by Radin and Vunnam in an attempt to find a compound suitable for the treatment of Gaucher disease.[13] This ceramide analog was quickly identified as a potent inhibitor of cellular growth and has served as a useful probe for the study of sphingolipid metabolism and function. In this chapter we review the use of PDMP and related homologues in more recent studies on sphingolipid function including the heat shock response.

INHIBITORS OF CERAMIDE GLUCOSYLATION BLOCK CELL GROWTH

The simplest glycosphingolipid, glucosylceramide, consists of three essential moieties. These are a sphingoid base backbone (usually sphingosine), a glucose group, and a fatty acid joined in amide linkage to the sphingoid base. PDMP, the prototypic glucosylceramide synthase inhibitor, has a morpholino group instead of a glucose group, a phenyl group instead of the long aliphatic chain characteristic of the long chain bases, and a fatty acid in amide linkage. In addition, PDMP is a different enantiomeric form (D-threo) than naturally occurring glucosylceramide (D-erythro) (Fig. 7.1).

PDMP inhibits the growth of a wide range of cultured cells.[14] When mice or rats are treated with the inhibitor, PDMP distributes preferentially to the kidney where it also blocks organ growth.[15]

The growth inhibitory effects of PDMP are temporally associated both with a decrease in glucosylceramide and other glucosylceramide based glycolipids and with an increase in cellular ceramide content.[16] In MDCK cells[17] and cultured keratinocytes,[18] both ceramide accumulation and glucosylceramide depletion are probably impor-

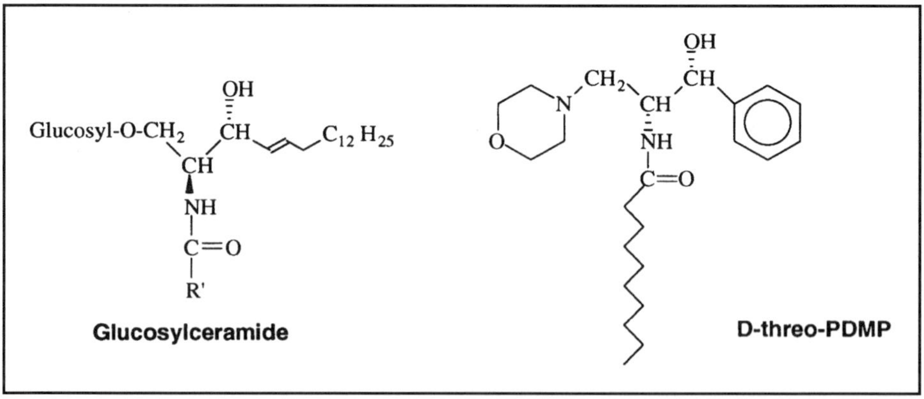

Fig. 7.1. Structural comparison between glucosylceramide and D-threo-1-phenyl-2-decanoylamino-3-morpholino-1-propanol.

tant in the growth effects. The addition of conduritol B epoxide, a β-glucosidase inhibitor, causes accumulation of glucosylceramide and growth stimulation. In other systems, e.g., renal epithelial MCT cells,[19] conduritol B epoxide is without effect.

The basis for PDMP mediated growth inhibition is not well understood. Based on reports by Bremer and Hakomori demonstrating a role for glycosphingolipids in the activity of receptor tyrosine kinases, it was postulated that glycosphingolipid depletion would inhibit cell growth by blocking an early event associated with growth factor signaling.[20] In NIH 3T3 cells stably transfected with the IGF1 receptor, however, PDMP did not block receptor tyrosine kinase activity, epidermal growth factor receptor kinase (ERK) activity, or phosphatidylinositol 3 kinase activity.[21]

On the contrary the PDMP effect correlated with cell cycle arrest. A block in G1/S and G2/M cell cycle progression was observed with corresponding changes in cyclin dependent kinase activities. Phosphorylation of the retinoblastoma protein, a necessary requirement for G1/S progression, is inhibited by ceramide.[22] Protein phosphatase 2a activity, a mediator of G2 arrest, is stimulated by ceramide.[23] Determining whether these effects are the basis for the growth arrest will require further investigation.

Inhibitors of Ceramide Glucosylation Have at Least Two Sites of Action

Based on PDMP as a parent compound, structural homologues were synthesized in order to identify more active glucosylceramide synthase inhibitors. Acyl substitutions for the decanoyl group revealed that palmitoyl analogs were significantly more potent as inhibitors of the cerebroside synthase.[24] A series of compounds replacing the morpholine ring with other cyclic tertiary amines was also created. A pyrrolidine substitution (1-phenyl-2-palmitoyl-3-pyrrolidino-1-propanol, P4) was discovered to be particularly active, with an ID_{50} of 2 μM (Fig. 7.2).[25]

To our surprise, when the DL-erythro and DL-threo diastereomers of the pyrrolidino substituted inhibitor were tested for cell growth inhibition, both compounds were observed to inhibit cell proliferation at equivalent concentrations. However, only the threo-P4 inhibited glucosylceramide formation. Upon further study it was discovered that both compounds raised cell ceramide levels equally. These data suggested that an alternative mechanism for ceramide accumulation exists.

The conclusion that PDMP and the acyl and morpholino substituted homologues had an alternative site of action was further substantiated when compounds

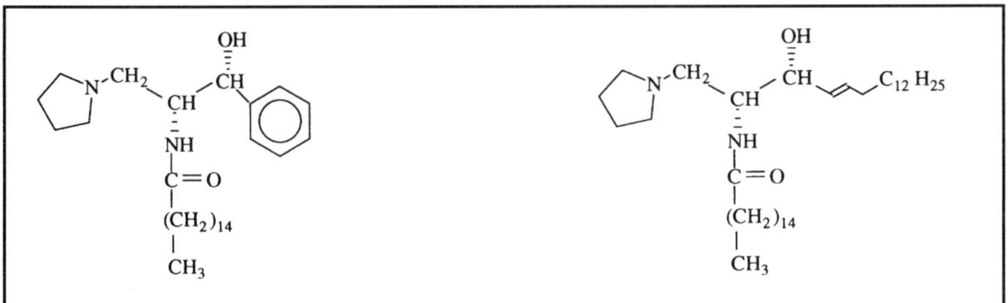

Fig. 7.2. Enantiomeric structures of two PDMP analogs.

with aliphatic substitutions for the phenyl group were synthesized and tested. 1-Pyrrolidino-1-deoxyceramide (Fig. 7.2) was observed to block glucosylceramide synthesis without any effect on ceramide levels or cell growth.[26] In addition, C2 ceramide (N-acetylsphingosine) itself raises cellular levels of both normal chain length ceramides and glucosylceramide. This effect is probably due to both the direct conversion of short chain ceramide to long chain ceramide based sphingolipids and the action of short chain ceramides as competitive inhibitors of ceramide metabolism. Comparing the effects of the P4 diastereoisomers and the D-threo-deoxyceramide enantiomers on cell growth, ceramide and glucosylceramide levels, we can conclude that ceramide accumulation and not glucosylceramide depletion correlates best with the growth inhibitory effects of these compounds (Table 7.1).

Several ceramide utilizing enzymes were assayed in an attempt to identify the basis for the ceramide accumulation in cells treated with the pyrrolidine homologues. Sphingomyelinase, sphingomyelin synthase, ceramidase, and ceramide synthase activities were not substantially inhibited or stimulated by the P4 diastereomers.

Although not every ceramide-utilizing enzyme was assayed, the failure to identify a second site of action for the P4 compounds raised the possibility that a novel ceramide pathway was involved. In order to assess this possibility, MDCK cells were incubated with tritiated C2 ceramide. Previous studies utilizing radiolabeled N-acetylsphingosine utilized ceramides labeled in the acetyl group. If deacetylated by a cellular ceramidase, the acetyl group would be only slightly metabolized to other labeled sphingolipids. MDCK cells rapidly metabolized N-acetyl[^{3}H]-sphingosine to free tritiated sphingosine, C_2-sphingomyelin, C2-glucosylceramide, and normal chain length ceramides, sphingomyelins, and glycosphingolipids (Fig. 7.3).[27]

When the cellular lipids were extracted without alkaline methanolysis and separated by thin layer chromatography, a rapidly migrating radiolabeled product was observed (Fig. 7.3). When developed with a chromatography solvent consisting of chloroform:methanol:acetic acid (90:1:9) this product migrated faster than ceramide,

Table 7.1. Effects of PDMP, some structurally related homologues, and N-acetylsphingosine on sphingolipid levels and cell growth

Compound	Ceramide	Glucosylceramide	Cell Growth
D-threo-1-Phenyl-2-decanoylamino-3-morpholino-propanol (D-threo-PDMP)	increased	decreased	decreased
DL-threo-1-Phenyl-2-palmitoyl-3-pyrrilidino-propanol (threo-P4)	increased	decreased	decreased
DL-erythro-1-Phenyl-2-palmitoyl-3-pyrrilidino-propanol (erythro-P4)	increased	no change	decreased
1-Pyrrilidino-1-deoxyceramide	no change	decreased	no change
N-Acetylsphingosine (C2-ceramide)	increased	increased	decreased

consistent with the formation of a less polar sphingolipid. Alkaline methanolysis of this lipid resulted in the reformation of N-acetylsphingosine, consistent with the presence of a fatty acid in ester linkage.

In order to determine the position of the O-acyl group, the radiolabeled unknown product was treated with 2,3-dichloro-5,6-dicyanobenzoquinone. This reagent oxidizes α,β-unsaturated alcohols to corresponding ketones. Under these conditions, if the C3 position of the sphingosyl group contains a free hydroxyl, it is oxidized to a ketone group. If the C3 position contains a substituted ester, then it is unreactive. Following 2,3-dichloro-5,6-dicyanobenzyoquinone treatment most of the product was recovered in a still higher position on the thin layer chromatography plate. This new product corresponded to the position of synthetic O-palmitoyl-N-acetyl-3-keto-sphingosine.

In 1977 Okabe and Kishimoto reported the existence of 1-O-acylceramide in rat brain.[28] However, the enzymatic pathway for the formation of acylated ceramides had not subsequently been elucidated. We observed that neither free fatty acid nor fatty acylCoA were suitable substrates for the formation of the 1-O-acylceramide. On the other hand phospholipids, particularly phosphatidylethanolamine and phosphatidylcholine, could serve as donors of fatty acyl groups from their sn-2 positions. Arachidonoylphosphatidylethanolamine was an excellent substrate for the enzyme. In the absence of ceramide, free arachidonic acid was liberated from phosphatidylethanolamine. These data are consistent with the view that the transacylase is a phospholipase A2 having dual competing activities: the hydrolytic release of arachidonate and the transfer of the arachidonate to ceramide.

The properties of the transacylase/phospholipase were further characterized. The activity was recovered in a 100,000 x g cell supernatant and had an apparent Km of 9.4 μM for N-acetylsphingosine. The enzyme had a pH optimum of 4.2 and was not inhibited by nonadecyltetraenyl trifluoromethyl ketone, a potent inhibitor of cytosolic phospholipase A2. The enzyme therefore appears to be a previously uncharacterized phospholipase. In cellular homogenates, product formation was limited over time and eventually decreased, consistent with the presence of an endogenous hydrolase. Subcellular fractionation showed that the

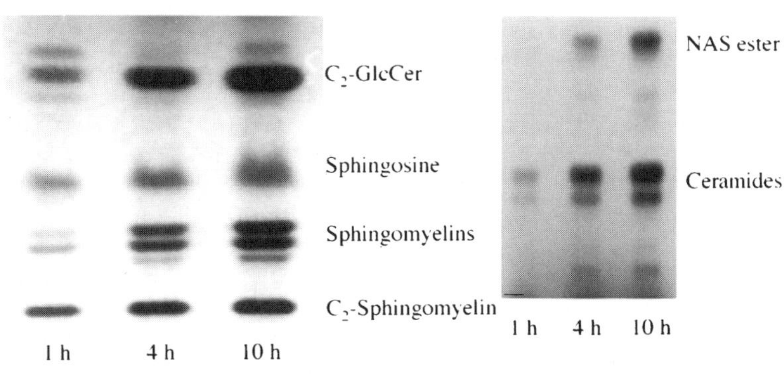

Fig. 7.3. Products of N-acetylsphingosine metabolism in MDCK cells. MDCK cells were incubated with 10 mM N-acetyl-[^{3}H]sphingosine for 1, 4, or 10 hours. The total lipids were extracted, separated by thin layer chromatography, and detected by autoradiography. NAS ester denotes 1-O-acyl-N-acetylsphingosine. Reprinted with permission from Abe A et al, J Biol Chem 1996; 271:14383-14389.

hydrolytic activity was absent from the cytosolic fraction.

We also know that the enzyme is inhibited 50% by 20 μM PDMP and is inhibited by both diastereomers of P4 at concentrations comparable to those that increase cell ceramide. Therefore the basis for the ceramide accumulation appears to be inhibition of a novel pathway, ceramide acyltransferase (Fig. 7.4).

Several important questions remain unanswered. First, what is the composition of endogenous 1-O-acylceramide? We observed that truncated ceramides were acylated at a greater rate than normal fatty acyl chain length ceramides. Although this may be an artifact due to the solubility differences between short and long chain ceramides, the recent discovery of naturally occurring N-acetylsphingosine[29] raises the possibility that short chain ceramides are acylated and exist in vivo. In addition, it is important to consider whether arachidonate is a preferred fatty acid for 1-O-acylation and whether the 1-O-acylceramide hydrolase liberates arachidonic acid.

Second, is the primary function of the acyltransferase to act as a phospholipase A2 or are there acceptors for the acyltransferase other than ceramide? In this regard it is important to determine the precise subcellular location of the ceramide acyltransferase.

Third, is the acyltransferase truly the mediator of PDMP-mediated cell growth arrest? If so, then PDMP homologues that inhibit the acyltransferase should follow the same rank order as those that block cell proliferation. In addition, ceramide or 1-O-acylceramide should colocalize in cells at the site of the acyltransferase.

IDENTIFICATION OF A ROLE FOR CERAMIDE IN THE HEAT SHOCK RESPONSE FOLLOWING AN ATTEMPT TO CLONE GLUCOSYLCERAMIDE SYNTHASE

In order to understand the specificity of these inhibitors further, we attempted to clone the glucosylceramide synthase. Our cloning strategy was based on two observations. The first observation was that the synthase was actually induced under conditions where cultured cells were incubated with PDMP and then the inhibitor diluted out prior to the assay.[30] The time-dependent, actinomycin D inhibitable increase in the glycosyltransferase was observed, consistent with new transcription of the synthase gene. The second observation was based on a consensus sequence defining a UDP-hexose binding domain for glycosyltransferases utilizing lipoidal substrates.[31] Using degenerate primers to this binding domain, a cDNA library was constructed and screened for PDMP inducibility.

An induced gene was identified in NIH 3T3 cells. Surprisingly, it was not glucosylceramide synthase, but rather a small heat shock protein, αB-crystallin. αB-Crystallin is a ubiquitous and multifunctional protein. The α-crystallins are members of a super-

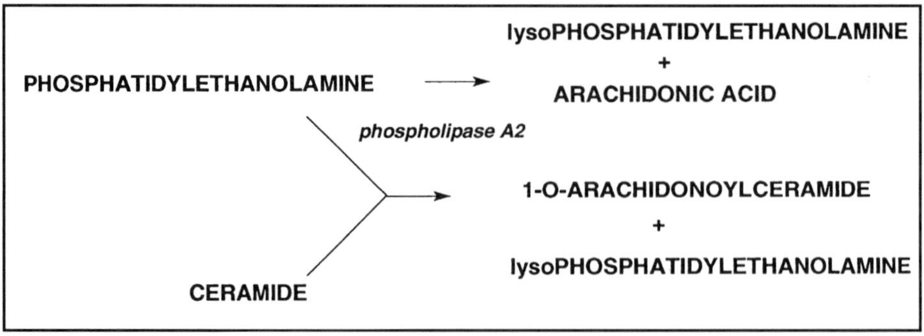

Fig. 7.4. Enzymatic pathway for 1-O-acylceramide formation.

family of small HSPs and share many of the structural and functional properties of the small HSPs.[32] They not only have chaperone activity, but they also possess autokinase activity, and confer thermostability on cells. αB-crystallin associates with the cytoskeleton and may inhibit the assembly of intermediate filaments. αB-Crystallin, like ceramide, may therefore play an active role in cellular morphogenesis.

To evaluate this finding further, we incubated the NIH 3T3 cells with a cell permeant short chain ceramide, N-acetylsphingosine. A time and concentration dependent increase in αB-crystallin transcription was observed. To determine whether endogenous cell ceramides containing fatty acyl moieties of normal chain length were also active in stimulating transcription, we incubated cells with either bacterial sphingomyelinase or PDMP. Sphingomyelinase treatment caused the rapid accumulation of ceramide. The αB-crystallin mRNA was detectable at one hour and no longer detectable by 12 hours of incubation. In contrast, PDMP treatment resulted in a more gradual accumulation of ceramide. Under these conditions the αB-crystallin mRNA peaked at 12 hours and was still detectable at 24 hours.

The ability of ceramide to induce an increase in αB-crystallin mRNA was structurally specific. The saturated form of ceramide, N-acetyldihydrosphingosine, had no effect on αB-crystallin mRNA levels. By contrast, N-acetylsphingosine induced αB-crystallin mRNA expression to levels comparable to those seen with heat shock.

The ceramide effect was specific for αB-crystallin. HSP25 is a structurally similar protein with a high degree of homology at the amino acid and gene levels. The exposure of cells to 42.5°C for 1 hour resulted in detectable increases in both αB-crystallin and HSP25 mRNA. However, only αB-crystallin and not HSP25 mRNA was detected following treatment with N-acetylsphingosine. The elevation of the cell temperature resulted in the formation of both αB-crystallin and HSP25 protein. But only the protein level of αB-crystallin was increased following N-acetylsphingosine treatment.

To ascertain whether endogenous ceramides would be formed as part of the heat shock response, 3T3 cells were incubated at 42.5°C for 1 hour and then at 37°C. Ceramide levels doubled following heat treatment and remained elevated for 6 hours following the return to 37°C. This finding has recently been confirmed in the U937 cell line.

Historically, establishing a biochemical nexus between heat shock and heat shock protein formation has been difficult. An important clue to the signaling basis for heat shock has emerged from work in sphingolipid deficient yeast. Lester and Dickson identified two mutations in sphingolipid deficient strains of *S. cerevisiae*.[33] One mutation is in the *lcb1* gene, resulting in inability to synthesize long chain bases. The other is a semidominant mutation which allows the yeast to grow without its particular long chain base. Further study revealed that while the sphingolipid deficient yeast grew normally under the usual culture conditions, they could not survive when stressed with low pH, an osmotic challenge, or elevated temperature.

A CAT REPORTER CONSTRUCT OF THE αB-CRYSTALLIN GENE IDENTIFIES CIS-ACTING SITES FOR CERAMIDE-INDUCED TRANSCRIPTION IN THE –661 TO +44 BASE PAIR REGION OF THE GENE

The transcriptional regulation of the αB-crystallin gene is complex, demonstrating both tissue specific control elements as well as altered expression to cell stress. Expression studies in transgenic mice utilizing a transgene comprised of a -661/+44 fragment of the αB-crystallin gene fused to the bacterial CAT gene revealed tissue specific expression in some but not all tissues where αB-crystallin is normally expressed.[32] These experiments suggested that the regulatory elements for αB-crystallin expression are located in the 5' flanking sequence for lens and skeletal muscle, but lie outside of this region for expression in kidney and brain.

An enhancer important for the expression of αB-crystallin in skeletal muscle cells has been identified between positions -427 and -259. Four functional elements were discovered in the enhancer by DNase I footprinting and established by additional studies. One element, MRF, appears to be muscle specific. The other elements, αBE-1, αBE-2, and αBE-3 are important for expression in other tissues.

In order to evaluate the role of sphingolipids in mediating αB-crystallin transcription, a CAT reporter construct containing the -661/+44 fragment of the αB-crystallin gene was obtained from Dr. Joram Piatigorsky. NIH 3T3 cells were cotransfected with β-galactosidase and the CAT construct. N-Acetylsphingosine (25 μM) treatment resulted in a level of CAT activity that was comparable to that produced by heat shock (Fig. 7.5); N-acetyldihydrosphingosine (50 μM) and sphingosine-1-phosphate (25 μM) had no significant effect on reporter activity.

These data correlate with the structure-activity profile observed on Northern analyses of αB-crystallin expression, suggesting that the "ceramide response element" is located within this -661/+44 fragment. Of interest is the presence of two AP1 sites (AGTCA) and one AP1-like site (CGTCA) within this region. Two other ceramide responsive genes, interleukin 6 and cyclooxygenase 2 also contain AP1 sites in their upstream noncoding regions. HSP25, the small heat shock protein with sequence homology to αB-crystallin, lacks comparable sites.

Recently, one form of mitogen activated protein kinase, the stress activated protein kinase (SAP kinase), has been shown to be activated by stress induced changes in cell ceramide levels.[34] It is tempting to postulate that SAP kinase stimulated AP1 activation is the proximate signal for αB-crystallin transcription. This hypothesis is currently being tested.

SPHINGOLIPID STRUCTURE: TEMPERATURE DEPENDENT EFFECTS ON PHASE TRANSITION

It is worth speculating as to why sphingolipid metabolism may be uniquely

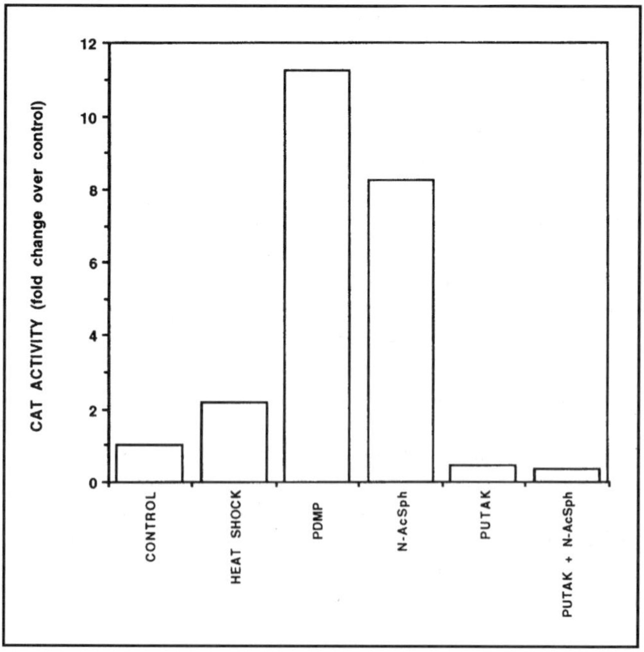

Fig. 7.5. Chloramphenicol acetyltransferase activity from NIH3T3 cells transfected with an αB-crystallin reporter construct in response to heat shock and sphingolipid treatment. Cells were transfected for 5 hours with the pD96B2 plasmid containing the -661/+44 base pair region of the 5' untranslated region of the αB-crystallin gene or with a neutral CAT promoter (PUTAK3). Cotransfections were performed with a pSV-β-galactosidase vector driving the transcription of the lacZ gene for monitoring of transfection efficiency.

sensitive to heat shock. One clue may lie within the structural uniqueness of ceramide based sphingolipids in comparison to diglyceride based glycerolipids. For example, sphingomyelin, like phosphatidylcholine, is a lipid sharing the common hydrophilic headgroup, phosphorylcholine. Both lipids are amphipathic molecules. However, the nonpolar lipoidal moieties are distinctly different. First, the fatty acid in the sn2 position of phosphatidylcholine is often unsaturated. In contrast, the fatty acid in amide linkage in sphingomyelin is usually saturated and is of 16-24 carbons in length. More than 50% of the fatty acyl groups are greater then 20 carbons in length. The nonpolar region of the sphingoid base is only 13-15 carbons in length. Thus the two hydrophobic chains of sphingomyelin may differ in length by as many as 7 methylene residues.

Second, the hydroxyl at carbon three and the amide group provide hydrogen bond donor capability not present in the glycerol backbone of other phospholipids. This difference in structure suggests that the hydrogen bonding between sphingolipids and cholesterol and sphingolipids and proteins will be stronger than between glycerolipids and like molecules.[35] These structural differences also impart unique physical properties to sphingolipids. Both the interdigitation of fatty acyl groups of sphingolipids on opposite leaflets of the plasma membrane and the increased hydrogen bonding increase the rigidity of membranes that are high in sphingolipid content.

One consequence of these physical properties is that the temperature dependent phase behavior of sphingomyelin is distinctly different from that of phosphatidylcholine. When studied in artificial liposomes, sphingomyelin undergoes a transition from a liquid crystal to a liquid gel at ~41°C. When reconstituted with sphingomyelinase, the lipase activity is markedly increased as this phase transition is reached.[36]

Another consequence of these physical properties is the tendency of sphingolipids to self aggregate in membranes. Recently specialized cell structures termed caveolae have been shown to be enriched in glycosphingolipids. These caveolae are enriched in phosphatidylinositol linked proteins as well as proteins governing cell growth responses (PDGF receptor) and transporters. Interleukin 1 stimulated sphingomyelin hydrolysis is reported to occur in the caveolar sphingomyelin pool.

ESTABLISHING PROOF OF PRINCIPLE FOR CERAMIDE DEPENDENCE IN HEAT SHOCK RESPONSES

As with other cellular functional responses, the role of ceramide in mediating the heat shock response is correlational and not necessarily causal. However, since much is known about the heat shock response, a number of predictions can be tested to evaluate the role of sphingomyelin in the heat shock response. Four such predictions are offered.

First, other forms of cellular stress including osmotic stress, heavy metals (cadmium), and sodium arsenite exposure induce αB-crystallin formation. If sphingomyelin hydrolysis is a common factor in these other forms of cell stress, then ceramide levels should be elevated in these conditions as well.

Second, cells which are tolerized to heat will survive heat shock compared to those not undergoing thermotolerance. Ceramide pretreatment at concentrations below those which induce apoptosis should replicate this thermotolerance.

Third, those agonists which cause αB-crystallin formation in the absence of HSP25 formation should produce ceramide. TNFα is one such agonist. Goldman and co-workers have already demonstrated its specificity for αB-crystallin in a neuronal cell line.[37]

Finally, as discussed above, defined ceramide responsive elements in the αB-crystallin gene should be identifiable. Several elements have already been defined within the 5' region of the αB-crystallin gene. These include an AP1 site which is conceivably a site for SAPK/JNK induced c-jun binding and which is present in other ceramide responsive genes but not present in the ceramide unresponsive HSP25 gene. Reporter construct studies are underway to evaluate potential ceramide-dependent transcription elements in the αB-crystallin gene.

CONCLUSION

In summary, the pharmacological manipulation of cell sphingolipid levels is a potentially fruitful approach to the understanding of sphingolipid signaling events. A new generation of compounds modeled after the cerebroside synthase inhibitor PDMP have been designed and tested. These compounds have led to the identification of a new pathway for ceramide metabolism, the formation of 1-O-acylceramide. This lipid is formed by an apparently novel phospholipase A2 with transacylase activity. Cerebroside synthase inhibitors have also resulted in the unexpected discovery that signaling through the heat shock response may be mediated by ceramide generation.

These studies provide pharmacological support to the view that ceramides represent important cellular messengers for a diverse range of cellular responses. In addition, cerebroside synthase and ceramide acyltransferase inhibitors may have important therapeutic potential in the manipulation of cell growth and stress responses.

REFERENCES

1. Kim MY, Linardic C, Obeid L et al. Identification of sphingomyelin turnover as an effector mechanism for the action of tumor necrosis factor α and γ-interferon. J Biol Chem 1991; 266:484-489.
2. Ballou LR, Chao CP, Holness MA et al. Interleukin-1-mediated PGE_2 production and sphingomyelin metabolism. J Biol Chem 1992; 267:20044-20050.
3. Cifone MG, De Maria R, Roncaioli P et al. Apoptotic signaling through CD95 (Fas/Apo-1) activates an acidic sphingomyelinase. J Exp Med 1994; 80: 1547-1552.
4. Dobrowsky RT, Jenkins GM, Hannun YA. Neurotrophins induce sphingomyelin hydrolysis. Modulation by co-expression of p75NTR with Trk receptors. J Biol Chem 1995; 270:22135-22142.
5. Okazaki T, Bell RM, Hannun YA. Sphingomyelin turnover induced by vitamin D_3 in HL-60 cells. J Biol Chem 1989; 264:19076-19080.
6. Nakamura T, Abe A, Balazovich K, Wu D et al. Ceramide regulates oxidant release in adherent human neutrophils. J Biol Chem 1994; 269:18384-18389.
7. Niculescu F, Rus H, Shin S et al. Generation of diacylglycerol and ceramide during homologous complement activation. J Immunol 1993; 150:214-224.
8. Zhang H, Desai NN, Olivera A et al. Sphingosine-1-phosphate, a novel lipid involved in cellular proliferation. J Biol Chem 1990; 265:76-81.
9. Igarashi Y, Kitamura K, Toyokuni T et al. A specific enhancing effect of N,N-dimethylsphingosine on epidermal growth factor receptor autophosphorylation. J Biol Chem 1990; 265:5385-5389.
10. Wang E, Norred WP, Bacon CW et al. Inhibition of sphingolipid biosynthesis by fumonisins. Implications for diseases associated with Fusarium moniliforme. J Biol Chem 1991; 266:14486-14490.
11. Grabowski GA, Osiecki-Newman K, Dinur T et al. Human acid beta-glucosidase. Use of conduritol B epoxide derivatives to investigate the catalytically active normal and Gaucher disease enzymes. J Biol Chem 1986; 261:8263-9.
12. Medlock KA, Merrill AH Jr. Inhibition of serine palmitoyltransferase in vitro and long-chain base biosynthesis in intact Chinese hamster ovary cells by β-chloroalanine. Biochemistry 1988; 27:7079-7084.
13. Radin NS, Vunnam RR. Inhibitors of cerebroside metabolism. Methods Enzymol 1981; 72:673-84.
14. Radin NS, Shayman JA, Inokuchi J. Metabolic effects of inhibiting glucosylceramide synthesis with PDMP and other substances. Adv Lipid Res 1993; 26:183-213.
15. Shukla A, Radin NS. Metabolism of D-[^{3}H]threo-1-phenyl-2-decanoylamino-3-morpholino-1-propanol, an inhibitor of glucosylceramide synthesis, and the synergistic action of an inhibitor of microso-

mal monooxygenase. J Lipid Res 1991; 32:713-22.
16. Shayman JA, Mahdiyoun S, Deshmukh G et al. Glucosphingolipid dependence of hormone-stimulated inositol trisphosphate formation. J Biol Chem 1990; 265:12135-12138.
17. Shayman JA, Deshmukh GD, Mahdiyoun S et al. Modulation of renal epithelial cell growth by glucosylceramide: association with protein kinase C, sphingosine, and diacylglyceride. J Biol Chem 1991; 266: 22968-22974.
18. Holleran WM. Lipid modulators of epidermal proliferation and differentiation. Adv Lipid Res 1991; 24:119-139.
19. El-Khatib M, Radin NS, Shayman JA. A role for de novo glycosphingolipid synthesis in the proliferative response of renal proximal tubule cells to hyperglycemia. Am J Physiol 1996; 270:F476-F484.
20. Bremer EG, Schlessinger J, Hakomori S-i. Ganglioside-mediated modulation of cell growth. J Biol Chem 1986; 261:2434-2440.
21. Rani CS, Abe A, Chang Y et al. Cell cycle arrest induced by an inhibitor of glucosylceramide synthase—correlation with cyclin-dependent kinases. J Biol Chem 1995; 270:2859-2867.
22. Dbaibo GS, Pushkareva MY, Jayadev S et al. Retinoblastoma gene product as a downstream target for a ceramide-dependent pathway of growth arrest. Proc Natl Acad Sci 1995; 92:1347-1351.
23. Dobrowsky RT, Kamibayashi C, Mumby MC et al. Ceramide activates heterotrimeric protein phosphatase 2A. J Biol Chem 1993; 268:15523-15530.
24. Abe A, Inokuchi J, Jimbo M et al. Improved inhibitors of glucosylceramide synthase. J Biochem (Tokyo) 1992; 111: 191-196.
25. Abe A, Radin NS, Shayman JA et al. Structural and stereochemical studies of potent inhibitors of glucosylceramide synthase and tumor cell growth. J Lipid Res 1995; 36:611-621.
26. Carson KG, Ganem B, Radin NS et al. Studies on morpholinosphingolipids: potent inhibitors of glucosylceramide synthase. Tetrahedron Lett 1994; 35: 2659-2662.
27. Chang Y, Abe A, Shayman JA. Ceramide formation during heat shock: a potential mediator of αB-crystallin transcription. Proc Natl Acad Sci 1995; 92:12275-12279.
28. Okabe H, Kishimoto Y. In vivo metabolism of ceramides in rat brain. Fatty acid replacement and esterification of ceramide. J Biol Chem 1977; 252:7068-7073.
29. Lee TC, Ou MC, Shinozaki K et al. Biosynthesis of N-acetylsphingosine by platelet-activating factor: sphingosine CoA-independent transacetylase in HL-60 cells. J Biol Chem 1996; 271:209-17.
30. Abe A, Radin NS, Shayman JA. Induction of glucosylceramide synthase by ceramide and synthase inhibitors. Biochim Biophys Acta 1996; 1299:333-341.
31. Schulte S, Stoffel W. Ceramide UDP-galactosyltransferase from myelinating rat brain: purification, cloning, and expression. Proc Natl Acad Sci 1993; 90:10265-10269.
32. Sax CM, Piatigorsky J. Expression of the α-crystallin/small heat-shock protein/ molecular chaperone genes in the lens and other tissues. Adv Enz Related Areas 1994; 69:155-201.
33. Nagiec MM, Baltisberger JA, Wells GB et al. The LCB2 gene of Saccharomyces and the related LCB1 gene encode subunits of serine palmitoyltransferase, the initial enzyme in sphingolipid synthesis. Proc Natl Acad Sci 1994; 91:7899-7902.
34. Verheij M, Bose R, Lin XH et al. Requirement for ceramide-initiated SAPK/ JNK signalling in stress-induced apoptosis. Nature 1996; 380:75-9.
35. Pascher I. Molecular arrangements in sphingolipids. Conformation and hydrogen bonding of ceramide and their implication on membrane stability and permeability. Biochim Biophys Acta 1976; 455:433-451.
36. Cohen R, Barenholz Y. Correlation between the thermotropic behavior of sphingomyelin liposomes and sphingomyelin hydrolysis by sphingomyelinase of Staphylococcous aureus. Biochim Biophys Acta 1978; 509:181-187.
37. Head MW, Corbin E, Goldman JE. Coordinate and independent regulation of alpha B-crystallin and hsp27 expression in response to physiological stress. J Cell Physiol 1994; 159:41-50.

CHAPTER 8

Role of Sphingolipids in Regulating the Phospholipase D Pathway and Cell Division

Antonio Gómez-Muñoz, Abdelkarim Abousalham,
Yutaka Kikuchi, David W. Waggoner and David N. Brindley

INTRODUCTION

This article deals with the interactions of some bioactive glycero-lipids and sphingolipids that control signal transduction, cell activation and cell division. In particular, the review will concentrate on the consequences of agonist-induced hydrolysis of phosphatidylcholine and sphingomyelin. These two phospholipids are related to each other structurally and essentially differ in that phosphatidylcholine has a glycerolipid backbone whereas sphingomyelin contains an N-acylated fatty acid linked to sphingosine (Table 8.1). The hydrolysis of these two lipids gives rise to bioactive lipid analogs such that diacylglycerol is related to ceramide, lysophosphatidate to sphingosine 1-phosphate and monoacylglycerol to sphingosine. The discussion will emphasize the interaction or "cross-talk" between the glycerolipid and sphingolipid signaling pathways.

CELL SIGNALING BY PHOSPHATIDATE, LYSOPHOSPHATIDATE AND DIACYLGLYCEROL

The turnover of lipids in cell membranes plays a critical role in signal transduction and cell regulation. The interaction of extracellular agonists with specific membrane receptors results in the activation of specific phospholipases which generate intracellular second messengers from structural lipid precursors. A well known example is the activation of the phosphoinositide-specific phospholipase C that degrades phosphatidylinositol 4,5-bisphosphate to generate inositol 1,4,5-trisphosphate and diacylglycerol (Fig. 8.1). Both products play critical roles in the transduction of the extracellular signals: inositol trisphosphate causes the release of stored Ca^{2+} from the endoplasmic reticulum which increases the intracellular Ca^{2+} concentration in the cytosol.[1] Ca^{2+} can then interact with specific Ca^{2+}/calmodulin kinases and thus regulate a variety of cellular functions. The diacylglycerol

Sphingolipid-Mediated Signal Transduction, edited by Yusuf A. Hannun.

Table 8.1. Structures of sphingolipids

Chemical Structure	Sphingolipid	Glycerolipid analogue
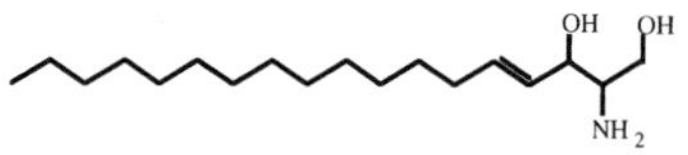	Sphingosine	Monoacylglycerol
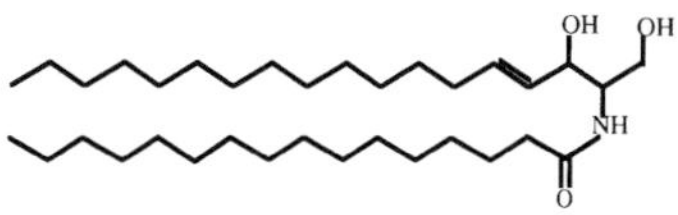	Ceramide	Diacylglycerol
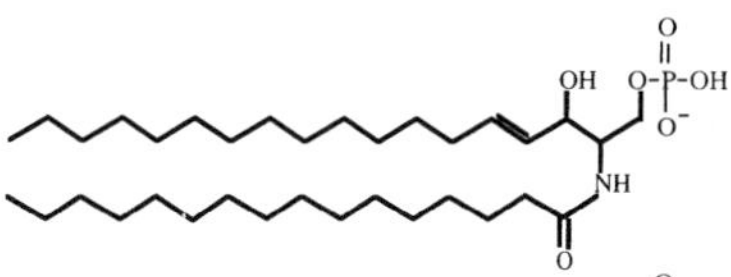	Ceramide 1-phosphate	Phosphatidate
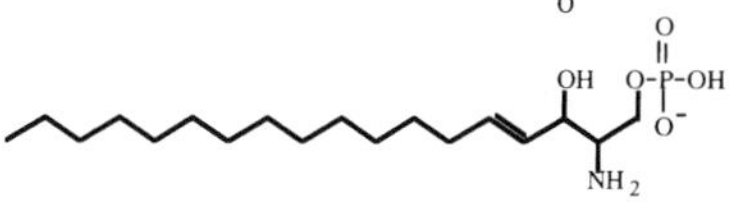	Sphingosine 1-phosphate	Lysophosphatidate
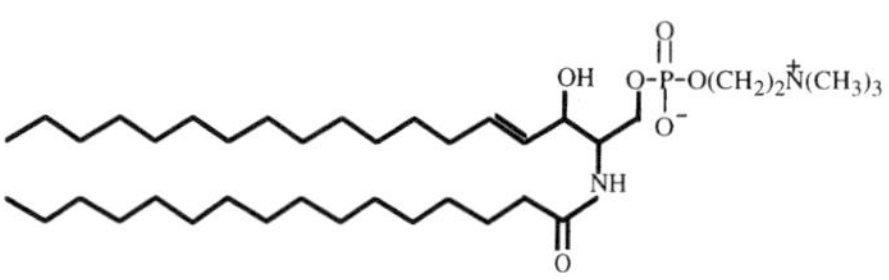	Sphingomyelin	Phosphatidylcholine

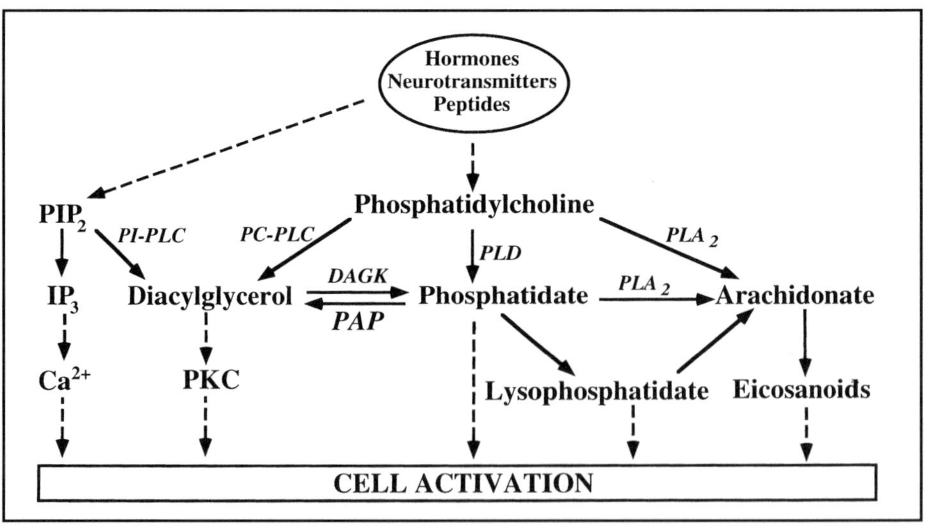

Fig. 8.1. Production of bioactive glycerolipids following agonist-stimulated activation of phospholipases. Abbreviations are as follows: DAGK, diacylglycerol kinase; IP_3, inositol 1,4,5-trisphosphate; PAP, phosphatidate phosphohydrolase; PIP_2, phosphatidylinositol 4,5-bisphosphate; PI-PLC, phosphatidylinositol-specific phospholipase C; PKC, protein kinase C; PLA_2, phospholipase A_2; PC-PLC, phosphatidylcholine-specific phospholipase C; PLD, phospholipase D. The figure is modified from Martin et al.[3]

together with Ca^{2+} activates protein kinase C (α, β, and γ isoforms) and this aspect of signal transduction is involved in the regulation of cell division. The increase in diacylglycerol concentrations that results from the hydrolysis of phosphatidylinositol 4,5-bisphosphate commonly lasts about 1 minute and it is followed by a second and more sustained production of diacylglycerol that is derived from other phospholipids, mainly phosphatidylcholine.[2,3] This production of diacylglycerol results directly from the activation of a phosphatidylcholine-specific phospholipase C, or indirectly via the sequential activation of phospholipase D to produce phosphatidate and phosphatidate phosphohydrolase to yield diacylglycerol (Fig. 8.1). This second phase of diacylglycerol production occurs without an accompanying release of Ca^{2+} and so might be related to the activation of the Ca^{2+}-independent isoforms of protein kinase C.

The stimulation of phospholipase D normally occurs in response to the activation of tyrosine kinases by growth factors, e.g., epidermal growth factor, or following the stimulation of protein kinase C.[4,5] These events may be linked in some cases, since the activation of phospholipase C-γ by growth factors through the stimulation of tyrosine kinases can then produce the Ca^{2+} and diacylglycerol signals that subsequently activate protein kinase C.

Another dimension of the phospholipase D signaling pathway involves phosphatidate, a bioactive lipid (Table 8.2) that

Table 8.2. Some signaling effects of diacylglycerol, phosphatidate and lysophosphatidate

Lipid	Effect	Reference
Diacylglycerol	Activates protein kinase C	1-3
Phosphatidate	Activates phospholipase D	3,20
	Activates phospholipase C-γ	11
	Stimulates superoxide formation	6,7
	Stimulates serine/threonine kinase activities	8,9
	Stimulates phosphatydylinositol 4-kinase	10
	Stimulates monoacylglycerol acyltransferase	12
	Increases GTP-Ras	13
	Activates Raf	14
	Stimulates MAP kinase activities	32
	Stimulates cell proliferation	3,19,20,32
	Stimulates stress fiber formation	21,22
	Decreases cAMP levels	3,18,20,23
Lysophosphatidate	Activates phospholipase D	3,15,20
	Stimulates tyrosine kinase activity	15
	Stimulates Ca^{2+} mobilization	15,18
	Stimulates arachidonate release	15,18
	Activates phospholipase C-γ	18
	Stimulates stress fiber formation	15
	Stimulates cell proliferation	3,15,18,20,26,71
	Decreases cAMP levels	15,18,20
	Competitive inhibitor of phosphatidate phosphohydrolase	34,70,71

is involved in H_2O_2 formation in neutrophils.[6,7] Phosphatidate also stimulates the activities of protein kinases,[8] protein kinase C-ζ,[9] phosphatidylinositol 4-kinase,[10] phospholipase C-γ,[11] monoacylglycerol acyltransferase[12] and increases the GTP-bound form of Ras.[13] The latter effect would increase indirectly the activation of mitogen activated protein (MAP) kinase via stimulation of Raf. A further role for the phospholipase D pathway and phosphatidate in activating Raf is provided by the observation that phosphatidate binds to Raf and enables it to translocate to membranes where it becomes physiologically active.[14] The effect of phorbol esters in causing the translocation of Raf to membranes was decreased by the addition of ethanol to MDCK cells.[14] This effect correlated with the diversion of the phospholipase D reaction to the formation of phosphatidylethanol rather than the production of phosphatidate.

A further function of the phospholipase D pathway is the conversion of phosphatidate to lysophosphatidate via a phospholipase A type reaction. If this were to involve phospholipase A_2, then this might lead to the formation of arachidonic acid and a subsequent eicosanoid signal (Fig. 8.1). Lysophosphatidate is thought to be a paracrine or autocrine signal. For example, lysophosphatidate is released by activated platelets and this has been postulated as a means of stimulating local wound repair.[15] Eichholtz et al[16] have demonstrated that approximately 90% of the newly formed lysophosphatidate in thrombin activated platelets is released into the medium. In addition, production of lysophosphatidate through activation of secretory phospholipase A_2 on microvesicles may represent a pro-inflammatory pathway.[17] Lysophosphatidate, or phosphatidate, when added to cells elicits a variety of signal transducing processes (Table 8.2). Exogenous phosphatidate and lysophosphatidate increase the activity of tyrosine kinases and activate the Ras-Raf-MAP kinase pathway[18] and they are potent mitogens in fibroblasts.[15,19,20] They also activate focal adhesion kinase and the formation of stress fibers involving Rho.[15,21,22] Lysophosphatidate can activate phospholipase C-γ, Ca^{2+} mobilization in cells and arachidonate release.[18] Lysophosphatidate and phosphatidate both activate phospholipase D, thus leading to the further generation of phosphatidate.[15,20] Finally, both phosphatidate and lysophosphatidate decrease the activity of adenylate cyclase through pertussis-toxin sensitive mechanisms and thereby lower cAMP concentrations.[18,20,23] This latter effect may be involved in the stimulation of the Ras-Raf-MAP kinase pathway since low concentrations of cAMP facilitate the activation of Raf by Ras.[24,25] These effects therefore contribute to the mitogenic actions of phosphatidate and lysophosphatidate that are seen in cultured fibroblasts.[16,18,20] In other cells, such as Sp2/0-Ag14 myeloma cells, lysophosphatidate is antimitogenic. This action is not affected by pertussis toxin and involves an increase rather than a decrease in cAMP concentrations.[26] Furthermore in Swiss 3T3 fibroblasts, cAMP is a positive effector of mitogenesis.[27]

CELL SIGNALING BY SPHINGOLIPIDS

Sphingolipids have long been considered as structural units of the plasma membrane of cells. However, in the last decade it has become evident that sphingolipids have multiple biological activities (Table 8.3). They are involved in the transformation, differentiation and proliferation of cells, and many have been described as potent second messengers and regulators of cell activation. The stimulation of sphingomyelinases by various agonists has been described in chapter 2 and this reaction leads to the formation of ceramides (Fig. 8.2). Ceramides (Table 8.3) unlike their glycerolipid analogs, diacylglycerols, inhibit cell division and lead to apoptosis (chapters 2 and 6). Ceramides can be metabolized to sphingosine by ceramidases that change the essential nature of the cell signal (chapter 11). Sphingosine itself, or following its conversion to sphingosine 1-phosphate, is responsible for stimulating cell division (chapters 9 and 10).

Table 8.3. Some signaling effects of ceramides, sphingosine, sphingosine-1-phosphate and ceramide-1-phosphate

Lipid	Effect	Reference
Ceramides	Induces cell differentiation	41,92,93
	Inhibits cell division and DNA synthesis	20,33,41,92,93
	Induces apoptosis	41,92,93
	Inhibits vesicle movement and endocytosis	68
	Induces interleukin-2 secretion	94
	Stimulates serine/threonine kinase activity	40,41
	Stimulates protein kinase C-ζ	95,96
	Activates Raf and MAP kinase	35
	Stimulates phosphoprotein phosphatase activity	42,43,92
	Inhibits phospholipase D activity	20,33,34,38,39, 63,66,67
	Stimulates degradation of phosphatidate, lysophosphatidate, sphingosine-1-phosphate, and ceramide-1-phosphate	20,32,33,34,71
Sphingosine	Inhibits protein kinase C	75
	Inhibits phosphatidate phosphohydrolases 1 and 2	3,71,76-83
	Stimulates phospholipase D	88
	Stimulates Ca^{2+} mobilization	74
	Stimulates diacylglycerol kinase	86,87
	Stimulates cell proliferation	33,75
Sphingosine 1-phosphate	Stimulates phospholipase D	27,33,90
	Stimulates Ca^{2+} mobilization	27,90
	Stimulates kinase activity	97,98
	Competitive inhibitor of phosphatidate phosphohydrolase	34,70,71
	Stimulates MAP kinase, AP-1 and cell proliferation	25,90,97-101
	Prevents ceramide-induced apoptosis	102
Ceramide 1-phosphate	Stimulates DNA synthesis and cell proliferation	32
	Competitive inhibitor of phosphatidate phosphohydrolase	34,70,71

Kolesnick and co-workers identified a novel pathway in which ceramide originating from the action of neutral sphingomyelinase, but not glucosylceramidase, is converted to ceramide 1-phosphate by a Ca^{2+}-dependent kinase.[28] Shingal et al[29] identified a ceramide 1-phosphate phosphatase in rat brain suggesting that ceramide 1-phosphate might regulate some aspects of synaptic vesicle function. Boudker and Futerman[30] characterized a phosphatase that specifically hydrolyzes ceramide-1-phosphate in the plasma membrane. These results suggest that ceramide 1-phosphate may play an important role in cell activation, but little is known about its physiological effects. The pathological and toxic effects of ceramide 1-phosphate have been recognized since a sphingomyelinase D was identified as the active principle of the venom of the brown recluse spider, *Loxosceles reclusa*.[31] Sphingomyelinase D is also produced by

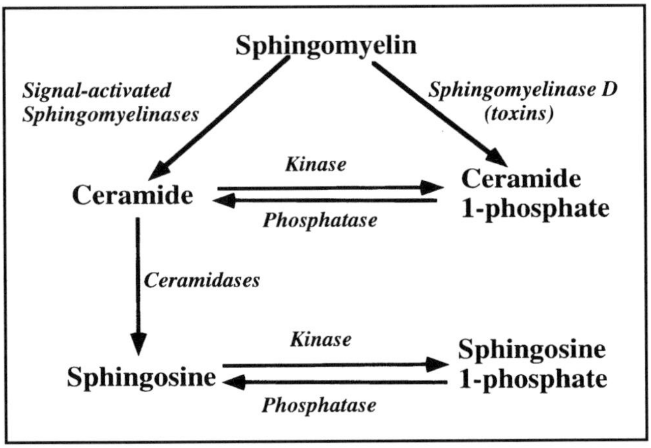

Fig. 8.2. Production of bioactive sphingolipids following the action of sphingomyelinases. The figure shows the metabolism of bioactive sphingolipids following the activation of agonist-stimulated sphingomyelinases. Also indicated is the action of the bacterial and arthropod sphingomyelinase D.

some bacteria including *Corynebacterium pseudotuberculosis* and *Vibrio damsela*.[31] Ceramide 1-phosphate can therefore be produced directly from sphingomyelin by the toxins of these organisms (Fig. 8.2) although there is as yet no evidence for this reaction occurring as a result of a mammalian enzyme. It will also become evident that ceramide 1-phosphates or derivatives might have application as novel pharmacological agents for altering cell signaling by mechanisms that differ from those employed by phosphatidate, lysophosphatidate and sphingosine 1-phosphate.[32]

CERAMIDES INHIBIT THE MITOGENIC EFFECTS OF PHOSPHATIDATE, LYSOPHOSPHATIDATE, SPHINGOSINE 1-PHOSPHATE AND CERAMIDE 1-PHOSPHATE

As discussed before (chapters 2 and 6) ceramides generally inhibit cell division and cause programmed cell death. By contrast phosphatidate, lysophosphatidate, sphingosine and sphingosine 1-phosphate stimulate DNA synthesis and cell division (Tables 8.2 and 8.3). We therefore investigated whether ceramides would inhibit the effects of the mitogenic lipids. In these experiments C_2, C_6 or C_8-ceramides (N-acetyl-, N-hexanoyl- or N-octanoyl-sphingosines, respectively) were used to mimic the effects of the natural long-chain ceramides. These cell-permeable ceramides inhibited the stimulation of DNA synthesis that was induced by phosphatidate, lysophosphatidate,[20] C_2- and C_8-ceramide 1-phosphates[32] and sphingosine 1-phosphate.[33] These effects of ceramides were also accompanied by a decrease in cell division.[32] The effects of the ceramides on the stimulation of DNA synthesis by the bioactive lipid phosphates were relatively specific since they were unable to block the induction of DNA synthesis by insulin, epidermal growth factor or platelet-derived growth factor.[33] Interestingly, relatively high concentrations of insulin potentiate the mitogenic effect of sphingosine 1-phosphate and protect the stimulation of DNA synthesis by sphingosine 1-phosphate against the inhibition by C_2-ceramide.[33] This observation indicates that insulin may be an important factor in regulating ceramide action, or vice versa. However, the concentrations of insulin used in these experiments were relatively high and it is not clear at present if this effect of insulin has physiological relevance.

It is important to define whether the action of ceramide was due to ceramide itself or to some potential byproduct of ceramide metabolism, especially sphingosine. Therefore, the effect of sphingosine on phosphatidate- or lysophosphatidate-induced DNA synthesis was evaluated. Rather than blocking the stimulation of DNA synthesis, sphingosine increased DNA synthesis further in the presence of phosphatidate or lysophosphatidate.[20] Although these effects were not strictly additive, these results indicate that sphingosine stimulated DNA synthesis by a mechanism that differed from that used by phosphatidate and lysophosphatidate.[20] This conclusion is supported by the observation that the increased DNA synthesis that is induced by sphingosine is only inhibited by about 16% by cell-permeable ceramides, whereas that stimulated by phosphatidate, lysophosphatidate, ceramide 1-phosphate and sphingosine 1-phosphate is inhibited almost completely.[20,32,33] It is therefore concluded that the inhibition of DNA synthesis by ceramide does not depend upon its metabolism.

One mechanism whereby ceramides inhibit the mitogenic effect of the bioactive lipid phosphates in rat fibroblasts might have been by inhibiting the decrease of cAMP accumulation. However, the cell-permeable ceramides were unable to reverse the effects of phosphatidate, lysophosphatidate, sphingosine 1-phosphate and sphingosine in lowering cAMP levels in the presence or absence of forskolin.[20,33] This implies that the mechanism by which ceramides inhibit the mitogenic effect of phosphatidate, lysophosphatidate, or sphingosine 1-phosphate is independent of interaction with adenylate cyclase. It was also significant to note that C_8-ceramide 1-phosphate, which was mitogenic for rat fibroblasts, did not change adenylate cyclase activity after 1 minute to 2 hours.[32]

Lysophosphatidate and phosphatidate also increase the activity of MAP kinase up to 5- to 6-fold in rat fibroblasts.[32] We therefore examined the interaction of C_2-ceramide on this activation to see whether this would explain the inhibition of DNA synthesis. Lysophosphatidate activated MAP kinase with a peak of activity at 5 minutes.[34] By contrast, C_2-ceramide alone produced an initial increase in MAP kinase activity of about 1.2- to 2-fold after 10-20 minutes. This was followed by a second increase in MAP kinase activity which was observed after 2 hours.[34] The activation of MAP kinase by ceramide is probably caused by a ceramide-induced phosphorylation of Raf.[35] There was no significant interaction in the rat fibroblast system between lysophosphatidate and C_2-ceramide in modifying the stimulation of MAP kinase activity.[34] Consequently, the inhibition of lysophosphatidate-induced DNA synthesis by ceramide could not be explained by changes in MAP kinase activity alone.

Unlike with phosphatidate, lysophosphatidate and sphingosine 1-phosphate (Tables 8.2 and 8.3) there was no change in MAP kinase activity after 1 minute to 2 hours when C_8-ceramide 1-phosphate was added to rat fibroblasts.[32] These results demonstrate that exogenous phosphatidate and ceramide 1-phosphate, which are structurally related (Table 8.1), exhibit different effects on signal transduction via cAMP and MAP kinase despite their apparently similar effects on DNA synthesis and cell division.

CERAMIDES INHIBIT THE ACTIVATION OF PHOSPHOLIPASE D

As mentioned before, phosphatidate is a potent bioactive lipid and mitogen (Table 8.2). Therefore the control of phospholipase D activity may be of critical importance for the regulation of cell activation. However, the generation of phosphatidate through activation of phospholipase D by different agonists need not be related directly to their mitogenic effects.[20,36] For example, endothelin is more efficient than lysophosphatidate in generating phosphatidate from phosphatidylcholine via phospholipase D, but it is a poor mitogen.[37] It is therefore unlikely that the stimulation of the phospholipase D pathway alone provides a complete mitogenic signal and that other events such as the stimulation of tyrosine kinases, or a decrease in cAMP concentrations, may be necessary for cell division to occur.

Lysophosphatidate, phosphatidate and sphingosine 1-phosphate stimulate the activity of phospholipase D when added to cells and sphingosine 1-phosphate is about 1000-fold more potent in molar terms than phosphatidate and lysophosphatidate.[20,33] Although ceramide 1-phosphate is an analog of phosphatidate (Table 8.1), it did not increase phospholipase D activity significantly[32] which provides further evidence that the bioactive lipid phosphates have different effects on signal transduction. The stimulation of phospholipase D by lysophosphatidate, phosphatidate and sphingosine 1-phosphate are all inhibited by C_2- and C_6-ceramides.[20,33] Similarly, the stimulation of phospholipase D by thrombin or serum is also blocked by cell-permeable ceramides. The same ceramide-induced inhibition of phospholipase D can be seen in permeabilized fibroblasts that were treated with GTPγS.[20] Ceramides have also been shown to inhibit the activation of phospholipase D in other cell systems. C_2-ceramide and sphingoid bases inhibited diradylglycerol formation by the phospholipase D pathway in neutrophils,[38] and C_6-ceramide inhibited the serum-stimulated accumulation of diacylglycerol and phospholipase D activation in senescent cells.[39]

Cell-permeable ceramides are able to stimulate the generation of long-chain ceramides in cells and these ceramides probably augment the action of the exogenous short-chain analogues.[33] Some effects of ceramides on cell signaling are caused by changes in the serine/threonine phosphorylation state of target proteins caused by activation of kinase activity,[40,41] or by increasing phosphoprotein phosphatases.[42,43] The latter effect can be blocked by okadaic acid but the addition of this compound to the fibroblasts did not reverse the ceramide-induced inhibition of phospholipase D.[33] Okadaic acid at 0.5 and 1.0 μM, in fact, decreased the activation of phospholipase D by sphingosine 1-phosphate.

In cell-free systems phospholipase D activity is dependent on the presence of both membrane and cytosolic components. These latter consist of small molecular weight G-proteins of the Ras super-family, such as ADP-ribosylation factor (ARF) and Rho.[44-47] ARF was first identified as a cofactor necessary for the ADP-ribosylation of the α-subunit of heterotrimeric G-proteins, i.e., G_s, by cholera toxin.[48] ARF is implicated in vesicular transport in the Golgi[49] and in endocytosis.[50] ARF-stimulated-PLD has been partially purified from HL-60 cell membranes and this stimulation was dependent on the presence of phosphatidylinositol 4,5-bisphosphate.[44] Subsequently, ARF-stimulated phospholipase D was separated from oleate-stimulated phospholipase D after solubilization from brain membranes, and its activation by class I, II, and III mammalian ARFs was demonstrated.[51] Rho proteins regulate the assembly of focal adhesion complexes and actin stress fibers in fibroblasts,[52] they inhibit phorbol ester-induced and integrin-dependent aggregation in lymphocytes[53] and they play a critical role in coupling of G-protein-linked chemoattractant receptors to integrin-mediated adhesion in leukocytes.[54] The exact mechanisms by which ARF and Rho A are involved in the PLD activation are still unclear.

Hammond et al[55] identified human phospholipase D cDNA, which defines a new and highly conserved gene family and described the critical role for phosphatidylinositol 4,5-bisphosphate in ARF-stimulated phospholipase D. In cell-free assays, ARF translocation correlates with potentiation of GTPγS-stimulated PLD activity in membrane fractions of HL-60 cells.[56] Several studies show that ARF acts synergistically with members of the Rho family,[57] and with other cytosolic factors from human neutrophils[58] and HL-60 cells.[59] In these studies, the cytosolic stimulating fraction prepared by gel filtration chromatography was estimated to be 50 kDa. Singer et al[60] recently reported that ARF and Rho A functions are synergistic with protein kinase C-α in the activation of phospholipase D partially purified from membranes of porcine brain. This effect of protein kinase C-α is independent of ATP. By contrast, Ohguchi et al[61] demonstrated that the synergism between protein kinase C-α and RhoA in the activation of membrane-bound phospholipase D was absolutely dependent on ATP in HL-60 cells. Although Rho can activate phospholipase D in cell-free systems, its physiological involvement has been questioned, and the major stimulation by G-proteins was concluded to be mediated by ARF.[62]

The mechanism for the effects of ceramides in inhibiting the activation of phospholipase D was examined further by using a reconstituted assay system. Phospholipase D activity in membranes from HL-60 cells was measured by following the release of [^{3}H]choline from phosphatidyl[^{3}H]choline.[63] This activity depended upon the presence of phosphatidylinositol 4,5-bisphosphate, GTPγS and the cytosolic fraction as expected. The addition of C_2-, C_8- and long-chain ceramides to the lipid substrates used in this assay resulted in a decrease in phospholipase D activity (Table 8.4). By contrast, dihydro-C_2-ceramide had no significant effect and dipalmitoylglycerol stimulated slightly.[63] These results demonstrated the specificity of the ceramide action and indicated that the inhibition was not caused by a nonspecific action of the lipids on phospholipase D. C_2-ceramide also decreased the activation of phospholipase D that was stimulated by recombinant ARF and Rho.[63]

These effects were examined further by reisolating the membranes and cytosolic fractions after incubation in the reconstituted phospholipase D assay. Addition of GTPγS to the incubation stimulated the association of the G-proteins, ARF and Rho, with the membrane fraction. C_2- and C_8-ceramide displaced both ARF and Rho from the membrane fraction into the cytosol. This effect was not seen with dihydro-C_2-ceramide.[63] Consequently, the inhibition of phospholipase D by ceramides can be explained by an inhibition of G-protein translocation to the membrane fraction. These results in the cell-free system were not dependent on the stimulation of kinases or phosphatases by ceramides since the effect was not modified significantly by the additions of either apyrase to destroy any endogenous ATP, or by okadaic acid to inhibit phosphoprotein phosphatases.[63]

The translocation of ARF to the plasma membrane has been observed in differentiated HL-60 cells that have been stimulated with *N*-formylmethionylleucylphenylalanine[56] and our results demonstrate that C_2-ceramide can reverse this effect on the subcellular distribution of ARF and Rho.[63] In these experiments there was also a ceramide-induced decrease in the amount of protein kinase C-α in the membrane fraction.

Inhibition of phospholipase D has been ascribed to the effects of ceramides in inhibiting the translocation of protein kinase C-α to the membrane fraction[64] since this protein can activate phospholipase D by ATP-dependent[47] and ATP-independent mechanisms.[48,65] An inhibitory effect of C_2-ceramide on phospholipase D activation

Table 8.4. Effects of ceramides and 1,2-dipalmitoyl-sn-glycerol on phospholipase D activity

Additions	Relative PLD Activity
None	100
+C_2-ceramide (N-acetylsphingosine)	53 ± 10
+C_8-ceramide (N-octanoylsphingosine)	63 ± 3
+long-chain ceramide	48 ± 3
+dihydro-C_2-ceramide	93 ± 4
+dipalmitoylglycerol	107 ± 4

The membrane plus cytosol fractions from HL-60 cells were assayed for PLD activity using a mixed liposome of PIP_2, phosphatidylethanolamine and phosphatidyl[^{3}H-methyl]choline.[44] The effects of 100 μM concentrations of the ceramides and diacylglycerol on PLD activity are given as means ± SD for three independent experiments. The results are taken from Abousalham, Liossis and Brindley.[63]

was also observed in rat basophilic leukemia (RBL-2H3) cells.[66] In the same work, the authors demonstrated that the translocation of protein kinase C-α, β_1, β_2, but not ε and δ isozymes, from cytosol to membrane fraction was specifically prevented during C_2-ceramide inhibition. In other work the inhibition of phorbol ester-induced activation of phospholipase D by ceramide was also observed, but this effect was not accompanied by a decrease in protein kinase C in the membrane fraction.[67] Ceramides did inhibit phospholipase D activation in a reconstituted cell-free assay system containing GTPγS. This phospholipase D activity was enhanced by adding protein kinase C activators, and ceramides specifically inhibited this effect upstream of stimulation of phospholipase D.[67]

The literature so far is consistent in finding that inhibition of phospholipase D is a target for the action of ceramides. There may be a variety of mechanisms that mediate this action via the regulation of low molecular weight G-proteins and protein kinase C. Furthermore, the effect of ceramides on the association of G-proteins with membranes could have far-ranging implications for the regulation of cell activation, cell motility and vesicle trafficking. As mentioned above, ARF is involved in vesicle movement and Rho is implicated in the organization of the cytoskeleton. Ceramides inhibit vesicle transport[68] and an interaction of ceramides on the subcellular distribution of small molecular weight G-proteins could contribute to this effect.

EFFECTS OF CERAMIDES ON THE METABOLISM OF PHOSPHATIDATE, LYSOPHOSPHATIDATE, SPHINGOSINE 1-PHOSPHATE AND CERAMIDE 1-PHOSPHATE

In addition to an effect on phospholipase D, ceramides can also modify the metabolism of phosphatidate, lysophosphatidate, sphingosine 1-phosphate and ceramide 1-phosphate and this could contribute to the antagonism of ceramides towards the mitogenic effects of the bioactive lipid phosphates. This was first shown using rat fibroblasts, where pretreatment with cell-permeable ceramides stimulated the degradation of exogenous phosphatidate and lysophosphatidate and decreased the interaction of these lipids with the cells.[20] Therefore, the inhibition of the interaction of the exogenous phospholipids with the cell membrane by ceramides and the enhanced degradation of phosphatidate and lysophosphatidate could be important steps in decreasing the mitogenic signal of these phospholipids. These changes were related to an increased activity of the plasma membrane phosphatidate phosphohydrolase.[20] At present the mechanism for this ceramide effect on the phosphohydrolase is not known.

Subsequent work showed that ceramides also stimulate the dephosphorylation of sphingosine 1-phosphate and ceramide 1-phosphate when these agonists are added externally to cells.[32,33] Thus, the action of ceramides in these latter cases is to promote the degradation of these two mitogenic lipid phosphates and to increase their conversion to ceramide. This action decreases the relative concentrations of the mitogenic lipids relative to ceramides. We provided preliminary evidence that the phosphohydrolase that degrades sphingosine 1-phosphate is identical to phosphatidate phosphohydrolase.[33] This prediction was confirmed in subsequent work with the Mg^{2+}-independent- and *N*-ethylmaleimide-insensitive phosphatidate phosphohydrolase that was purified from plasma membranes.[69] This enzyme degraded phosphatidate, lysophosphatidate, ceramide 1-phosphate and sphingosine 1-phosphate. Furthermore, these substrates were mutually competitive, indicating that they were degraded by the same active site on a common phosphohydrolase.[70,71] Glycerol phosphate was not a substrate for this phosphatidate phosphohydrolase, nor was it an inhibitor. Consequently, the plasma membrane phosphatidate phosphohydrolase appears to be a relatively nonspecific lipid phosphomonoesterase. There is, however, some degree of specificity since dolicholmonophosphate was a relatively poor substrate and also a poor inhibitor of phosphatidate hydrolysis.[70]

With regard to signal transduction, the Mg^{2+}-independent phosphatidate phosphohydrolase may terminate the signals from phosphatidate, lysophosphatidate, ceramide 1-phosphate and sphingosine 1-phosphate. In doing so it will also generate diacylglycerol, ceramide and sphingosine which have other signaling properties (Tables 8.2 and 8.3). Phosphatidate phosphohydrolase could therefore play an important role in cell signaling by modifying the balance of the bioactive lipids in cell activation and in controlling cell growth. Furthermore, the mutual competition between the lipid phosphomonoesters could be an additional site of "cross talk" between the glycerolipid and sphingolipid signaling pathways (Fig. 8.3).[70,71]

It is also significant that the activity of the plasma membrane phosphatidate phosphohydrolase is decreased in *ras*-transformed fibroblasts compared to nontransformed cells. This is accompanied by an increase in the production of phosphatidate relative to diacylglycerol following activation of phospholipase D.[72] This increased formation of phosphatidate may be important in: (1) producing the low cAMP concentrations

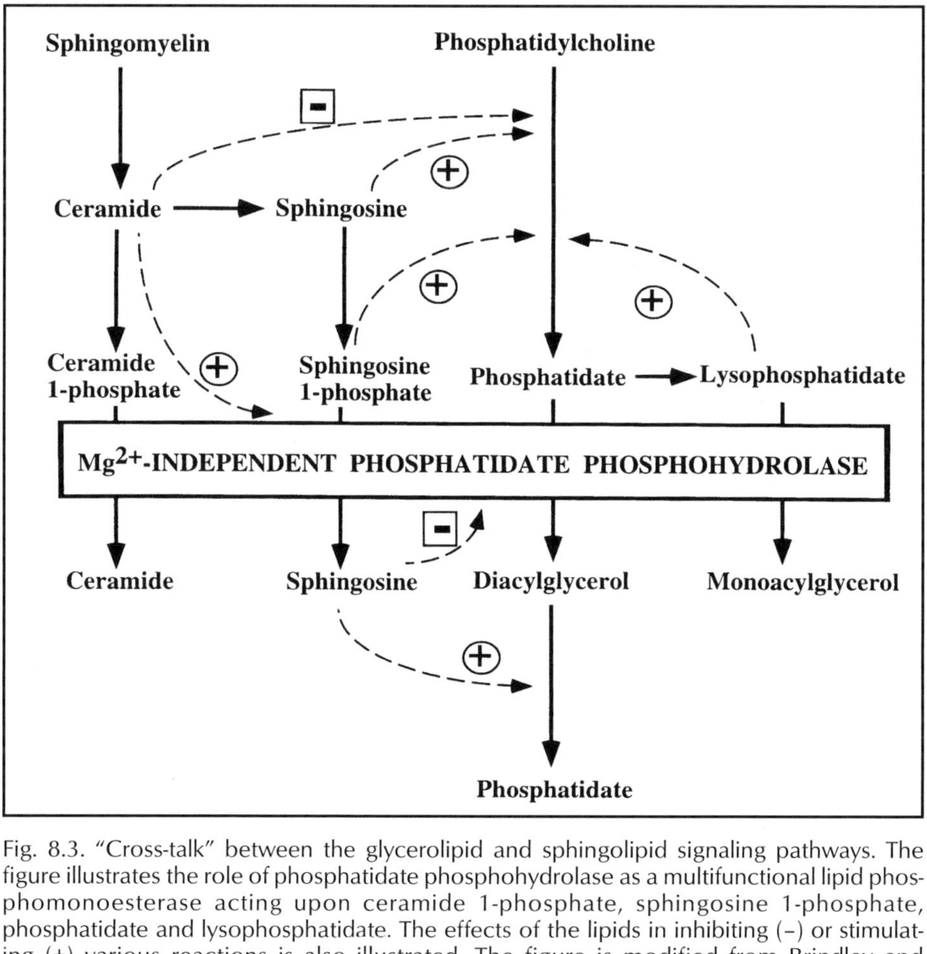

Fig. 8.3. "Cross-talk" between the glycerolipid and sphingolipid signaling pathways. The figure illustrates the role of phosphatidate phosphohydrolase as a multifunctional lipid phosphomonoesterase acting upon ceramide 1-phosphate, sphingosine 1-phosphate, phosphatidate and lysophosphatidate. The effects of the lipids in inhibiting (–) or stimulating (+) various reactions is also illustrated. The figure is modified from Brindley and Waggoner.[71]

that are observed in the transformed fibroblasts[73,74] and which may facilitate the activation of Raf by Ras[24,25] and (2) for increasing the attachment of Raf to the plasma membrane.[14] Such effects might be part of the mechanism for enabling the transformed cells to continue to divide, whereas the nontransformed fibroblasts attain growth arrest when they reach confluence.

EFFECTS OF SPHINGOSINE ON THE METABOLISM OF PHOSPHATIDATE AND DIACYLGLYCEROL

Interest in long-chain sphingoid bases increased considerably after the discovery that sphingosine inhibits protein kinase C activity.[75] Protein kinase C is regulated by diacylglycerol and it was shown subsequently that sphingosine can decrease the formation of diacylglycerol by inhibiting both the Mg^{2+}-dependent and Mg^{2+}-independent phosphatidate phosphohydrolase activities.[76-82] The former enzyme is involved in glycerolipid synthesis,[76] whereas the latter is thought to participate in signal transduction at the plasma membrane.[71,78] Studies in human neutrophils confirmed that sphingosine inhibits phosphatidate phosphohydrolase activities directly and not indirectly by an effect on protein kinase C.[79] It was also concluded from studies in yeast that the inhibition of phosphatidate phosphohydrolase by sphingoid bases is a physiologically relevant event.[82] Consequently, the inhibition of the protein kinase C via decreased diacylglycerol production may be particularly important compared to the direct action on protein kinase C itself.[3,71,77,79]

The ability of sphingosine to inhibit phosphatidate phosphohydrolase is shared by a family of amphiphilic amines including propranolol and chlorpromazine.[76-78,83] It was postulated for the Mg^{2+}-dependent phosphohydrolase that the inhibition was of a competitive type in which an inhibitor-substrate complex was formed.[76,84] The amphiphilic amine interacted with the membrane containing the phosphatidate through its hydrophobic region and the positively charged amine bound to the negative charge on the phosphate.[84] This model was compatible with the kinetic properties of the inhibition and the observation that the potency of the amphiphilic amines was related to their ability to bind to the phosphatidate substrate.[84] Addition of chlorpromazine to cultured hepatocytes resulted in the displacement of the Mg^{2+}-dependent phosphatidate phosphohydrolase from the membrane compartment (where phosphatidate is synthesized) to the cytosol.[85] This separation of the enzyme from its substrate caused a build up of phosphatidate and an strong inhibition of diacylglycerol production leading to a decreased synthesis of phosphatidylcholine and triacylglycerol.[85]

Our preliminary studies with the purified Mg^{2+}-independent phosphatidate phosphohydrolase indicate that sphingoid bases are more potent inhibitors than propranolol and in addition to exhibiting competitive inhibition also decrease the Vmax (Waggoner DW, Dewald J and Brindley DN, unpublished work). This form of phosphatidate phosphohydrolase is more likely to be involved in the phospholipase D pathway and signal transduction.

In addition to inhibiting phosphatidate phosphohydrolase, sphingosine is also able to modify diacylglycerol concentrations by stimulating the activity of diacylglycerol kinase[86,87] and thereby further increase the ratio of phosphatidate to diacylglycerol (Fig. 8.3). The increase in phosphatidate is also exaggerated because sphingosine can activate phospholipase D rapidly in some cells.[88] However, in NG108-15 neural-derived cells[89] and in rat fibroblasts[33] this process requires several hours to occur. This may indicate that conversion of sphingosine to sphingosine 1-phosphate is required in these cells for the activation, since the latter compound is a very potent stimulator of phospholipase D.[33,90] These actions of sphingosine and sphingosine 1-phosphate contrast with those of ceramides which decrease phospholipase D activation and increase the conversion of phosphatidate to diacylglycerol as described above (Table 8.3).

CONCLUSIONS

This review has emphasized the "cross talk" that occurs between sphingolipids and glycerolipids in controlling cell signaling particularly in relation to controlling the mitogenic action of the bioactive lipids. For example, ceramides block the mitogenic effects of phosphatidate, lysophosphatidate, ceramide 1-phosphate and sphingosine 1-phosphate. One site is suggested for this action in that the ceramides appear to increase the degradation of the lipid phosphomonoesters via the action of a common phosphatidate phosphohydrolase (Fig. 8.3). Ceramides also block the formation of phosphatidate and lysophosphatidate by inhibiting the activation of phospholipase D. Conversely, sphingosine (and/or sphingosine 1-phosphate) stimulates phospholipase D and inhibits the degradation of phosphatidate by phosphatidate phosphohydrolase. These actions increase the ratio of phosphatidate to diacylglycerol. It is therefore evident that the signal that a cell receives depends upon the balance of the production of the bioactive lipids.[33,34,71,91] For example, activation of sphingomyelinase to form ceramide may yield an apoptotic signal, whereas conversion of ceramide to sphingosine or sphingosine 1-phosphate may stimulate cell division. Similarly, the message received from activation of the phospholipase D pathway will depend upon the balance between the production of phosphatidate and lysophosphatidate versus diacylglycerol. As illustrated in Figure 8.3 the balance in the production of the bioactive glycerolipids and sphingolipids themselves is one of the factors that modifies the metabolism of the lipids that are produced by the phospholipase D and sphingomyelinase pathways. These complex inter-relationships and the balance of the "cross-talk" can provide for a variety of biological responses to different combinations of agonists that may be cell type specific.

ACKNOWLEDGMENTS

Our own work in this area was supported by an operating grant from the Medical Research Council of Canada, and by salary support to DNB from the Alberta Heritage Foundation for Medical Research and to DWW from the Canadian Diabetes Foundation.

REFERENCES

1. Berridge MJ. Inositol phosphate and calcium signaling. Nature 1991; 361: 315-325.
2. Exton JH. Signaling through phosphatidylcholine breakdown. J Biol Chem 1990; 265:1-4.
3. Martin A, Gómez-Muñoz A, Duffy PA et al. Phosphatidate phosphohydrolase: the regulation of signal transduction by phosphatidate and diacylglycerol. In: Liscovitch M, ed. Signal-Activated Phospholipases. Austin: RG Landes Co., 1994: 139-164.
4. Liscovitch M, Cantley LC. Lipid second messengers. Cell 1994; 77:329-334.
5. Liscovitch M, Chalifa V. Signal-activated phospholipase D. In: Liscovitch M, ed. Signal-Activated Phospholipases. Austin: RG Landes Co., 1994:31-63.
6. Rossi F, Grzeskowiak M, Della Bianca V et al. Phosphatidic acid and not diacylglycerol generated by phospholipase D is functionally linked to the activation of the NADPH oxidase by FMLP in human neutrophils. Biochem Biophys Res Comm 1990; 168:320-327.
7. Agwu DE, McPhail LC, Sozzani S et al. Phosphatidic acid as a second messenger in human polymorphonuclear leukocytes. J Clin Invest 1991; 88:531-539.
8. Bocckino SD, Wilson PB, Exton JH. Phosphatidate-dependent protein phosphorylation. Proc Natl Acad Sci USA 1991; 88:6210-6213.
9. Lamatola C, Schaap D, Moolenaar WH et al. Phosphatidic acid activation of protein kinase C-ζ overexpressed in COS cells: comparison with other C-protein kinase isotypes and other acid lipids. Biochem J 1994; 304:1001-1008.
10. Moritz A, De Graan PNE, Gispen WH et al. 1992. Phosphatidic acid is a specific activator of phosphatidylinositol-4-phosphate kinase. J Biol Chem 1992; 267:7207-7210.
11. Jones GA, Carpenter G. The regulation of phospholipase C-γ1 by phosphatidic acid. J Biol Chem 1993; 268:20845-20850.

12. Bhat BG, Wang P, Coleman RA. Hepatic monoacylglycerol acyltransferase is regulated by sn-1,2-diacylglycerol and by specific lipids in Triton X-100/phospholipid mixed micelles. J Biol Chem 1994; 269: 13172-13178.
13. Tsai M, Yu CL, Stacey DW. A cytoplasmic protein inhibits the GTPase activity of H-Ras in a phospholipid-dependent manner. Science 1990; 250:982-985.
14. Ghosh S, Strum JC, Sciorra VA et al. Raf-1 kinase possesses distinct binding domains for phosphatidylserine and phosphatidic acid. J Biol Chem 1996; 271:8472-8480.
15. Moolenaar WH. Lysophosphatidic acid, a multifunctional phospholipid messenger. J Biol Chem 1995; 270:12949-12952.
16. Eichholtz T, Jalink K, Fahrenfort I et al. The bioactive phospholipid lysophosphatidic acid is released from activated platelets. Biochem J 1993; 291:677-680.
17. Fourcade O, Simon MF, Viodé C et al. Secretory phospholipase A_2 generates the novel lipid mediator lysophosphatidic acid in membrane microvesicles shed from activated cells. Cell 1995; 80:919-927.
18. van Corven EJ, Groenink A, Jalink K et al. Lysophosphatidate-induced cell proliferation: identification and dissection of signaling pathways mediated by G proteins. Cell 1989; 59:45-54.
19. van Corven EJ, van Rijswijk A, Jalink K et al. Mitogenic action of lysophosphatidic acid and phosphatidic acid on fibroblasts. Biochem J 1992; 281:163-169.
20. Gómez-Muñoz A, Martin A, O'Brien L et al. Cell-permeable ceramides inhibit the stimulation of DNA synthesis and phospholipase D activity by phosphatidate and lysophosphatidate in rat fibroblasts. J Biol Chem 1994; 269:8937-8943.
21. Ha KS, Exton JH. Activation of actin polymerization of phosphatidic acid derived from phosphatidylcholine. J Cell Biol 1993; 123:1789-1796.
22. Cross MJ, Roberts S, Ridley AJ et al. Stimulation of actin stress fibre formation by activation of phospholipase D. Curr Biol 1996; 6:588-596.
23. Murayama T, Ui M. Phosphatidic acid may stimulate membrane receptors mediating adenylate cyclase inhibition and phospholipid breakdown in 3T3 fibroblasts. J Biol Chem 1987; 262:5522-5529.
24. Cook SJ, McCormick F. Inhibition by cAMP of Ras-dependent activation of Raf. Science 1993; 262:1069-1072.
25. Hordijk PL, Verlaan I, Jalink K et al. cAMP abrogrates the $p21^{ras}$-mitogen-activated protein kinase pathway in fibroblasts. J Biol Chem 1994; 269:3534-3538.
26. Tigyi G, Dyer DL, Miledi R. Lysophosphatidic acid possesses dual action in cell proliferation. Proc Natl Acad Sci USA 1994; 91:1908-1912.
27. Spiegel S, Olivera A, Carlson RO. The role of sphingosine in cell growth regulation and transmembrane signaling. Adv Lipid Res 1993; 25:105-127.
28. Dressler KA, Kolesnick RN. Ceramide-1-phosphate, a novel phospholipid in human leukemia (HL-60) cells: synthesis via ceramide from sphingomyelin. J Biol Chem 1990; 265:14917-14921.
29. Shinghal R, Cheller RH, Bajalieh SM. Ceramide-1-phosphate phosphatase activity in brain. J Neurochem 1993; 61: 2279-2285.
30. Boudker O, Futerman AH. Detection and characterization of ceramide-1-phosphate phosphatase activity in rat liver plasma membrane. J Biol Chem 1993; 268: 22150-22155.
31. Truett AP, King LE Jr. Sphingomyelinase D: A pathogenic agent produced by bacteria and arthropods. Adv Lipid Res 1993; 26:275-289.
32. Gómez-Muñoz A, Duffy PA, Martin A et al. Short-chain ceramide-1-phosphates are novel stimulators of DNA synthesis and cell division: antagonism by cell-permeable ceramides. Mol Pharmacol 1995; 47:883-889.
33. Gómez-Muñoz A, Waggoner DW, O'Brien L et al. Interaction of ceramides, sphingosine, and sphingosine 1-phosphate in regulating DNA synthesis and phospholipase D activity. J Biol Chem 1995; 270:26318-26325.
34. Brindley DN, Abousalham, A, Kikuchi Y et al. "Cross talk" between the bioactive glycerolipids and sphingolipids in signal transduction. Biochem Cell Biol 1996; 74:469-476.
35. Yao S, Zhang Y, Delikat S et al. Phos-

phorylation of Raf by ceramide-activated protein kinase. Nature 1995; 378: 307-310.

36. Paul A, Plevin R. Evidence against a role for phospholipase D in mitogenesis. TiPS 1994; 15:174.
37. van der Bend RL, de Widt J, van Corven E et al. The biologically active phospholipid, lysophosphatidic acid, induces phosphatidylcholine breakdown in fibroblasts via activation of phospholipase D. Biochem J 1992; 285:235-240.
38. Nakamura T, Abe A, Balazovich K J et al. Ceramide regulates oxidant release in adherent neutrophils. J Biol Chem 1994; 269: 18394-18389.
39. Venable MK, Blobe GC, Obeid LM. Identification of a defect in the phospholipase D/diacylglycerol pathway in cellular senesence. J Biol Chem 1994; 269:26040-26044.
40. Mathias S, Dressler K, Kolesnick R. Characterization of a ceramide-activated protein kinase: stimulation by tumor necrosis factor-α. Proc Natl Acad Sci 1991; 88:10009-10013.
41. Kolesnick R, Golde DW. The sphingomyelin pathway in tumor necrosis factor and interleukin-1 signaling. Cell 1994; 77:325-328.
42. Dobrowsky RT, Hannun YA. Ceramide stimulates a cytosolic protein phosphatase. J Biol Chem 1992; 267:5048-5051.
43. Fishbein JD, Dobrowsky RT, Bielawska A et al. Ceramide-mediated growth inhibition and CAPP are conserved in *Saccharomyces cerevisiae*. J Biol Chem 1993; 268:9255-9261.
44. Brown HA, Gutowski S, Moomaw CR et al. ADP-ribosylation factor, a small GTP-dependent regulatory protein, stimulates phospholipase D activity. Cell 1993; 75:1137-1144.
45. Bowman EP, Uhlinger DJ, Lambeth JD. Neutrophil phospholipase D is activated by a membrane-associated Rho family small molecular weight GTP-binding protein. J Biol Chem 1993; 268:21509-21512.
46. Cockcroft S, Thomas GMH, Fensome A et al. Phospholipase D: a downstream effector of ARF in granulocytes. Science 1994; 263: 523-526.
47. Malcolm C, Ross AH, Qui RG et al. Activation of rat liver phospholipase D by the small GTP-binding protein RhoA. J Biol Chem 1994; 269:25951-25954.
48. Kahn RA, Gilman AG. Purification of a protein cofactor required for ADP-ribosylation of the stimulatory regulatory component of adenylate cyclase by cholera toxin. J Biol Chem 1984; 259:6228-6234.
49. Rothman JE, Orci L. Molecular dissection of the secretory pathway. Nature 1992; 355:409-415.
50. D'Souza-Schorey C, Li G, Colombo MI et al. A regulatory role for ARF6 in receptor-mediated endocytosis. Science 1995; 267:1175-1178.
51. Massenburg D, Han JS, Liyanage M et al. Activation of rat brain phospholipase D by ADP-ribosylation factors 1, 5 and 6: separation of ADP-ribosylation factor-dependent and oleate-dependent enzymes. Proc Natl Acad Sci 1994; 91: 11718-11722.
52. Ridley AJ, Hall A. The small GTP-binding protein rho regulates the assembly of focal adhesion and stress fibers in response to growth factors. Cell 1992; 70:389-399.
53. Tominaga T, Sugie K, Hirata M et al. Inhibition of PMA induced, LFA-1-dependent by lymphocyte aggregation by ADP ribosylation of the small molecular weight GTP binding protein, *rho*. J Cell Biol 1993; 120:1529-1535.
54. Laudanna C, Campbell JJ, Butcher EC. Role of Rho in chemoattractant-activated leukocyte adhesion through integrins. Science 1996; 271:981-983.
55. Hammond S, Altshuller YM, Sung TC et al. Human ADP-ribosylation factor activated phosphatidylcholine-specific phospholipase D defines a new and highly conserved gene family. J Biol Chem 1995; 270:29640-29643.
56. Houle MG, Kahn RA, Naccache PH et al. ADP-ribosylation factor translocation correlates with potentiation of GTPγS-stimulated phospholipase D activity in membrane fractions of LH-60 cells. J Biol Chem 1995; 270:22795-22800.
57. Siddiqi AR, Smith JL, Ross AH et al.

Regulation of phospholipase D in HL60 cells. Evidence for a cytosolic phospholipase D. J Biol Chem 1995; 270: 8466-8473.
58. Lambeth JD, Kwak JY, Bowman EP et al. ADP-ribosylation factor functions synergistically with a 50-kDa cytosolic factor in cell-free activation of human neutrophil phospholipase D. J Biol Chem 1995; 270:2431-2434.
59. Bourgoin S, Harbour D, Desmarais Y et al. Low molecular weight GTP-binding proteins in HL-60 granulocytes. Assessment of the role of ARF and of a 50-kDa cytosolic protein in phospholipase D activation. J Biol Chem 1995; 270:3172-3178.
60. Singer WD, Brown A, Jiang X et al. Regulation of phospholipase D by protein kinase C is synergistic with ADP-ribosylation factor and independent of protein kinase activity. J Biol Chem 1996; 271: 4504-4510.
61. Ohguchi K, Banno Y, Nakashima S et al. Regulation of membrane-bound phospholipase D by protein kinase C in HL60 cells. Synergistic action of small GTP-binding protein RhoA. J Biol Chem 1996; 271:4366-4372.
62. Martin A, Brown F, Hodgkins MN et al. Activation of phospholipase D and phosphatidylinositol 4-phosphate 5-kinase in HL60 membranes is mediated by endogenous Arf but not Rho. J Biol Chem 1996; 271:(in press).
63. Abousalham A, Liossis C, O'Brien L, Brindley DN. Cell-permeable ceramides prevent the activation of phospholipase D by ADP-ribosylation factor and Rho. J Biol Chem 1996; 272:1069-1075.
64. Jones MJ, Murray AW. Evidence that ceramide selectively inhibits protein kinase C-α translocation and modulates bradykinin activation of phospholipase D. J Biol Chem 1995; 270:5007-5013.
65. Conricode KM, Brewer KA, Exton JH. Activation of phospholipase D by protein kinase C. J Biol Chem 1992; 267:7199-7202.
66. Nakamura Y, Nakashima S, Ojio K et al. Ceramide inhibits IgE-mediated activation of phospholipase D, but not of phospholipase C, in rat basophilic leukemia (RBL-2H3) cells. J Immunol 1995; 156:256-262.
67. Venable ME, Bielawsha A, Obeid LM. Ceramide inhibits phospholipase D. J Biol Chem 271:24800-24805.
68. Chen CS, Rosenwald AG, Pagano RE. Ceramide as a modulator of endocytosis. J Biol Chem 1995; 270:13291-13297.
69. Waggoner DW, Martin A, Dewald J et al. Purification and characterization of a novel plasma membrane phosphatidate phosphohydrolase from rat liver. J Biol Chem 1996; 270:19422-19429.
70. Waggoner DW, Gómez-Muñoz A, Dewald J et al. Phosphatidate phosphohydrolase catalyzes the hydrolysis of lysophosphatidate, ceramide 1-phosphate, and sphingosine 1-phosphate. J Biol Chem 1996; 271:16506-16509.
71. Brindley DN, Waggoner DW. Phosphatidate phosphohydrolase and signal transduction. Chem Phys Lipids 1996; 80:45-57.
72. Martin A, Goméz-Muñoz A, Waggoner DW et al. Decreased activities of phosphatidate phosphohydrolase and phospholipase D in *ras* and tyrosine kinase *(fps)* transformed fibroblasts. J Biol Chem 1993; 268:23924-23932.
73. Tagliaferri P, Katsaros D, Clair T et al. Reverse transformation of Harvey murine sarcoma virus-transformed NIH/3T3 cells by site-selective cyclic AMP analogs. J Biol Chem 1988; 263:409-416.
74. Davies SA, Houslay MD, Wakelam MJO. The effects of $p21^{N-ras}$ expression in NIH-3T3 cells upon cyclic AMP metabolism. Biochim Biophys Acta 1989; 1013: 173-179.
75. Merrill AH Jr, Stevens VL. Modulation of protein kinase C and diverse cell functions by sphingosine—a pharmacologically interesting compound linking sphingolipids and signal transduction. Biochim Biophys Acta 1989; 1010: 131-139.
76. Brindley DN. Phosphatidate phosphohydrolase: Its role in glycerolipid synthesis. In: Brindley DN, ed. CRC Series in Enzyme Biology. Boca Raton: CRC Press, 1988; 1:21-77.
77. Lavie Y, Piterman O, Liscovitch M. Inhibition of phosphatidic acid phos-

phohydrolase activity by sphingosine. Dual action of sphingosine in diacylglycerol signal termination. FEBS Lett 1990; 277:7-10.
78. Jamal Z, Martin A, Gómez-Muñoz A et al. Plasma membrane fractions from rat liver contain a phosphatidate phosphohydrolase distinct from that in the endoplasmic reticulum and cytosol. J Biol Chem 1991; 266:2988-2996.
79. Mullman TJ, Siegel MI, Egan RW et al. Sphingosine inhibits phosphatidate phosphohydrolase in human neutrophils by a protein kinase C-independent mechanism. J Biol Chem 1991; 266:2013-2016.
80. Gómez-Muñoz A, Hamza EH, Brindley DN. Effects of sphingosine, albumin and unsaturated fatty acids on the activation and translocation of phosphatidate phosphohydrolases in rat hepatocytes. Biochim Biophys Acta 1992; 1127:49-56.
81. Perry DK, Hand WL, Edmondson DE et al. Role of phospholipase D-derived diradylglycerol in the activation of the human neutrophil respiratory burst oxidase. Inhibition of phosphatidic acid phosphohydrolase inhibitors. J Immunol 1992; 149:2749-2758.
82. Wu WI, Lin YP, Wang E et al. Regulation of phosphatidate phosphatase activity from the yeast *Saccharomyces cerevisiae* by sphingoid bases. J Biol Chem 1993; 268:13830-13837.
83. Liscovitch M, Lavie Y. Sphingoid bases as endogenous cationic amphiphilic "drugs". Biochem Pharmacol 1991; 42: 2071-2075.
84. Bowley M, Cooling J, Burditt SL et al. The effects of amphiphilic cationic drugs and inorganic cations on the activity of phosphatidate phosphohydrolase. Biochem J 1977; 165:447-454.
85. Martin A, Hopewell R, Martin-Sanz P et al. Relationship between the displacement of phosphatidate phosphohydrolase from the membrane-associated compartment by chlorpromazine and the inhibition of the synthesis of triacylglycerol and phosphatidylcholine. Biochim Biophys Acta 1986; 876:581-591.
86. Sakane F, Yamada K, Kanoh H. Different effects of sphingosine, R59022 and amionic amphiphiles on two diacylglycerol kinase isozymes purified from porcine thymus cytosol. FEBS Letts 1989; 255:409-413.
87. Yamada K, Sakane F, Imai S-I et al. Sphingosine activates cellular diacylglycerol kinase in intact Kurkat calls, a human cell line. Biochim Biophys Acta 1993; 1169:217-224.
88. Natarajan V, Jayaram HN, Scribner WM et al. Activation of endothelial cell phospholipase D by sphingosine and sphingosine 1-phosphate. Am J Respir Cell Mol Biol 1994; 11:221-229.
89. Lavie Y, Liscovitch M. Activation of phospholipase D by sphingoid bases in NG108-15 neural-derived cells. J Biol Chem 1990; 265:3868-3872.
90. Desai NN, Zhang H, Olivera A, Mattie ME, Spiegel S. Sphingosine-1-phosphate, a metabolite of sphingosine, increases phosphatidic acid levels by phospholipase D activation. J Biol Chem 1992; 267:23122-23128.
91. Coroneos E, Martinez M, McKenna, S et al. Differential regulation of sphingomyelinase and ceramidase activities by growth factors and cytokines. J Biol Chem 1995; 27:23305-23309.
92. Hannun YA. The sphingomyelin cycle and second messenger function of ceramide. J Biol Chem 1994; 269:3125-3128.
93. Heller RA, Krönke M. Tumor necrosis factor receptor-mediated signaling pathways. J Cell Biol 1994; 126:5-9.
94. Mathias S, Younes Y, Kan CC et al. Activation of the sphingomyelin signaling pathway on intact EL4 cells and in a cell-free system by IL-1 beta. Science 1993; 259: 519-522.
95. Lozano J, Berra E, Municio MM et al. Protein kinase C ζ isoform is critical for κB-dependent promoter activation by sphingomyelinase. J Biol Chem 1994; 269:19200-19202.
96. Müller G, Ayoub M, Storz P et al. PKC ζ is a molecular switch in signal transduction of TNF-α, bifunctionally regulated by ceramide and arachidonic acid. EMBO J 1995; 14:1961-1969.
97. Wu J, Spiegel S, Sturgill TW. Sphingosine 1-phosphate rapidly activates the mitogen-activated protein kinase pathway by a G protein-dependent mechanism. J Biol

Chem 1995; 270:11484-11488.
98. Pyne S, Pyne NJ. The differential regulation of cAMP by sphingomyelin-derived lipids and the modulation of sphingolipid-stimulated extracellular signal regulated kinase-2 in airway smooth muscle. Biochem J 1996; 315:917-923.
99. Olivera A, Spiegel S. Sphingosine-1-phosphate as a second messenger in cell proliferation induced by PDGF and FCS mitogens. Nature 1993; 365:557-560.
100. Goodmote KA, Mattie ME, Berger A et al. Involvement of a pertussis toxin-senstive G protein in the mitogenic signaling pathway of sphingosine 1-phosphate. J Biol Chem 1995; 270:10272-10277.
101. Su Y, Rosenthal D, Smulson M et al. Sphingosine 1-phosphate, a novel molecule, stimulates DNA binding activity of AP-1 in quiescent Swiss 3T3 fibroblasts. J Biol Chem 1994; 269:16512-16517.
102. Cuvillier O, Pirianov G, Kleuser B et al. Suppression of ceramide-mediated programmed cell death by sphingosine-1-phosphate. Nature 1996; 381:800-803.

CHAPTER 9

Sphingosine-1-Phosphate: Member of a New Class of Lipid Second Messengers

Sarah Spiegel, Olivier Cuvillier, Elena Fuior and Sheldon Milstien

INTRODUCTION

Homeostasis of multicellular organisms as well as their normal development depends on the balance between cellular proliferation, differentiation, and cell death or apoptosis. Ceramide, sphingosine, and sphingosine-1-phosphate (SPP), metabolites of sphingolipids, and ubiquitous components of eukaryotic cell membranes, have recently emerged as members of a new class of signaling molecules regulating these diverse cellular processes.[1-4] Sphingolipid metabolism involves removal of their polar headgroups; for example, phosphorylcholine from sphingomyelin by acidic or neutral sphingomyelinases to produce ceramide,[5] which can then be cleaved by ceramidases to release fatty acid and the free long-chain base (sphingosine or sphinganine).[6] Sphingosine can be phosphorylated to SPP by sphingosine kinase,[7] reacylated to ceramide, or methylated.[8] SPP in turn can undergo dephosphorylation to sphingosine,[9] or cleavage to ethanolamine phosphate and trans-2-hexadecenal by a pyridoxal phosphate-dependent lyase.[10,11] Although all of these sphingolipid metabolites may play important roles in cell regulation, this review is focused on current knowledge regarding the second messenger role of SPP in regulating the fate of the cell.

SPHINGOSINE-1-PHOSPHATE IN CELL GROWTH REGULATION

Branching pathways of sphingolipid metabolism can mediate either mitogenic or apoptotic effects depending on the cell type and the nature of the stimulus.[1-4] Ceramide has emerged as a critical regulatory component of apoptosis induced by ionizing radiation and by members of the tumor necrosis factor (TNF) superfamily, TNFα and Fas ligand.[1-3,12-15] The binding of these ligands to their receptors results in stimulation of sphingomyelinase and increases ceramide which in turn mediates apoptosis (reviewed in refs. 1, 3, 12). The observation that exogenous ceramide mimics the hallmarks of apoptosis, cell growth arrest and DNA fragmentation, whereas closely related analogs are ineffective also supports its

Sphingolipid-Mediated Signal Transduction, edited by Yusuf A. Hannun.

critical role.[1,16] On the other hand, further metabolites of ceramide, sphingosine, produced from ceramide by ceramidase, and its metabolite, sphingosine-1-phosphate (SPP), have been shown to mediate mitogenesis in several mammalian cell lines.[17-24]

Sphingosine stimulated the proliferation of quiescent Swiss 3T3 fibroblasts,[17,22] Rat-1 fibroblasts[20] and reinitiated DNA synthesis in HL-60 cells after phorbol ester-induced cell growth arrest.[25] In contrast to the lack of stereospecificity in inhibition of protein kinase C, the mitogenic effect of sphingosine was stereospecific as only the D(+)-*erythro* enantiomer stimulated cell growth.[21] Increased endogenous levels of sphingosine and sphinganine by Fumonisin B_1, a fungal toxin that inhibits sphinganine-N-acyltransferase, also stimulates DNA synthesis in Swiss 3T3 fibroblasts.[22] However, in some cell types sphingosine inhibits cell growth and thus has bimodal effects[27] depending on cell type and nature of the stimuli.[4,26,28] The growth-arresting effect of sphingosine could result partly from cytotoxicity due to its detergent properties,[26,29] inhibition of protein kinase C[26] or by increased dephosphorylation of the retinoblastoma gene product resulting in loss of its effect on gene transcription.[30] In addition, sphingosine has also been implicated as an endogenous modulator of apoptosis during phorbol ester-induced differentiation of HL-60 cells[31] and TNFα-induced apoptosis in human neutrophils.[32] Because ceramide and sphingosine are interconverted by simple acylation/deacylation reactions, further studies are needed to determine whether one or both are the active species or whether these are cell type specific responses. As a further complication, we[33] and others[34,35] have observed that treatment with exogenous sphingomyelinase, as well as cell-permeable analogs of ceramide, increased proliferation of quiescent Swiss 3T3 fibroblasts, suggesting that ceramide may not only play a role in differentiation and apoptosis, but also may be involved in cellular proliferation. In agreement, it has been recently found that CD28, a costimulatory signal for the T cell receptor, stimulates sphingomyelin hydrolysis in spleen cells and exogenous sphingomyelinase synergized with CD3 to induce a proliferative response in these cells.[36] Although the explanation for these contrasting effects of both ceramide and sphingosine on cell growth is still not clear, it is possible that they have different cellular targets in different cell types, or alternatively, they may be metabolized or sorted to specific compartments in different cell types. Moreover, the pleiotropic effects of ceramide and sphingosine may depend on the context of other modulatory signals.[2,37-39]

The mitogenic effects of exogenous sphingosine may be mediated via its conversion to SPP, as sphingosine is rapidly taken up by cells and phosphorylated on its primary hydroxyl group to SPP.[18] Furthermore, SPP is a more potent mitogenic agent than sphingosine itself.[18] Finally, inhibitors of sphingosine kinase block the mitogenic effect of sphingosine.[40] In contrast to the bimodal effects of ceramide and sphingosine, SPP has only been shown to be a positive regulator of cell growth. SPP stimulated proliferation of quiescent Swiss 3T3 fibroblasts,[18] Rat-1 fibroblasts,[41] and increased DNA synthesis in airway smooth muscle cells,[42] and in arterial smooth muscle cells.[43]

The steady-state level of SPP in cells is very low and increases rapidly and transiently in response to various stimuli, such as platelet-derived growth factor (PDGF) in Swiss 3T3 fibroblasts,[40] arterial smooth muscle cells[43] and in airway smooth muscle cells.[42] PDGF also increased levels of sphingosine, the precursor of SPP in Swiss 3T3 fibroblasts,[40] vascular smooth muscle cells[44] and glomerular mesangial cells[45] (Table 9.1). These responses are specific for certain growth promoting agents, since other potent mitogens, including EGF and endothelin-1 did not induce significant changes.[40] The level of SPP in cells is determined by the relative contributions of its formation, mediated by sphingosine kinase,[7] and its degradation, catalyzed by a pyridoxal phosphate-dependent lyase located in the endoplasmic reticulum,[11] and by a plasma-membrane associated SPP-phosphatase.[9] Further studies have revealed that growth factor-in-

Table 9.1. Various stimuli increase cellular levels of sphingosine and sphingosine-1-phosphate

Stimuli	Sphingolipid	Cell Type
Serum/plasma	Sphingosine	Human neutrophils
Serum/PDGF/TPA	Sphingosine/SPP	Swiss 3T3 fibroblasts
PDGF	Sphingosine	Vascular smooth muscle
PDGF	Sphingosine	Mesangial cells
Fumonisin B_1	Sphingosine Sphinganine	Many cell types, cattle, horses
Dexamethasone/ Phorbol dibutyrate	Sphingosine	Epstein Barr transformed human B lymphocytes
PDGF	SPP	Arterial smooth muscle cells
PDGF	SPP	Airway smooth muscle cells
TPA	SPP	Balb/c 3T3 fibroblasts
TPA	SPP	Swiss 3T3 fibroblasts
TPA	SPP	HL-60, U937, Jurkat T cells
Thrombin	SPP	Human platelets
Antigen clustering of FcεRI	SPP	Rat mast-cell line

duced elevations in intracellular sphingosine are not the sole regulatory factor influencing SPP levels, as PDGF and serum also activated cytosolic sphingosine kinase.[40] Similarly, other mitogens such as the B subunit of cholera toxin and the phorbol ester 12-O-tetradecanoyl phorbol 13 acetate (TPA) significantly increase sphingosine kinase activity,[46,47] whereas some mitogens, including bombesin, bradykinin, insulin and EGF have little or no effect. Further support for the role of SPP in cellular proliferation arose from the use of a competitive inhibitor of sphingosine kinase, D,L-*threo*-dihydrosphingosine, which completely eliminates formation of SPP and selectively blocks cellular proliferation induced by PDGF but not by EGF.[40] In agreement, L-cycloserine, an inhibitor of the initial reaction in sphingolipid biosynthesis, and N-oleoylethanolamine, an inhibitor of ceramidase activity, inhibited PDGF-stimulated DNA synthesis in vascular smooth muscle cells and in rat glomerular mesangial cells, respectively.[44,45] Moreover, cell growth inhibition induced by ISP-1/myriocin, a potent immunosuppressant which inhibits serine palmitoyltransferase and sphingolipid metabolism, was completely abolished by the addition of sphingosine or SPP.[23] Thus, sphingosine kinase is apparently an important regulatory enzyme in signal transduction pathways which regulate cell growth. Sphingosine kinase activity has been detected in yeast,[48] *Tetrahymena pyriformis*,[49] rat liver, kidney and brain,[50] bovine brain,[51] and human and porcine platelets.[7,50] Although it has not yet been purified to homogeneity or cloned, we have recently purified rat kidney sphingosine kinase approximately 30,000-fold (Olivera et al, unpublished). Cloning of sphingosine kinase is an important task which is necessary to fully understand the role of SPP in cellular proliferation.

SPHINGOSINE-1-PHOSPHATE INHIBITS CERAMIDE-MEDIATED PROGRAMMED CELL DEATH

Paradoxically, key regulatory components that play crucial roles in cell growth, such as p53, Rb-1, cyclin D1, c-fos, c-myc and $p34^{cdc2}$ kinase, have also been implicated in programmed cell death.[52,53] Furthermore, as apoptosis exhibits numerous overall features reminiscent of mitosis, it has been suggested that the pathways of apoptosis and mitosis may be mechanistically related or tightly coupled.[53] Because ceramide and SPP have opposite effects on cell growth and are potentially interconvertable sphingolipid metabolites, it was of interest to determine whether SPP counteracts apoptosis mediated by the sphingomyelinase/ceramide pathway. Recently, it was demonstrated that SPP acts similarly to the diacylglycerol/protein kinase C pathway[16,39,54] and prevents programmed cell death induced by the related sphingolipid ceramide.[47] In human monoblastic leukemia U937 cells, human promyelocytic HL-60 cells, and human Jurkat T cells, SPP suppresses the appearance of the hallmarks of apoptosis, intranucleosomal DNA fragmentation and morphological changes, resulting from elevations of ceramide induced by TNFα and Fas ligation, sphingomyelinase treatment, or by cell permeable ceramide analogs. The concentration of SPP required for these cytoprotective effects is similar to that previously shown to stimulate proliferation of various cell types and to activate the transcription factor activator protein-1.[18,19] However, in different cell types undergoing apoptosis it might be anticipated that the effective concentration of SPP could be different depending on basal levels of ceramide, the increases in ceramide induced by external stimuli, and the rate of degradation of SPP. Furthermore, high concentrations of SPP might be able to induce apoptosis in some cell types due to conversion to sphingosine, which might induce apoptosis,[31] potentiate the lethal actions of ceramide[55] or by reacylation of sphingosine to ceramide.

It is well established that activation of protein kinase C by TPA or diacylglycerol (DAG) prevents DNA fragmentation induced by TNFα, Fas ligation, ionizing radiation, and cell permeable ceramide analogs, indicating that protein kinase C can counteract the ceramide-mediated apoptotic pathway.[14,16,54,56] Although numerous targets of protein kinase C have been identified, for several reasons it appears that one important target of protein kinase C is the activation of sphingosine kinase. Firstly, activation of protein kinase C in diverse cell types stimulates sphingosine kinase activity leading to increased cellular levels of SPP. Secondly, inhibition of protein kinase C, as well as inhibition of sphingosine kinase by dimethylsphingosine leads to apoptosis, which can be overcome by addition of SPP.[47]

Although neither SPP nor ceramide have direct effects on protein kinase C, both have been shown to cross-talk with signaling pathways modulating the levels of DAG. Ceramide inhibits DAG generation through inhibition of phospholipase D (PLD) (see below) and it has been suggested that the growth arresting effects of ceramide as well as the defects in mitogenic signaling pathways in cellular senescence are due to this inhibition.[20,57] It is interesting to note that in a wide variety of cells, SPP uniformly has a stimulatory effect on phospholipase D.[58] Thus, SPP and ceramide act in feedback regulatory loops to regulate protein kinase C either positively or negatively, respectively. Since activation of protein kinase C results in increased SPP and decreased ceramide levels,[47] these feedback loops amplify the effects of protein kinase C on the balance between the concentrations of ceramide and SPP.

It appears that the relative intracellular levels of ceramide and SPP are a determining factor in the fate of the cell. This relationship may be analogous to the Bcl-2/Bax rheostat switch in which the signal must reach a critical threshold for a response and the extent of the response is dictated by a set point.[59] In a similar manner, the balance between cellular levels of ceramide that favor death to levels of SPP that inhibit death is critical. Although the ceramide/SPP rheostat may be an inherent characteristic of each cell type, external stimuli can reset this ratio. Dissociation of growth factor-induced mitogenesis from cytokine-mediated apoptosis is a consequence of distinct sphingolipid-derived second messengers. Growth promoting agents such as PDGF and TPA increased SPP levels, and TPA also blocked TNFα-induced increases in ceramide levels. Conversely, although TNFα not only increased ceramide levels, it markedly decreased the amount of intracellular SPP.[47] Thus, TPA and TNFα exert opposing effects on intracellular levels of both ceramide and SPP and therefore have a marked impact on the ratio of their concentrations (see Fig. 9.1). Similarly, in rat glomerular mesangial cells, PDGF, but not cytokines, mediates proliferation in part through ceramidase-regulated sphingosine formation, whereas inflammatory cytokines stimulate sphingomyelinase but not ceramidase activity leading to accumulation of ceramide.[45]

Little is yet known of the mechanism of the cellular death pathway but recent attention has been focused on the involvement of intracellular cysteine proteases belonging to the ICE/CED-3 family (for review see refs. 60, 61). Overexpression of ICE or CED-3 in mammalian cells induces apoptosis, suggesting that ICE, or a related protease, is an essential component.[62] Recent studies have shown that ceramide production requires an ICE-like protease.[63] Another member of this emerging family, CPP32/apopain, cleaves the DNA repair enzyme poly(ADP-ribose) polymerase (PARP) at the onset of apoptosis[64,65] and recently this process was shown to be stimulated by ceramide.[66] Moreover, Bcl-2, a protein known to spare cells from apoptosis triggered by

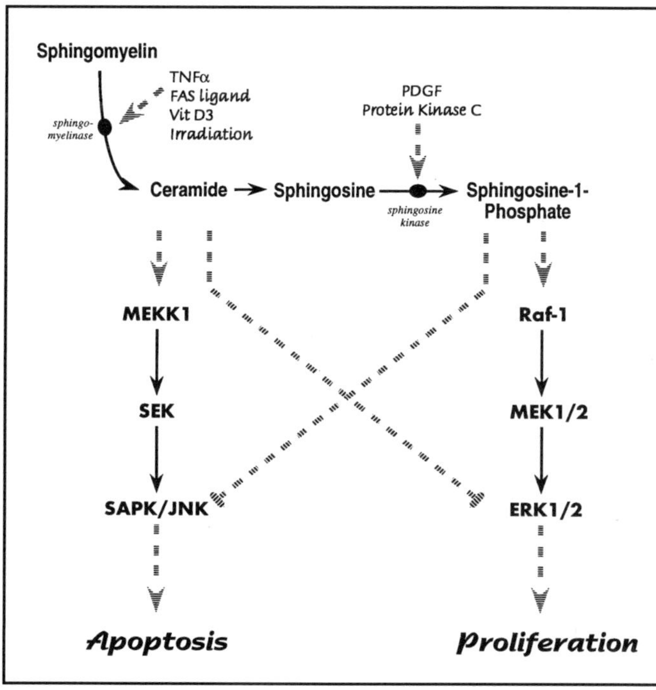

Fig. 9.1. The effects of growth factors and cytokines on sphingolipid metabolism and consequent regulation of distinct members of the mitogen activated protein kinase superfamily, culminating in either cellular proliferation or apoptosis.

ERK1/2 kinases are activated by dual specificity MAP kinase kinases (MEKs or MKKs), which in turn are regulated by Raf-1-catalyzed serine phosphorylation. Activation of JNK/SAPK is mediated through a kinase cascade in which MEKK1 phosphorylates and activates a dual-specificity threonine-tyrosine protein kinase SEK1 (also known as JUNKK1) which then phosphorylates and activates JNK/SAPK. Certain extracellular ligands, such as TNFα and Fas, stimulate sphingomyelin hydrolysis to form ceramide, an activator of the JNK/SAPK signaling pathway leading to apoptosis and an inhibitor of the ERK signaling pathway of cell growth. On the other hand, PDGF or activation of protein kinase C (PKC) stimulates sphingosine kinase leading to the formation of SPP, a stimulator of the ERK signaling pathway of cell growth and inhibitor of ceramide signaling pathway leading to apoptosis. Solid lines indicate pathways that have been established. Dashed lines refer to effects whose mechanisms have not yet been elucidated.

various stimuli, has been shown to inhibit both apoptosis and PARP proteolysis induced by ceramide, but not the ceramide generation, suggesting that Bcl-2 functions at a point downstream of ceramide.[66,67] The mechanism of action of Bcl-2 protein, which can counteract some, but not all, insults known to induce apoptosis, is still not known. This 26 kDa integral membrane protein has been localized to the endoplasmic reticulum, mitochondria, and continuous perinuclear membrane[68] and it has been suggested that Bcl-2 represses apoptosis by regulating endoplasmic reticulum-associated calcium fluxes.[69] Calcium plays an important role in apoptosis, and several groups have characterized nucleus-associated Ca^{2+}-dependent endonucleases that may be involved in internucleosomal DNA digestion typical of apoptotic cells. Interestingly, SPP generated in endoplasmic reticulum membranes has recently been shown to activate release of stored calcium in a previously undescribed manner, independent of inositol 1,4,5-trisphosphate ($InsP_3$) (see below).[70,71] Thus, a possible relationship between Bcl-2 and SPP makes an attractive hypothesis to explain their similar preventive effects on programmed cell death.

INTRACELLULAR ACTIVITIES OF SPHINGOSINE-1-PHOSPHATE

In the following sections, potential downstream targets for the actions of SPP and possible mechanisms by which it may exert effects on cell growth and survival are discussed. These signaling pathways include calcium mobilization, differentially regulated extracellular signal-regulated kinase and stress-activated protein kinase cascades, activation of PLD, and regulation of cAMP levels. Little emphasis will be placed on the effects of SPP on PLD since this topic has recently been reviewed elsewhere.[58]

Calcium Mobilization and $InsP_3$

Sphingosine and SPP mobilize calcium from internal sources in many cell types.[18,70-73] Calcium release is the only early event in sphingosine-induced mitogenesis that has been demonstrated to be stereospecific.[21] In addition, calcium release by sphingosine is time- and temperature-dependent, implicating an intervening enzymatic step, consistent with the conversion of sphingosine to SPP.[72] In agreement, inhibitors of sphingosine kinase block sphingosine-induced calcium mobilization.[21] Although in some cell types, both sphingosine and SPP also stimulate phospholipase C leading to an increase in $InsP_3$ formation, which is known to mobilize calcium from internal sources, there are several lines of evidence indicating that SPP can mobilize calcium by a unique mechanism. SPP stimulated a rapid release of intracellular calcium in human skin fibroblasts without affecting phosphoinositol turnover.[73] Moreover, complete inhibition of $InsP_3$ formation by brief treatment of Swiss 3T3 fibroblasts with TPA did not inhibit SPP-mediated calcium responses, although SPP did increase $InsP_3$ levels.[71] Furthermore, heparin, an $InsP_3$ antagonist, inhibited calcium release induced by exogenous $InsP_3$ but did not affect calcium release induced by SPP in permeabilized cells,[71] or in endoplasmic reticulum from smooth muscle cells.[70] Thus, SPP appears to be able to mobilize calcium by an $InsP_3$-independent pathway. The rapidity, reversibility, and specificity of the effect of SPP all closely resemble the calcium-releasing effect of $InsP_3$ which mediates direct calcium channel activation. The endoplasmic reticulum membrane is known to contain sphingosine kinase activity which can convert sphingosine to SPP which may then interact directly with calcium channels within these membranes, perhaps constituting a novel calcium signaling pathway.[70] An intracellular sphingosinephosphorylcholine (SPC)-gated calcium channel with unique pharmacological and electrophysiological properties has recently been characterized.[74] In a recent report, a SPC-responsive endoplasmic reticulum channel was cloned and characterized by selection in *Xenopus* oocytes.[75] This 181-amino acid polypeptide, named SCaMPER (sphingolipid Ca release-mediating protein of ER) is a novel protein, with no sequence homology to known proteins. Hydrophobicity plots revealed two potential transmembrane domains, as well as a single N-glycosylation consensus site, suggesting that the C- and N-termini are cytosolic, with the domain in between the two transmembrane domains positioned extracellularly.[75] However, it has not yet been determined if this is a SPP-gated calcium channel.

In contrast to SPP, sphingosine may have differential effects on different types of calcium channels. Opposite to its mobilizing effects in many cell types, sphingosine inhibited the release of calcium from skeletal muscle sarcoplasmic reticulum and directly inhibited sarcoplasmic reticulum ryanodine receptors.[76] Differential modulation of PDGF-induced calcium signaling by intracellular levels of sphingosine and SPP was found in transformed oligodendrocytes.[77] SPP induced a rapid, nonoscillatory calcium response, whereas the response to sphingosine was delayed and oscillatory. These responses could result from different kinetics of production of sphingosine and SPP. Hence, the balance between SPP and sphingosine could control calcium mobilization both spatially and temporally. Recently, it was observed that calcium mobilization following FcεR1 engagement in mast cells is mediated by SPP and not by $InsP_3$ formation.[78] An inhibitor of sphingosine kinase, DL-*threo*-dihydrosphingosine, blocks the FcεR1-mediated calcium signal and partially inhibits histamine release. These findings suggest an exciting area in the development of SPP inhibitors as potential new antihistamine drugs.

In contrast to this wealth of information suggesting that SPP may regulate calcium levels by direct interaction with putative calcium permeable channels on the endoplasmic reticulum, other studies indicate that SPP activates a G protein-coupled receptor in the plasma membrane sensitive to pertussis toxin or a G protein-coupled to phospholipase C leading to calcium mobilization.[79-81] For example, in HL-60 cells, calcium release requires $InsP_3$ generation, since a specific phospholipase C inhibitor, U 73122, abolished SPP-induced calcium-mobilization.[82] There is also some evidence for a high-affinity, G protein-coupled SPP receptor in the plasma membrane, different from the receptor for the closely structurally related lipid, lysophosphatidic acid.[71,81-83] Moreover, in atrial myocytes, SPP and SPC activate muscarinic-like inwardly rectifying K-channels with an EC_{50} of 1.3 and 1.5 nM, respectively.[84] Activation of these channels occurred when SPP was applied at the extracellular face of the atrial myocyte plasma membrane as measured in cell-attached and inside-out patch clamp current recordings.[84] Sphingosine was completely inactive, indicating the importance of phosphorylation at the C-1 position for interaction with the receptor. Complete homologous and heterologous desensitization with each other, but not with acetylcholine, indicates that both SPP and SPC act on the same receptor and this activation was completely abolished in both cases by pertussis toxin. Furthermore, other sphingolipids were ineffective. Similarly, in MC3T3-E1 preosteoblasts, no elevation in $[Ca^{2+}]_i$ was detected in response to the application of N-acetyl sphingosine, but psychosine (1-galactosyl sphingosine) also caused calcium release from intracellular stores.[85] Sequential additions of combinations of sphingosine, SPP, SPC and psychosine revealed that all of these sphingolipids employ the same receptor.[85] In contrast, based on the different response of various cell types to SPP and/or SPC, it was suggested that these sphingolipids might act through several subtypes of G protein-coupled plasma membrane receptors.[81] However, it has not yet been shown that SPP and SPC are physiological ligands for the K-channels (responsible for the vagal slowing of cardiac frequency). Moreover, the fact that pertussis toxin inhibits but does not abolish SPP-induced calcium release,[79,82] suggests that SPP, in addition to its interaction with a high affinity Gi protein-coupled receptor also has a direct action on intra–cellular calcium stores.

Cytoskeletal Remodeling

Because calcium affects polymerization/depolymerization of actin filaments, the involvement of SPP in cytoskeletal remodeling, cellular motility and metastatic invasiveness is of great interest. In arterial smooth muscle cells, SPP inhibits PDGF-induced chemotaxis by favoring actin filaments disassembly and inhibiting focal adhesion formation.[43] In contrast to its effects on Swiss 3T3 fibroblasts, SPP induces more sustained PIP_2 hydrolysis in smooth muscle cells, stronger calcium mobilization, with subsequent cAMP increases, and protein kinase A activation. These responses result in the disruption of the equilibrium of assembly/disassembly of actin filaments, inhibiting cell spreading, extension of the leading lamellae, and chemotaxis toward PDGF. SPP also inhibits the motility of mouse melanoma cells by decreasing F-actin and interfering with actin nucleation and pseudopodium formation.[86] However, in Swiss 3T3 fibroblasts, SPP induced rapid reorganization of the actin cytoskeleton resulting in stress fiber formation which was mediated by the small GTP-binding protein $p21^{rho}$ (Rho).[87] In estrogen-independent breast cancer MDA-MB-231 cells, SPP not only markedly reduced motility, it also decreased invasion and proteolytic activity leading to Matrigel degradation.[88] Thus SPP may be an attractive compound for inhibition of metastasis. Exogenously added, but not microinjected, SPP can also induce calcium-independent actin cytoskeleton remodeling mediated by Rho in N1E-115 neuronal cells, resulting in neurite retraction and rounding of the cell body.[83] Lysophosphatidic acid was 100 times less potent than SPP and the responses to SPP and lysophosphatidic acid did not show cross-desensitization, suggesting that SPP activates its own high affinity receptor to trigger Rho-regulated cytoskeletal events.[83]

Focal adhesion kinase (FAK), a cytosolic tyrosine kinase, and its substrate paxillin are concentrated in focal adhesions, regions of dense cellular structures which link integrin receptors attached to the extracellular matrix to actin filament bundles. Although the precise cellular function of FAK is still not completely understood, its subcellular localization and regulations by growth factors and integrins suggests that it may participate in the organization of focal adhesions and actin stress fibers and cell motility. In Swiss 3T3 fibroblasts, sphingosine[89] and SPP[87] induce stress-fiber formation, with concomitant focal adhesion assembly and tyrosine phosphorylation of FAK and paxillin. Cytochalasin D, which disrupts the network of actin microfilaments, completely inhibits sphingosine-induced tyrosine phosphorylation of FAK and paxillin.[89] The exoenzyme C3 transferase, which inactivates Rho by ADP-ribosylation, inhibits both protein tyrosine-phosphorylation and stress fiber formation[87,89] indicating a novel, Rho-mediated, signaling pathway.

Phospholipase D and Phosphatidic Acid

Sphingosine and SPP uniformly have a stimulatory effect on phosphatidic acid levels in many cell types (reviewed in ref. 58). Sphingosine and SPP have been shown to stimulate PLD activity in several cell types.[58] However, sphingosine in contrast to SPP also inhibited phosphatidic acid phosphohydrolase.[90,91] Similar results were obtained with bovine pulmonary artery endothelial cells in which SPP stimulated PLD but had no effect on membrane-associated phosphatidic acid phosphohydrolase activity.[92] Sphingosine, but not SPP, also positively regulates diacylglycerol kinase[93] and negatively regulates monoacylglycerol acyltransferase,[94] leading to decreases in DAG levels. These observations may explain the effect of sphingosine on protein kinase C in vivo. In contrast to sphingosine and SPP, ceramide inhibits PLD activation in many cell types.[57,95] Increased levels of ceramide may exhibit growth arresting effects by blocking the PLD signaling pathway, while the mitogenic effects of sphingosine and SPP may be due to stimulation of PLD.[41] The regulation of phosphatidic acid levels by sphingolipid metabolites indicates that the sphingolipid cycle and the glycerophospholipid cycle could regulate each other.

Protein Kinase A and c-AMP

Both sphingosine[96] and SPP,[79] rapidly and drastically decrease cellular cAMP levels in Swiss 3T3 fibroblasts. Pertussis toxin treatment prevents the inhibitory effect of sphingosine and SPP on cAMP accumulation, suggesting that a pertussis toxin-sensitive G_i protein may be involved. In agreement, SPP suppresses cAMP accumulation in airway smooth muscle cells,[42] whereas sphingosine and ceramide stimulate cAMP formation, a negative modulator of cell growth in these cells. In contrast, in human smooth muscle cells, SPP increases cAMP levels leading to activation of cAMP-dependent protein kinase (protein kinase A), as a consequence of the marked increase in cytosolic free calcium induced by SPP.[43] The relevance of changes in cAMP levels to mitogenic effects of SPP is still unclear because in some cell types cAMP is a positive regulator of growth while it is a negative growth regulator in others, and the mitogenic effects of SPP do not correspond to the effects of cAMP.

Mitogen-Activated Protein Kinases

The mitogen-activated protein kinase (MAPK) signaling pathway is a network of interacting proteins that translocates information from activated plasma-membrane receptors to initiate nuclear transcriptional events. Although the role of MAP kinases (also known as extracellular signal regulated kinases, ERK-1 and ERK-2) in cellular proliferation and differentiation has been well established,[97] more recently novel classes of mammalian enzymes closely related to MAPKs have been identified whose roles in stress-activated processes are now being explored. Stress-activated protein kinases (SAPKs) represent a family of closely related enzymes activated by various forms of environmental stress and cytokines[98] which have been independently identified by virtue of their c-Jun amino terminal kinase activity, hence also termed JNKs.[99] While growth factors, such as PDGF, stimulate MAPK, SAPKs are activated by cytokines, such as TNFα and IL-1, and also by sphingomyelinase.[98] Recently, it has been suggested that activation of SAPK/JNK and concurrent inhibition of ERK are critical for promotion of cell death in rat pheochromcytoma PC-12 cells and conversely, activation of ERK signals and suppression of SAPK/JNK leads to the promotion of cell survival.[100] Therefore, it has been proposed that the dynamic balance between growth factor-activated ERK and stress-activated SAPK/JNK pathways, may be important in determining whether a cell survives or undergoes apoptosis.[100] It is interesting to note that ceramide induces activation of SAPK/JNK in HepG2 cells[98] and in HL-60 cells,[101] while SPP stimulates ERK in 3T3 fibroblasts.[102] Although the mechanism by which SPP stimulates the Raf/MEK/MAPK pathway is not known, the ability of SPP to stimulate PLD leading to an increase in phosphatidic acid might be relevant, since recently phosphatidic acid formation has been shown to regulate Raf activity.[103] Specific activations of these different signaling pathways might explain the opposite effects of ceramide and SPP on cell survival and cell death. Indeed, activation of the SAPK/JNK cascade has been shown to be required for ceramide-induced apoptosis,[104] whereas SPP not only stimulates ERK-1 and ERK-2, it also inhibits SAPK/JNK activity stimulated by TNFα or ceramide.[47] Similarly, in rat glomerular mesangial cells, growth factors and sphingosine activate ERK but not SAPK/JNK, whereas inflammatory cytokines or cell-permeable ceramide analogs activate SAPK/JNK but not ERK.[105] Ceramide can also suppress growth factor- or sphingosine-induced ERK activation as well as proliferation.[105] In agreement, short-chain ceramides are poor stimulators of ERKs in airway smooth-muscle cells, but are powerful activators of SAPK/JNK. In contrast, SPP activates ERK-2, potentiates growth-factor-stimulated DNA synthesis and fails to activate SAPK/JNK. These results in many different cellular systems imply that the relative intracellular levels of ceramide and SPP, and their consequent activation or inhibition of distinct members of the MAP kinase cascades are important factors in determining the fate of cells (Fig. 9.1).

CONCLUDING REMARKS

Much evidence has been accumulated that suggests that the sphingolipid metabolite SPP is an additional member of a new and growing class of lipid second messengers. The level of SPP can be regulated by growth factors and other stimuli. SPP triggers diverse and complex cellular responses, including release of calcium from internal sources, regulation of actin assembly, blockade of cell death, and regulation of the MAP kinase signaling pathway leading to gene activation. It appears that the relative intracellular levels of ceramide and SPP and consequent regulation of different family members of mitogen-activated protein kinases is an important factor determining whether cells will grow, divide, or differentiate. Thus, important steps leading towards the understanding of the regulation of enzymes involved in production and metabolism of these sphingolipids, sphingomyelinase, ceramidase, and sphingosine kinase by growth factors and cytokines are their purification and cloning and a detailed study of their mechanism of action and interaction with other cellular components of signal transduction pathways. An additional crucial question that remains to be addressed is the possibility that SPP may also function as an extracellular ligand for a specific cell surface receptor. If confirmed, this would be one of the first examples of a second messenger that can also act as a first messenger.

The beneficial effects of SPP in preventing apoptosis induced by ceramide might be important for those diseases in which ceramide may contribute to the pathology, including AIDS, neurodegenerative diseases, and ischemic stroke. Ceramide levels are markedly elevated in human immunodeficiency virus (HIV)-infected T lymphocytes,[106] which are known to undergo apoptosis, and ceramide also induces HIV replication in chronically infected HL-60 cells,[107] suggesting a role for ceramide in the progression of AIDS.[1,107] Recent data suggests that markedly increased levels of ceramide in peripheral blood lymphocytes are associated with the progressive course of HIV infection.[108] We are now developing SPP mimetics which may be useful therapeutic agents to counteract or prevent the effects of ceramide.

References

1. Hannun YA. The sphingomyelin cycle and the second messenger function of ceramide. J Biol Chem 1994; 269:3125-3128.
2. Hannun YA, Obeid LM. Ceramide: an intracellular signal for apoptosis. Trends Biochem Sci 1995; 20:73-77.
3. Kolesnick R, Golde DW. The sphingomyelin pathway in tumor necrosis factor and interleukin-1 signaling. Cell 1994; 77:325-328.
4. Spiegel S, Milstien S. Sphingolipid metabolites: members of a new class of lipid second messengers. J Membr Biol 1995; 146:225-237.
5. Spence MW. Sphingomyelinases. Adv Lipid Res 1993; 26:3-23.
6. Hassler DG, Bell RM. Ceramidases: Enzymology and metabolic roles. Adv Lipid Res 1993; 26:49-57.
7. Stoffel W, Hellenbroich B, Heimann G. Properties and specificities of sphingosine kinase from blood platelets. Hoppe-Seyler's Z Physiol Chem 1973; 354: 1311-1316.
8. Igarashi Y, Kitamura K, Toyokuni T et al. A specific enhancing effect of N,N-dimethylsphingosine on epidermal growth factor receptor autophosphorylation. Demonstration of its endogenous occurrence (and the virtual absence of unsubstituted sphingosine) in human epidermoid carcinoma A431 cells. J Biol Chem 1990; 265:5385-5389.
9. Van Veldhoven PP, Mannaerts GP. Sphinganine 1-phosphate metabolism in cultured skin fibroblasts: evidence for the existence of a sphingosine phosphatase. Biochem J 1994; 299:597-601.
10. Stoffel W, Assmann G. Enzymatic degradation of 4t-sphingenine 1-phosphate (sphingosine-1-phosphate) to 2t-hexadecen-1-al and ethanolamine phosphate. Hoppe-Seyler's Z Physiol Chem 1970; 351:1041-1049.

11. Van Veldhoven PP, Mannaerts GP. Subcellular localization and membrane topology of sphingosine-1-phosphate lyase in rat liver. J Biol Chem 1991; 266: 12502-12507.
12. Heller RA, Kronke M. Tumor necrosis factor receptor-mediated signaling pathways. J Cell Biol 1994; 126:5-9.
13. Cifone MG, De Maria R, Roncaioli P et al. Apoptotic signaling through CD95 (Fas/Apo-1) activates an acidic sphingomyelinase. J Exp Med 1993; 177: 1547-1552.
14. Haimovitz-Friedman A, Kan CC, Ehleiter D et al. Ionizing radiation acts on cellular membranes to generate ceramide and initiate apoptosis. J Exp Med 1994; 180:525-535.
15. Gulbins E, Bissonnette R, Mahboubi A et al. FAS-induced apoptosis is mediated via a ceramide-initiated RAS signaling pathway. Immunity 1995; 2:341-351.
16. Obeid LM, Linardic CM, Karolak LA et al. Programmed cell death induced by ceramide. Science 1993; 259:1769-1771.
17. Zhang H, Buckley NE, Gibson K et al. Sphingosine stimulates cellular proliferation via a protein kinase C-independent pathway. J Biol Chem 1990; 265:76-81.
18. Zhang H, Desai NN, Olivera A et al. Sphingosine-1-phosphate, a novel lipid, involved in cellular proliferation. J Cell Biol 1991; 114:155-167.
19. Su Y, Rosenthal D, Smulson M et al. Sphingosine 1-phosphate, a novel signaling molecule, stimulates DNA binding activity of AP-1 in quiescent 3T3 fibroblasts. J Biol Chem 1994; 269:16512-16517.
20. Gomez MA, Martin A, O'Brien L et al. Cell-permeable ceramides inhibit the stimulation of DNA synthesis and phospholipase D activity by phosphatidate and lysophosphatidate in rat fibroblasts. J Biol Chem 1994; 269:8937-8943.
21. Olivera A, Zhang H, Carlson RO et al. Stereospecificity of sphingosine-induced intracellular calcium mobilization and cellular proliferation. J Biol Chem 1994; 289:17924-17930.
22. Schroeder JJ, Crane HM, Xia J et al. Disruption of sphingolipid metabolism and stimulation of DNA synthesis by fumonisin B1. A molecular mechanism for carcinogenesis associated with Fusarium moniliforme. J Biol Chem 1994; 269: 3475-3481.
23. Miyake Y, Kozutsumi Y, Nakamura S et al. Serine palmitoyltransferase is the primary target of a sphingosine-like immunosuppressant, ISP-1/myriocin. Biochem Biophys Res Commun 1995; 211: 396-403.
24. Pyne S, Pyne NJ. The differential regulation of cyclic AMP by sphingomyelin-derived lipids and the modulation of sphingolipid-stimulated extracellular signal regulated kinase-2 in airway smooth muscle. Biochem J 1996; 315:917-923.
25. Merrill AH, Sereni AM, Stevens VL et al. Inhibition of phorbol ester-dependent differentiation of human promyelocytic leukemic (HL-60) cells by sphinganine and other long-chain bases. J Biol Chem 1986; 261:12610-12615.
26. Merrill AH, Stevens VL. Modulation of protein kinase C and diverse cell functions by sphingosine—a pharmacologically interesting compound linking sphingolipids and signal transduction. Biochim Biophys Acta 1989; 1010:131-139.
27. Iwata M, Herrington J, Zanger RA. Sphingosine: a mediator of acute renal tubular injury and subsequent cytoresistance. Proc Natl Acad Sci USA 1995; 92:8970-8974.
28. Chao R, Khan W, Hannun YA. Retinoblastoma protein dephosphorylation induced by D-erythro-sphingosine. J Biol Chem 1992; 267:23459-23462.
29. Stoffel W, Bister K. Stereospecificities in the metabolic reactions of the four isomeric sphinganines (dihydrosphingosines) in rat liver. Hoppe-Seyler's Z Physiol Chem 1973; 354:169-181.
30. Pushkareva M, Chao R, Bielawska A et al. Stereoselectivity of induction of the retinoblastoma gene product (pRb) dephosphorylation by D-erythro-sphingosine supports a role for pRb in growth suppression by sphingosine. Biochemistry 1995; 34:1885-1892.
31. Ohta H, Sweeney EA, Masamune A et al. Induction of apoptosis by sphingosine in human leukemic HL-60 cells: a possible endogenous modulator of apoptotic

DNA fragmentation occurring during phorbol ester-induced differentiation. Cancer Res 1995; 55:691-697.
32. Ohta H, Yatomi Y, Sweeney EA et al. A possible role of sphingosine in induction of apoptosis by tumor necrosis factor-alpha in human neutrophils. FEBS Lett 1994; 355:267-270.
33. Olivera A, Spiegel S. Sphingomyelinase and cell-permeable ceramide analogs stimulate cellular proliferation in quiescent Swiss 3T3 fibroblasts. J Biol Chem 1992; 267:26121-26127.
34. Hauser JM, Buehrer BM, Bell RM. Role of ceramide in mitogenesis induced by exogenous sphingoid bases. J Biol Chem 1994; 269:6803-6809.
35. Sasaki T, Hazeki K, Hazeki O et al. Permissive effect of ceramide on growth factor-induced cell proliferation. Biochem J 1995; 311:829-834.
36. Boucher LM, Wiegmann K, Futterer A et al. CD28 signals through acidic sphingomyelinase. J Exp Med 1995; 181:2059-2068.
37. Kolesnick R, Fuks Z. Ceramide: a signal for apoptosis or mitogenesis? J Exp Med 1995; 181:1949-1952.
38. Jarvis WD, Kolesnick RN, Fornari FA et al. Induction of apoptotic DNA damage and cell death by activation of the sphingomyelin pathway. Proc Natl Acad Sci USA 1994; 91:73-77.
39. Jayadev S, Liu B, Bielawska AE et al. Role for ceramide in cell cycle arrest. J Biol Chem 1995; 270:2047-2052.
40. Olivera A, Spiegel S. Sphingosine-1-phosphate as second messenger in cell proliferation induced by PDGF and FCS mitogens. Nature 1993; 365:557-560.
41. Gomez-Munoz A, Waggoner DW, O'Brien L et al. Interaction of ceramides, sphingosine, and sphingosine 1-phosphate in regulating DNA synthesis and phospholipase D activity. J Biol Chem 1995; 270:26318-26325.
42. Pyne S, Chapman J, Steele L et al. Sphingomyelin-derived lipids differentially regulate the extracellular signal-regulated kinase 2 (ERK-2) and c-Jun N-terminal kinase (JNK) signal cascades in airway smooth muscle. Eur J Biochem 1996; 237:819-826.
43. Bornfeldt KE, Graves LM, Raines EW et al. Sphingosine-1-phosphate inhibits PDGF-induced chemotaxis of human arterial smooth muscle cells: spatial and temporal modulation of PDGF chemotactic signal transduction. J Cell Biol 1995; 130:193-206.
44. Jacobs LS, Kester M. Sphingolipids as mediators of effects of platelet-derived growth factor in vascular smooth muscle cells. Am J Physiol 1993; 265:c740-c747.
45. Coroneos E, Martinez M, McKenna S et al. Differential regulation of sphingomyelinase and ceramidase activity by growth factor and cytokines: implication for cellular proliferation and differentiation. J Biol Chem 1995; 270:23305-23309.
46. Mazurek N, Megidish T, Hakomori S-I et al. Regulatory effect of phorbol esters on sphingosine kinase in BALB/C 3T3 fibroblasts (variant A31): demonstration of cell type-specific response. Biochem Biophys Res Commun 1994; 198:1-9.
47. Cuvillier O, Pirianov G, Kleuser B et al. Suppression of ceramide-mediated programmed cell death by sphingosine-1-phosphate. Nature 1996; 381:800-803.
48. Stoffel W, Sticht G, LeKim D. Synthesis and degradation of sphingosine bases in *Hansenula ciferrii*. Hoppe-Seyler's Z Physiol Chem 1968; 349:1149-1156.
49. Keenan RW. Sphingolipid base phosphorylation by cell-free preparations from *Tetrahymena pyriformis*. Biochim Biophys Acta 1972; 270:383-396.
50. Buehrer BM, Bell RM. Inhibition of sphingosine kinase in vitro and in platelets. Implications for signal transduction pathways. J Biol Chem 1992; 267:3154-3159.
51. Louie DD, Kisic A, Schroepfer GJ. Sphingolipid base metabolism. Partial purification and properties of sphinganine kinase of brain. J Biol Chem 1976; 52: 4557-4564.
52. Shi L, Nishioka WK, Th'ng J et al. Premature p34cdc2 activation required for apoptosis. Science 1994; 263:1143-1145.
53. Harrington EA, Fanadi A, Evan GI. Oncogenes and cell death. Curr Biol 1994; 4:120-129.
54. Jarvis WD, Fornari FA, Browning JL et al. Attenuation of ceramide-induced

apoptosis by diglyceride in human myeloid leukemia cells. J Biol Chem 1994; 269:31685-31692.
55. Jarvis WD, Fornari FA, Traylor RS et al. Induction of apoptosis and potentiation of ceramide-mediated cytotoxicity by sphingoid bases in human myeloid leukemia cells. J Biol Chem 1996; 271: 8275-8284.
56. Tepper CG, Jayadev S, Liu B et al. Role for ceramide as an endogenous mediator of Fas-induced cytotoxicity. Proc Natl Acad Sci USA 1995; 92:8443-8447.
57. Venable ME, Blobe GC, Obeid LM. Identification of a defect in the phospholipase D/diacylglycerol pathway in cellular senescence. J Biol Chem 1994; 269: 26040-26044.
58. Spiegel S, Milstien S. Sphingoid bases and phospholipase D activation. Chem Physics Lipids 1996; 80:27-36.
59. Oltvai ZN, Korsmeyer SJ. Checkpoints of dueling dimers foil death wishes. Cell 1994; 79:189-192.
60. Martin SJ, Green DR. Protease activation during apoptosis: death by a thousand cuts? Cell 1995; 82:349-352.
61. Patel T, Gores GJ, Kaufmann SH. The role of proteases during apoptosis. FASEB J 1996; 10:587-597.
62. Miura M, Zhu H, Rotello R et al. Induction of apoptosis in fibroblasts by IL-1β-converting enzyme, a mammalian homolog of the C. elegans cell death gene *ced-3*. Cell 1993; 75:653-660.
63. Pronk GJ, Ramer K, Amiri P et al. Requirement of an ICE-like protease for induction of apoptosis and ceramide generation by REAPER. Science 1996; 271:808-810.
64. Tewari M, Quan LT, O'Rourke K et al. Yama/CPP32β, a mammalian homolog of CED-3, is a CrmA-inhibitable protease that cleaves the death substrate poly (ADP-ribose) polymerase. Cell 1995; 81:801-809.
65. Nicholson DW, All A, Thornberry NA et al. Identification and inhibition of the ICE/CED-3 protease necessary for mammalian apoptosis. Nature 1995; 376: 37-43.
66. Smyth MJ, Perry DK, Zhang J et al. prICE: a downstream target for ceramide-induced apoptosis and for the inhibitory action of Bcl-2. Biochem J 1996; 316:25-28.
67. Zhang J, Alter N, Reed JC et al. Bcl-2 interrupts the ceramide-mediated pathway of cell death. Proc Natl Acad Sci USA 1996; 93:5325-5328.
68. Reed JC. Bcl-2 and the regulation of programmed cell death. J Cell Biol 1994; 124:1-6.
69. Lam M, Dubyak G, Chen L et al. Evidence that BCL-2 represses apoptosis by regulating endoplasmic reticulum-associated Ca^{2+} fluxes. Proc Natl Acad Sci USA 1994; 91:6569-6573.
70. Ghosh TK, Bian J, Gill DL. Sphingosine-1-phosphate generated in the endoplasmic reticulum membrane activates release of stored calcium. J Biol Chem 1994; 269:22628-22635.
71. Mattie ME, Brooker G, Spiegel S. Sphingosine-1-phosphate, a putative second messenger, mobilizes calcium from internal stores via an inositol trisphosphate-independent pathway. J Biol Chem 1994; 269:3181-3188.
72. Ghosh TK, Bian J, Gill DL. Intracellular calcium release mediated by sphingosine derivatives generated in cells. Science 1990; 248:1653-1656.
73. Chao CP, Laulederkind SJF, Ballou LR. Sphingosine-mediated phosphatidylinositol metabolism and calcium mobilization. J Biol Chem 1994; 269:5849-5856.
74. Kindman LA, Kim S, McDonald TV et al. Characterization of a novel intracellular sphingolipid-gated Ca^{2+}-permeable channel from rat basophilic leukemia cells. J Biol Chem 1994; 269:13088-13091.
75. Mao C, Kim SH, Almenoff JS et al. Molecular cloning and characterization of SCaMPER, a sphingolipid Ca^{2+} release-mediating protein from endoplasmic reticulum. Proc Natl Acad Sci USA 1996; 93:1993-1996.
76. Sabbadini RA, Betto R, Teresi A et al. The effects of sphingosine on sarcoplasmic reticulum membrane calcium release. J Biol Chem 1992; 267:15475-15484.
77. Fatatis A, Miller RJ. Sphingosine and sphingosine-1-phosphate differentially modulate platelet-derived growth factor-BB-induced Ca^{2+} signaling in

transformed oligodendrocytes. J Biol Chem 1996; 271:295-301.
78. Choi OH, Kim JH, Kinet JP. Calcium mobilization via sphingosine kinase in signalling by the Fc epsilon RI antigen receptor. Nature 1996; 380:634-636.
79. Goodemote KA, Mattie ME, Berger A et al. Involvement of a pertussis toxin-sensitive G protein in the mitogenic signaling pathways of sphingosine 1-phosphate. J Biol Chem 1995; 270:10272-10277.
80. Sakano S, Takemura H, Yamada K et al. Ca^{2+} mobilizing action of sphingosine in Jurkat human leukemia T cells. Evidence that sphingosine releases Ca^{2+} from inositol trisphosphate- and phosphatidic acid-sensitive intracellular stores through a mechanism independent of inositol trisphosphate. J Biol Chem 1996; 271:11148-11155.
81. van Koppen CJ, Meyer-zu Heringdorf D, Zhang C et al. A distinct G(i) protein-coupled receptor for sphingosylphosphorylcholine in human leukemia HL-60 cells and human neutrophils. Mol Pharmacol 1996; 49:956-961.
82. Okajima F, Tomura H, Sho K et al. Involvement of pertussis toxin-sensitive GTP-binding proteins in sphingosine 1-phosphate-induced activation of phospholipase C-Ca^{2+} system in HL60 leukemia cells. FEBS Lett 1996; 379:260-264.
83. Postma FR, Jalink K, Hengeveld T et al. Sphingosine-1-phosphate rapidly induces Rho-dependent neurite retraction: action through a specific cell surface receptor. Embo J 1996; 15:2388-2392.
84. van Koppen C, Meyer-zu Heringdorf M, Laser KT et al. Activation of a high affinity Gi protein-coupled plasma membrane receptor by sphingosine-1-phosphate. J Biol Chem 1996; 271:2082-2087.
85. Liu R, Farach-Carson MC, Karin NJ. Effects of sphingosine derivatives on MC3T3-E1 pre-osteoblasts: psychosine elicits release of calcium from intracellular stores. Biochem Biophys Res Commun 1995; 214:676-684.
86. Yamamura S, Sadahira Y, Ruan F et al. Sphingosine-1-phosphate inhibits actin nucleation and pseudopodium formation to control cell motility of mouse melanoma cells. FEBS Lett 1996; 382:193-197.
87. Wang F, Nobes CD, Hall A et al. Sphingosine-1-phosphate stimulates Rho-mediated stress fiber formation and tyrosine phosphorylation of focal adhesion kinase and paxillin. Biochem J 1996; (submitted).
88. Spiegel S, Olivera A, Zhang H et al. Sphingosine-1-phosphate, a novel second messenger involved in cell growth regulation and signal transduction, affects growth and invasiveness of human breast cancer cells. Breast Cancer Res Treat 1994; 31:195-206.
89. Seufferlein T, Rozengurt E. Sphingosine induces p125FAK and paxillin tyrosine phosphorylation, actin stress fiber formation, and focal contact assembly in Swiss 3T3 cells. J Biol Chem 1994; 269:27610-27617.
90. Lavie Y, Piterman O, Liscovitch M. Inhibition of phosphatidic acid phosphohydrolase activity by sphingosine: Dual action of sphingosine in diacylglycerol signal termination. FEBS Lett 1990; 277:7-10.
91. Desai NN, Zhang H, Olivera A et al. Sphingosine-1-phosphate, a metabolite of sphingosine, increases phosphatidic acid levels by phospholipase D activation. J Biol Chem 1992; 267:23122-23128.
92. Natarajan V, Jayaram HN, Scribner WM et al. Activation of endothelial cell phospholipase D by sphingosine and sphingosine-1-phosphate. Am J Respir Cell Mol Biol 1994; 11:221-229.
93. Yamada K, Sakane F. The different effects of sphingosine on diacylglycerol kinase isozymes in Jurkat cells, a human T-cell line. Biochim Biophysica Acta 1993; 1169:211-216.
94. Bhat BG, Wang P, Coleman RA. Sphingosine inhibits rat hepatic monoacylglycerol acyltransferase in Triton X-100 mixed micelles and isolated hepatocytes. Biochemistry 1995; 34:11237-11244.
95. Gomez-Munoz A, Martin A, O'Brien L et al. Cell-permeable ceramides inhibit the stimulation of DNA synthesis and phospholipase D activity by phosphatidate and lysophosphatidate in rat fibroblasts. J Biol Chem 1994; 269: 8937-8943.
96. Zhang H, Desai NN, Murphey JM et al. Increases in phosphatidic acid levels ac-

company sphingosine-stimulated proliferation of quiescent Swiss 3T3 cells. J Biol Chem 1990; 265:21309-21316.
97. Marshall CJ. Specificity of receptor tyrosine kinase signaling transient versus sustained extracellular signal-regulated kinase activation. Cell 1995; 80:179-185.
98. Kyriakis JM, Banerjee P, Nikolakaki E et al. The stress-activated protein kinase subfamily of c-Jun kinases. Nature 1994; 369:156-160.
99. Derijard B, Raingeaud J, Barrett T et al. JNK1: a protein kinase stimulated by UV light and Ha-Ras that binds and phosphorylates the c-Jun activation domain. Cell 1995; 76:1025-1037.
100. Xia A, Dickens M, Raingeaud J et al. Opposing effects of ERK and JNK-p38 MAP kinases on apoptosis. Science 1995; 270:1326-1331.
101. Westwick JK, Bielawska AE, Dbaibo G et al. Ceramide activates the stress-activated protein kinases. J Biol Chem 1995; 270: 22689-22692.
102. Wu J, Spiegel S, Sturgill TW. Sphingosine-1-phosphate rapidly activates the MAP kinase pathway by a G-protein dependent mechanism. J Biol Chem 1995; 270:11484-11488.
103. Ghosh S, Strum JC, Sciorra VA et al. Raf-1 kinase possesses distinct binding domains for phosphatidylserine and phosphatidic acid. Phosphatidic acid regulates the translocation of Raf-1 in 12-O-tetradecanoylphorbol-13-acetate-stimulated Madin-Darby canine kidney cells. J Biol Chem 1996; 271:8472-8480.
104. Verheij M, Bose R, Lin X et al. Requirement for ceramide-initiated SAPK/JNK signalling in stress-induced apoptosis. Nature 1996; 380:75-79.
105. Coroneos E, Wang Y, Panuska JR et al. Sphingolipid metabolites differentially regulate extracellular signal-regulated kinase and stress-activated protein kinase cascades. Biochem J 1996; 316:13-7.
106. van Veldhoven PP, Matthews TJ, Bolognesi DP et al. Changes in bioactive lipids, alkylglycerol and ceramide, occur in HIV-infected cells. Biochem Biophys Res Commun 1992; 187:209-216.
107. Rivas CI, Golde DW, Vera JC et al. Involvement of the sphingomyelin pathway in TNF signaling for HIV production in chronically infected HL-60 cells. Blood 1993; 83:2191-2197.
108. De Simone C, Cifone MG, Roncaioli P et al. Ceramide, AIDS and long-term survivors. Immunol Today 1996; 17:48.

CHAPTER 10

Functional Roles of Glycosphingolipids and Sphingolipids in Signal Transduction

Sen-itiroh Hakomori

INTRODUCTION: EARLY STUDIES LEADING TO THE CONCEPT THAT GSLs AND SLs AFFECT TRANSMEMBRANE SIGNALING FOR CONTROL OF CELL GROWTH

The role of GSLs in the control of cell growth was initially suggested by dramatic changes of GSL composition observed in polyoma virus-transformed cells,[1] transformed vs. nontransformed cells with temperature-sensitive mutants of Rous sarcoma virus,[2] and transformed cells transfected with Ras-K oncogene.[3] Figure 10.1 summarizes the dramatic reduction of GM_3 and associated increase of LacCer, GlcCer, and/or Cer in transformed cells in these early studies. These observations are consistent with today's views on roles of these GSLs and SLs in control of signal transduction (see Fig. 1 legend). Subsequent studies on roles of GSLs in control of cell proliferation and signal transduction (as understood today) were based on: (a)exogenous addition of GSLs to transformed cells, which partially restored normal cell growth and extended the G1 phase of cell cycle;[4] (b)use of anti-GM_3 Fab antibodies which arrested cell growth at the G1 phase[5] and induced normal cell growth phenotype[6] (c)a close correlation between "cell contact response" of GSL synthesis and "contact inhibition" of cell growth.[7,8] Phenomenon (c) (illustrated in Fig. 10.2) is relevant to the current hypothesis that GM_3- or Gb3-dependent signal transduction is altered in association with cell contact and contact inhibition, and that these signaling mechanisms are lost in transformed cells.

Experiments using chemically-defined media (which first became available in the early 1980s) showed that BHK cell growth depends on FGF, and that FGF receptor function (internalization of ^{125}I-labeled FGF) is inhibited by exogenous GM_3, although GM_3 does not interact with FGF.[9] Consequently, we were able to demonstrate that various gangliosides affect Tyr kinase activity associated with growth factor receptors or hormone receptors.[10,11]

Sphingolipid-Mediated Signal Transduction, edited by Yusuf A. Hannun.

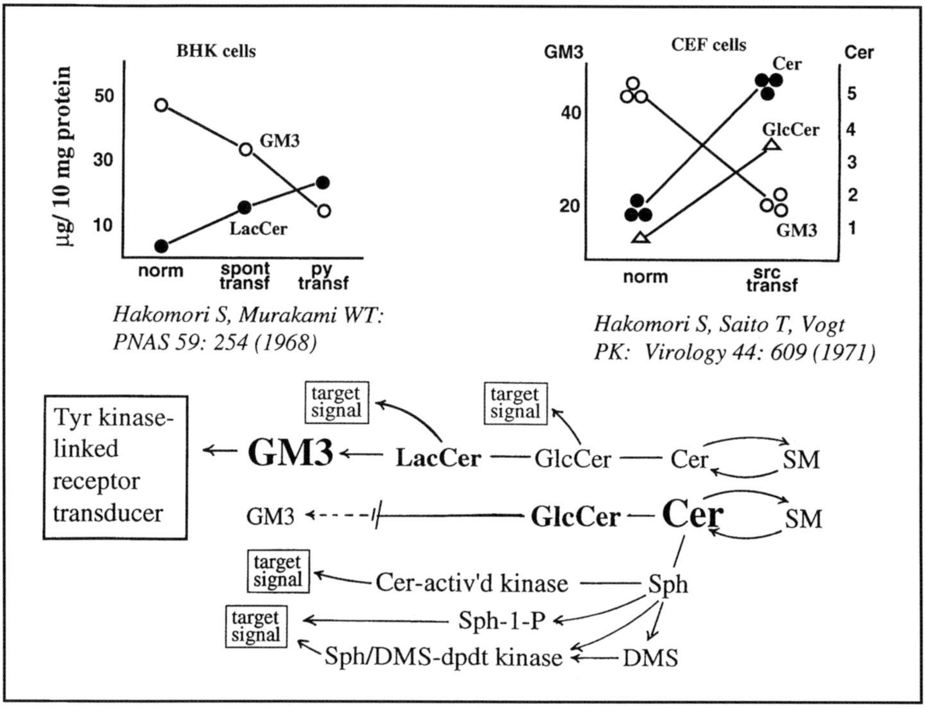

Fig. 10.1. Reciprocal relationships between quantity of GM3 vs. LacCer, or GM3 vs. GlcCer/Cer, in normal and transformed fibroblasts. *Upper left:* baby hamster kidney (BHK) cells transformed by polyoma virus (BHK-py). GM3 decreased to 1/4 of original level, while LacCer increased five-fold. *Upper right:* chick embryonic fibroblasts (CEF) transformed by Rous sarcoma virus (CEF-src). GM3 decreased to 1/3 of original level, while GlcCer and Cer increased five-fold. *Lower panel:* model of signal transduction changes associated with GSL and SL changes, based on data in upper panels. Bold letters indicate increased level. Unblocked GM3 synthesis from Cer through GlcCer and LacCer takes place in normal BHK and CEF cells. Accumulation of LacCer in BHK-py and of GlcCer/Cer in CEF-src results in activation of target signaling mechanisms, particularly Cer-activated kinase. GM3 inhibits Tyr kinase associated with growth factor receptor.

Two direct degradation products of gangliosides (lysogangliosides and de-N-acetylgangliosides; see below) were found to respectively inhibit and promote EGF-RK.[12-14]

During 1982-1987, while we were studying effects of gangliosides and their degradation products on RK activity, Bell, Hannun, and associates observed that Sph, the common backbone of GSLs and SLs, inhibits PKC.[15,16] This is analogous to findings that phospholipid breakdown products affect signal transduction, e.g., DAG strongly promotes PKC.[17] These observations encouraged many subsequent studies on roles of Sph, Sph derivatives, and Cer as second messengers or modifiers of signaling. In these early studies, the stereoisomer of Sph used was ambiguous. In follow-up studies, we showed that the N,Ndimethyl compound (DMS) with D-erythro configuration had a stronger inhibitory effect than other stereoisomers,[18] and that DMS is synthesized by methyltransferase in mouse brain.[19] Other studies indicate that Sph inhibits not only PKC but also a number of cellular components, including Na/K-ATPase,[20] c-src and v-src kinase.[18] On the other hand, Sph activates some protein kinases. For example, it enhances phosphorylation at Thr 669 (MAP kinase site) of EGF receptor through a

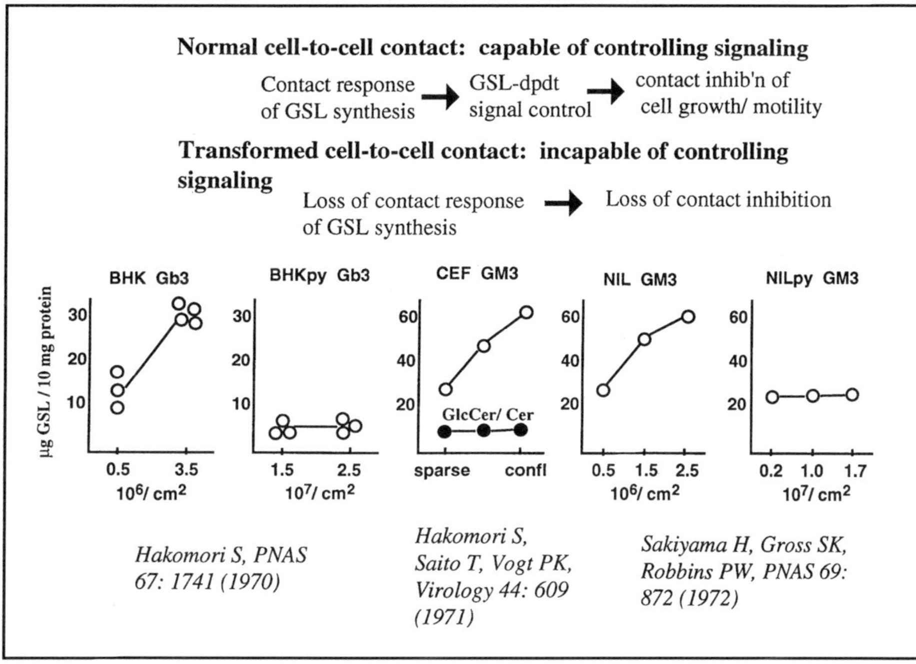

Fig. 10.2. Cell contact-dependent response of GSL synthesis, and its absence in transformed cells. Nontransformed fibroblasts (BHK, CEF, NIL) showed three-fold increase of Gb3 or GM3 synthesis when undergoing contact (and consequent inhibition of cell growth) at high density. Transformed cells (BHKpy, NILpy) do not show changes of Gb3 or GM3 synthesis. In CEF, only GM3 (not GlcCer/Cer) shows contact response of synthesis.

These observations, from over 20 years ago, are now interpreted as showing that nontransformed cells are susceptible to cell contact to initiate GSL synthesis, which induces signaling to stop proliferation, and that this GSL response and signaling are lost in transformed cells.

pathway which appears to be independent of PKC (see refs. 17, 21 for review of these early studies). Exogenous addition of Sph to culture inhibits cell growth in most cell types. However, it induces proliferation in Swiss 3T3 cells through its conversion into Sph-1-P, which opens up a new signaling pathway, possibly through intracellular Ca^{2+} mobilization.[22,23] Sph-1-P induces Ca^{2+} mobilization similarly to inositol triphosphate[24] (see ref. 25 for review of these early studies).

A significant amount of cell surface SM is hydrolyzed to Cer and phosphorylcholine by activation of neutral SMase in HL-60 cells stimulated by 1-α,25-dihydroxy vitamin D_3 to induce differentiation. During a lapse of 4 hours, levels of Cer and SM become normalized.[26] This observation and many subsequent studies suggested that Cer and Sph-1-P play important roles as second messengers.[27-29] However, roles of gangliosides and other GSLs in modulation of Tyr kinase linked receptors and other transducer molecules involved in signal transduction pathways have also become increasingly clear. Signaling systems and transducer molecules may be essentially the same among various cell types, except for small variations in the regulatory domain of transducer molecules, which may create differences in susceptibility to GSLs or SLs. Alternatively, GSL and SL patterns may vary extensively among different types of cells, leading to major differences in intracellular signaling.

EFFECTS OF GSLs AND SLs ON FIVE CLASSES OF RECEPTORS AND SIGNAL TRANSDUCERS

Five classes of receptors or signal transducers are involved in initiation of signal transduction pathways: ion channel linked, G-protein linked, Tyr kinase linked, PKC-dependent, and SMase linked (see Fig. 10.3). Effects of GSLs and SLs on function and subsequent signaling system for each type of receptor are briefly described below.

EFFECTS OF GSLs ON SIGNALING THROUGH TYR KINASES ASSOCIATED WITH GROWTH FACTOR RECEPTORS

Effects of gangliosides or GSLs have been most extensively studied for class III receptors, in which cytoplasmic Tyr kinases associated with or directly linked to each type of receptor are inhibited or promoted by GSLs or gangliosides (see Fig. 10.4). Tyr phosphate (Y-P) formed on class III receptors is recognized by various docking proteins, through which different signals are created. A typical example is shown in Fig. 5. In this signaling pathway, Tyr kinases are sensitive not only to specific gangliosides and GSLs (shown in Fig. 10.4), but also to lyso-GSLs or a mixture of GSL and lyso-PC. For example, EGF-RK is highly susceptible to GM_3 and lyso-GM_3, and the inhibitory effect of GM_3 is greatly enhanced in the presence of lyso-PC but not other lyso-phospholipids.[30]

Raf-K is also a common element in two major signaling pathways involving PKC and PKA. Raf-K is stimulated by PKC[31,32] but inhibited by PKA.[33,34] Activities of both PKC and Raf-K are modulated by chaperon protein 14-3-3, which is in turn highly susceptible to Sph and DMS (see section 2.5). This signaling pathway is the major and best known example, and effects of GSLs and SLs on key transducer molecules have been partially (but definitively) identified (Fig. 10.5).

ION-CHANNEL LINKED AND G-PROTEIN LINKED RECEPTORS

Ion-channel linked receptors (class I) include the Na^+/K^+ pump associated with Na^+/K^+-ATPase, synaptic signaling between electrically excitable cells, acetylcholine receptors at the neuromuscular junction, etc. In a classic study, Mg^{2+}-ATPase (but not Na^+/K^+-ATPase) in rat brain was activated by various gangliosides.[35] The observed susceptibility in kidney cells of ion transport modulator to GM1,[36] and susceptibility in rat brain membrane of Ca^{2+}-dependent kinase to GD1a[37] and polysialoganglioside-dependent myelin protein[38] suggests that ion-channel linked receptors in general are modulated by gangliosides.

Signaling via G-protein linked cell surface receptors (class II) is widely-occurring and closely associated with key signaling enzymes. There has been no systematic study of possible susceptibility to gangliosides, GSLs, or SLs of the events subsequent to G-protein activation, or of G-protein itself. Gangliosides were suggested to be cofactors of adenyl cyclase in an early study.[39] Extensive future studies along this line are needed.

Pertussis toxin is a functional uncoupler of G-protein from its receptor. This toxin ADP-ribosylates the α-subunit of G-protein (GTP-binding protein).[40,41] In similar fashion, exoenzyme C3 of *Clostridium botulinum* type C and D ADP-ribosylates the Rho family of GTP-binding proteins, which play essential roles in formation of actin filament networks in eukaryotic cells.[42] ADP-ribosyltransferase activities of both pertussis toxin and exoenzyme C3 are inhibitable by various gangliosides, most notably the novel ganglioside GQ1bα ($IV^3NeuAcIII^6NeuAcII^3NeuAc2$-8NeuAcGg4Cer).[43] However, the ganglioside concentrations required are high, and the physiological significance of the phenomenon remains unclear.

Sph-1-P-dependent activation of DNA synthesis in Swiss 3T3 cells is also inhibitable by pertussis toxin.[44] Together, these findings suggest that G-protein is involved in regulation of the mitogenic effect of Sph-1-P, and that ADP-ribosylation could be modulated by gangliosides.

SMase Associated Receptors

The receptors for TNFα, IL-1, and apoptosis ligand Fas have been shown recently to be closely associated with SMase activity through an unknown mechanism, and are tentatively termed class IV receptors. SMase activity is enhanced by TNFα, IL-1, or Fas bound to its receptor, leading to production of Cer and possibly Sph and Sph derivatives (see for review refs. 27-29). The close association of intracellular Cer and Sph levels with apoptosis and cell motility depends on class IV receptors.

Thus, the SM cycle is affected by activation of class IV receptors to enhance de novo Cer level in response to cell stimulation. However, a significant amount of Cer (5-20 μg per 100 mg cellular protein, equivalent to the amount of GlcCer) is present in essentially all animal cells. Cer created de novo in response to cell stimulation ranges from 30-300% of the quantity present prior to stimulation. We must assume that the de novo Cer has different in situ organization than the Cer previously present. Susceptibility of SMase to GSLs or SLs is totally unknown.

Signaling Through PKC

PKC, a ubiquitous transmembrane signal transducer, is activated by DAG, which is produced by hydrolysis of PIP2 as catalyzed by PL C-β and C-γ. In contrast, PKC is inhibited by Sph, DMS, and lyso-gangliosides.[16-18,21] Thus, numerous signaling pathways depending on PKC are controlled by GSLs and SLs. PKC promotes Raf kinase (MAPKKK), and this promotion is inhibited by the phosphorylated forms (α and δ isoforms) of 14-3-3,[45] a versatile "chaperon" protein which affects functions of many regulatory proteins.[46,47] Nonphosphorylated forms (β and ζ isoforms) of 14-3-3 interact with and activate Raf-K to maintain its activity.[48,49] 14-3-3 was recently shown to be one of the substrates of Sph-/DMS-dependent kinase. Only Sph and DMS (but not Cer, SPC, Sph-1-P, or 13 other lipids) are capable of inducing phosphorylation of 14-3-3.[50] Thus, SL-dependent regulation of PKC and Raf-K through 14-3-3 is a crucial event in control of various cellular phenotypes.

Signaling Through Changes in Level of Ca^{2+} or Other Ions

Cells use Ca^{2+} as a signaling molecule for stimulation of motility and growth. The resting cell must keep cytosolic [Ca^{2+}] low (10^{-7} M). Various stimuli lead to increased levels of IP_3 (the hydrolysis product of PIP2 by PL C-β and C-γ) and PA (hydrolysis product of PC by PL D). Both IP_3 and PA somehow open up Ca^{2+} channels at the endoplasmic reticulum and mitochondrial membrane, leading to increase of cytosolic [Ca^{2+}] up to 10^{-3} M level. Calmodulin, a versatile Ca^{2+} receptor/modulator, binds to ganglioside,[51] and calmodulin-dependent phosphodiesterase is inhibited by ganglioside.[52] Anti-GalCer antibody induces a sustained increase in Ca^{2+} level in primary cultures of oligodendroglia.[53] Exogenous addition of GM_3 specifically enhances intracellular Ca^{2+} in platelets.[54] We found that phosphorylation of both a Ca^{2+} regulatory protein (endoplasmin) and glucose-regulated protein (Grp; Mr 100 kDa) is stimulated by Sph and DMS, but not by Cer, Sph-1-P, or other SLs (Megidish T, Takio K, Titani K, Hakomori S, unpublished data; see below). Thus, the process of Ca^{2+} influx may be regulated by GSLs and SLs at various stages.

FUNCTION OF GM_3 AND ITS PRIMARY DEGRADATION PRODUCTS IN CONTROL OF TRANSMEMBRANE SIGNALING

Inhibitory Effect of GM_3 on Growth Factor RK, and Its Physiological Relevance

GM_3 inhibits EGF-RK and EGF-dependent cell growth in vitro. However, its physiological significance in vivo was unclear until studies were performed employing GM_3-expressing mutant cells. One example is an epimerase-less mutant of CHO cells (1a1D) transfected with an EGF receptor

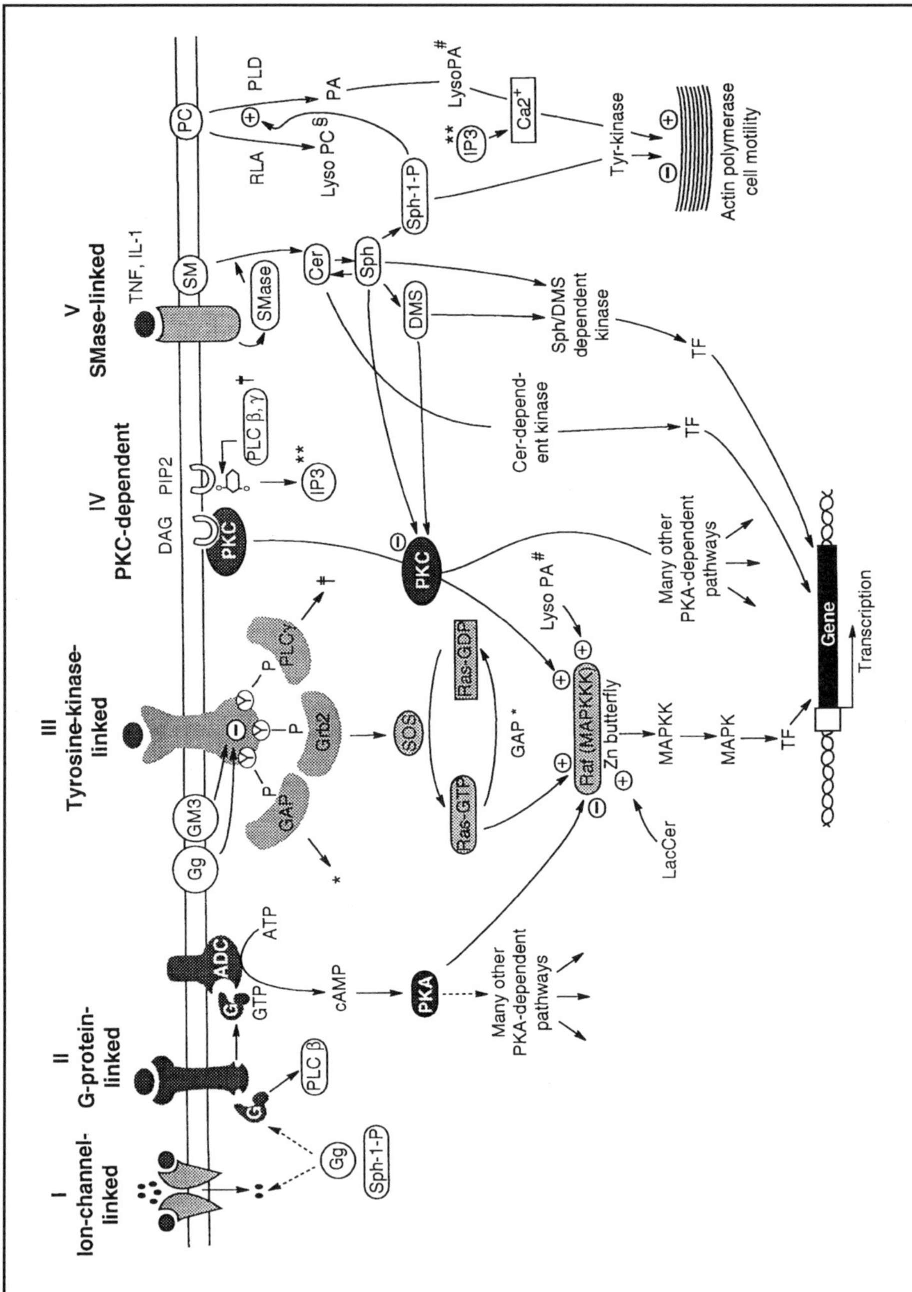
I
Ion-channel-linked
II
G-protein-linked
III
Tyrosine-kinase-linked
IV
PKC-dependent
V
SMase-linked
TNF, IL-1
Gg
Sph-1-P
PLC β
GTP
ADC
ATP
cAMP
PKA
Many other PKA-dependent pathways
Gg
GM3
GAP
Grb2
PLCγ
SOS
Ras-GTP
Ras-GDP
GAP *
Raf (MAPKKK)
Zn butterfly
LacCer
MAPKK
MAPK
TF
Gene
Transcription
DAG
PIP2
PKC
PLC β, γ
IP3
Lyso PA #
Many other PKA-dependent pathways
SM
SMase
Cer
Sph
DMS
Cer-dependent kinase
Sph/DMS dependent kinase
TF
PC
RLA
PLD
Lyso PC §
PA
LysoPA #
Sph-1-P
IP3
Ca2+
Tyr-kinase
Actin polymerase
cell motility

Fig. 10.3. Five classes of receptors or signal transducers which initiate transmembrane signaling. Abbreviations and conventions: Shaded and nonshaded compounds are proteins and lipids, respectively. ADC, adenyl cyclase. G,G-protein. GAP, GTP phosphatase activating protein. Gg, ganglioside. GTP, guanidine triphosphate. IP_3, inositol triphosphate. PIP_2, phosphatidylinositol diphosphate. PLA, phospholipase A. PLC, phospholipase C. PLD, phospholipase D. TF, transcription factor. Zn butterfly, a symmetric, cysteine-rich domain showing lipid-binding affinity.

Five classes of receptors or transducers are involved: ion channel linked (I), G-protein linked (II), Tyr kinase linked (III), PKC-dependent (IV), and SMase linked (V). Function of classes I and II is thought to be modulated by gangliosides, but no extensive studies have been performed (see text). G-protein ("G"), activated upon GTP binding, in turn activates PLCβ. On the other hand, PLCγ, which has SH domain (*src* homology domain 2,3), is capable of adapting to Tyr phosphate of receptor class III. Signaling through PLC is controlled by GM3 and other gangliosides. GTP-bound, activated G-protein also activates adenylate cyclase (ADC), which converts ATP to cAMP, which in turn activates PKA. Thus, many PKA-dependent pathways are opened. ADC is known to be susceptible to gangliosides,[39] but further extensive studies are needed. PC is converted to lyso-PC through action of PLA, and to PA by PLD. Activity of PLD is enhanced by Sph-1-P. Lyso-PC, in combination with GM3, strongly inhibits Tyr kinase linked receptor (class III). PA is converted to lyso-PA, which strongly enhances activity of Tyr kinase associated with podosomes, and also enhances actin polymerization and cell motility. PLCβ,γ, activated by G-protein or Tyr kinase receptor, activates PKC via DAG formation. This event plays a crucial role in signal transduction and opens many signaling pathways. In contrast to DAG, which promotes PKC, Sph and DMS strongly inhibit PKC. PKC activity is also inhibited by various lyso-GSLs and gangliosides. Tyr kinase linked receptors (class III) are inhibited by GM3, SPG, and other gangliosides. Activation of class III receptors by binding of various signaling molecules (growth factors) results in Tyr phosphorylation at cytoplasmic sites ("P-Y"). This creates "docking sites" for binding of adapter proteins having *src* homology domains (SH2 and SH3), such as Grb2, phospholipase Cγ, GTPase activating protein (GAP), etc. Binding of Grb2, for example, leads to activation of Sos, the GDP/GTP exchanger for Ras. This triggers a cascade of phosphorylation through Raf (MAP kinase kinase kinase), MAP kinase kinase, and finally MAPK, which phosphorylates one of the transcription factors (TF) for activation of gene transcription. One of the activated genes encodes MAPK phosphatase, which feeds back to inactivate MAPK (see Fig. 10.5). An alternative pathway leading to MAPK activation and mediated by PKC and PKA is also present. This pathway, regarded as the central event of signal transduction, is closely associated with two other well-established pathways which also depend on PKC and PKA. It has recently become clear that Raf kinase activity is stimulated by PKC[31,32] and inhibited by PKA.[33,34] Some types of receptors (e.g., IL-1, TNFα) show specific linkage to SMase, which induces hydrolysis of SM to give increased level of Cer. This triggers activation of Cer-dependent kinase and consequent activation of TF.[28,101]

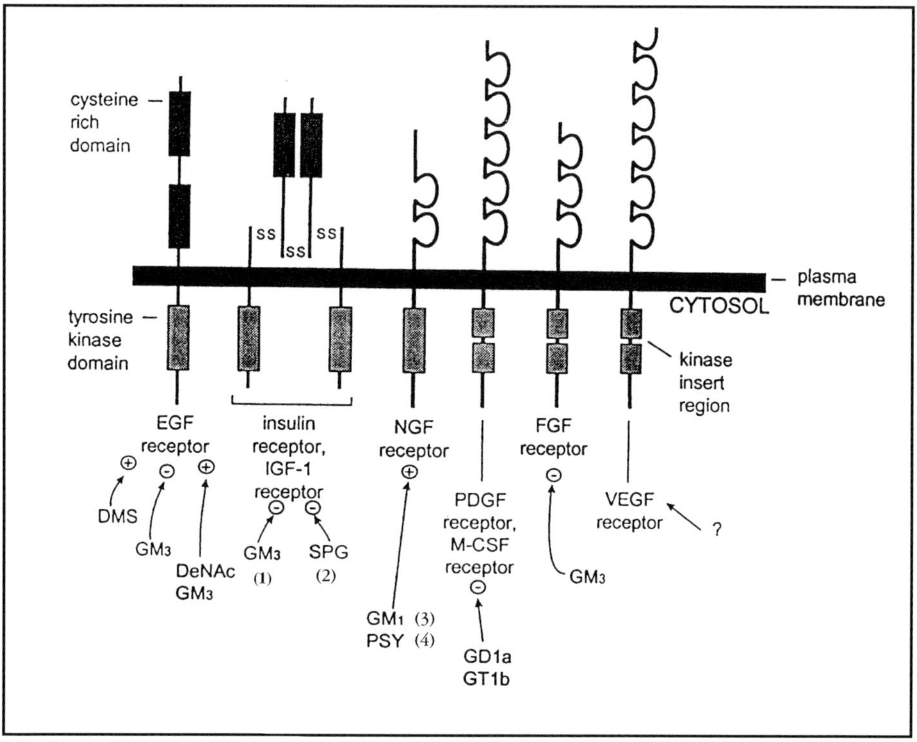

Fig. 10.4. Six subfamilies of growth factor/hormone receptor having tyrosine kinases, and their susceptibility to various types of gangliosides and GSLs. These receptors all have a cytoplasmic tyrosine kinase domain which is susceptible to GSL or sphingolipid. +, promotion. –, inhibition. DeNAcGM3, de-N-acetyl-GM3. PSY, plasmalopsychosine. VEGF, vascular endothelial growth factor. (1) rat and mouse insulin receptor susceptible to GM3. (2) human insulin receptor susceptible to SPG. (3) requires large quantity of GM1 (200 μM- 1 mM). (4) highly susceptible to PSY; requires only 1-10 μM. (?) unidentified; affects VEGF receptor. Reprinted and modified with permission from Hakomori S, Igarashi Y. J Biochem 1995; 118:1091-1103.

gene. Thus, the mutant expresses EGF receptor but is incapable of synthesizing LacCer or GM_3 unless Gal is added to the medium. In this mutant, EGF-dependent cell growth inhibition was only observed in association with the onset of GM_3 synthesis induced by addition of Gal to the medium.[55] In another example, GM_3-defective mouse mammary carcinoma mutant FM28-7, containing LacCer but not GM_3, showed strong insulin-dependent cell growth, which was completely inhibited by exogenous addition of GM_3.[56] Mouse insulin RK is highly susceptible to GM_3, whereas human insulin RK is susceptible to SPG.[57]

According to the Ullrich-Schlessinger model,[58] dimerization or multimerization of receptor upon binding of growth factor to its receptor is a prerequisite for internalization of receptor-growth factor complex and simultaneous activation of cytoplasmic Tyr kinase. In A431 cells, GM_3 inhibits Tyr kinase in a dose-dependent manner, but has no effect on quantity of receptor dimer yielded upon addition of EGF.[59] We conclude tentatively that the inhibitory effect of GM_3 on EGF-dependent cell growth and Tyr kinase is due to direct interaction of GM_3 with EGF-RK, rather than indirect effect of interference with receptor-receptor intera-

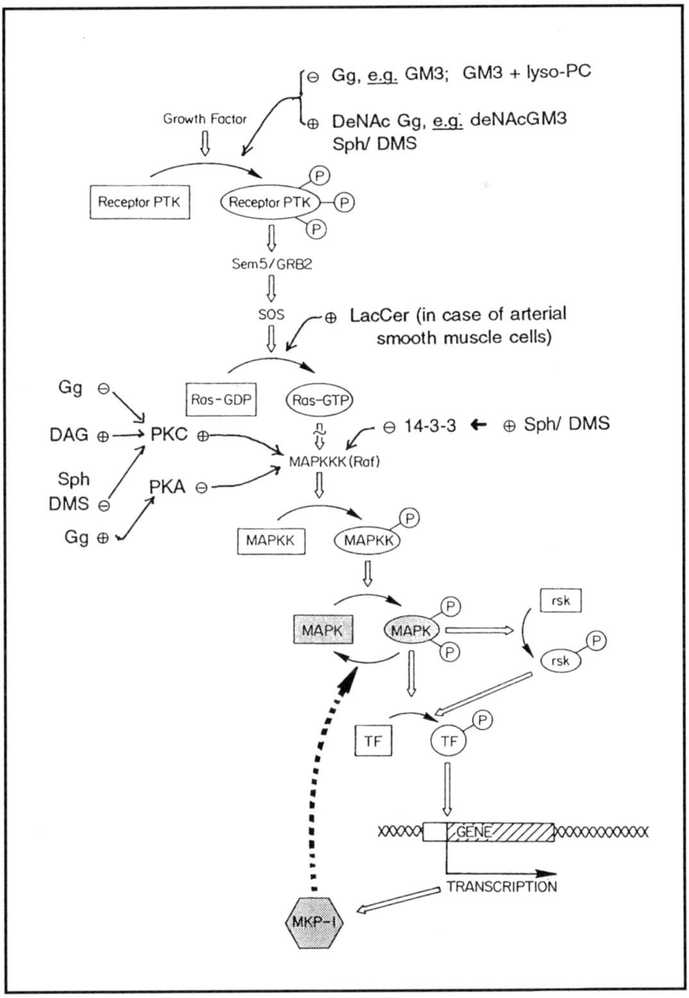

Fig. 10.5. A signaling pathway triggered by Tyr kinase linked growth factor receptor (PTK), and susceptibility of signaling to GSLs, SLs, or other modulators. This scheme (modified from ref. 102) summarizes current concepts of cellular response to growth factor receptor kinase (PTK) through binding to its ligand. Autophosphorylation of the receptor creates docking sites that recruit adapter protein Grb2, leading to activation of SOS, the GTP/GDP exchanger for Ras. A cascade of phosphorylation is triggered, leading to mitogen-activated protein kinase (MAPK). MAPK in turn phosphorylates transcription factors (TF), either directly or through activity of another Ser/Thr kinase, Rsk. This promotes transcription of genes requires for growth response. One of the genes induced by growth factor stimulation through the MAPK pathway encodes a MAPK phosphatase (MKP-1) which feeds back to the pathway by dephosphorylating and inactivating MAPK (dotted line).

The initial Tyr phosphorylation of receptor is inhibited by GM3, strongly inhibited by GM3 plus lyso-PC, and promoted by deNAcGM3, Sph, or DMS. Activity of MAPKKK (Raf) is promoted in the presence of nonphosphorylated 14-3-3. Phosphorylation of 14-3-3 is promoted by Sph and DMS; this process may inhibit MAPKKK.

PKC promotes MAPKKK activity. Since PKC activity is inhibited by Sph, DMS, and some gangliosides (Gg), these compounds indirectly inhibit MAPKKK activity. PKA inhibits MAPKKK activity. GDP loading reaction on Ras is promoted by LacCer in arterial smooth muscle cells.

ction.[59] Further extensive studies are necessary to confirm this idea.

Cooperative Inhibitory Effect of GM_3 and Lyso-PC on EGF-RK

The effect of GM_3 on cell growth through EGF-RK modulation was made clear by studies of GM_3 mutants (see above). However, essentially all cells which show growth factor-dependent growth contain some quantity of GM_3. How can one reconcile the consistent presence of GM_3 with the apparent fact that only a part of GM_3 may be used for regulatory effect on receptor function? This question is partially answered by the observation that the inhibitory effect of GM_3 on EGF-RK is greatly enhanced in the presence of lyso-PC. Lyso-PE, lyso-PS, and lyso-PI were ineffective. Lyso-PC is produced by hydrolysis of PC by PLA2. Therefore, it is possible that cell growth stimulation by growth factor may activate PLA2, produce lyso-PC surrounding GM_3, and that the lyso PC-GM_3 complex controls EGF-RK.[30]

Effect of Lyso-GM_3 and de-N-acetyl-GM_3 on EGF-RK

The modified gangliosides lyso-GM_3 and de-N-acetyl-GM_3 were identified in 1987-88 as possible modulators of transmembrane signaling through inhibitory or stimulatory effects on EGF-RK.[12-14] Lyso-GSLs (e.g., psychosine and lactosyl-Sph) were also shown to inhibit PKC.[16] Lyso-GM_3 strongly inhibits activity of not only EGF-RK but also PKC. In contrast, GM_3 inhibits EGF-RK but not PKC. Lyso-GM_3 was detected as a physiological component of A431 cells,[14] but the presence of lyso-GSLs in other types of cells remains to be explored. In human brain, psychosine was completely absent, while Sph and a major cationic GSL termed plasmalopsychosine were present (see section 5).

The novel ganglioside deNAcGM_3, in which the N-acetyl group of sialic acid is absent, was detected in various types of cells through positive reactivity with mAb DH5, which reacts specifically with deNAcGM_3 but not related structures. deNAcGM_3, in contrast to lyso-GM_3, strongly promotes EGF-RK, and promotes cell growth when added exogenously.[13] DeNAcGM_3 promotes Ser as well as Tyr phosphorylation of EGF-RK.[59] The physiological significance of this phenomenon remains to be studied. Structures of deNAcGM_3 and lyso-GM_3, and their relationships as signal modulators, are shown in Figure 10.6. Presence of de-N-acetyl sialic acid in ganglioside occurs in not only GM_3 but also other gangliosides such as GD3 and GD2.[60] De-N-acetylation of GD3 in human melanoma cells is induced by the Tyr kinase inhibitor Genistein. If de-N-acetyl gangliosides are involved in promotion of cell proliferation and cell cycle, it is important to identify the target molecules in this process. Either deNAcGM_3 or lyso-GM_3 can be formed by N-acetylase or GM_3:ceramidase in response to cell stimulation, but are immediately converted back to GM_3 by N-acetylation or N-fatty acylation. Proposed relationships between these compounds, and their functions, are shown in Fig. 10.7.

EFFECT OF SPG AND GM_3 ON INSULIN-DEPENDENT RK

Essentially all animal cells require insulin and its receptor for maintenance of growth in vitro.[61] The receptor consists of two insulin-binding subunits and two kinase subunits. In human cell lines, Tyr phosphorylation of the kinase subunit with Mr 97 kDa is highly susceptible to SPG. Exogenous addition of SPG (but not other gangliosides) inhibits insulin-dependent growth of human cells.[57] In mouse or rat cells, insulin RK is susceptible to GM_3. Insulin-dependent growth of these cells is inhibited by GM_3, as demonstrated with FM28-7 cells.[56] Type 2 diabetes may result from deficient insulin receptor function, which may result from aberrant synthesis of SPG or GM_3 in specific tissues (particularly muscle and fat, where insulin-dependent metabolism is high).

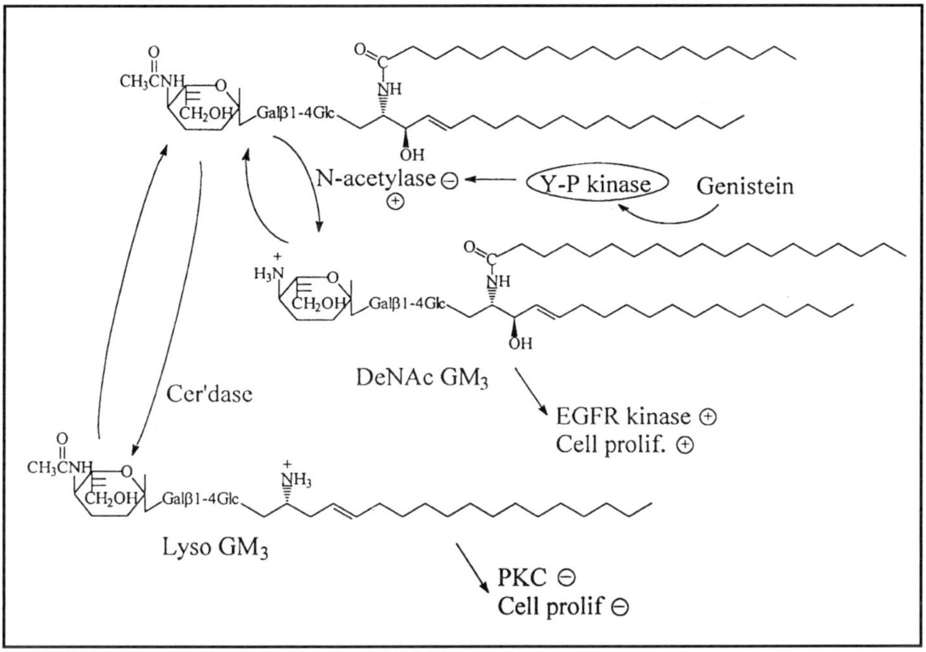

Fig. 10.6. Two catabolites of GM3 act as second messengers. DeNAcGM3 (yielded by de-N-acetylation of sialic acid moiety of GM3) enhances EGF receptor Tyr and Ser phosphorylation, and induces cell proliferation. The de-N-acetylation process is apparently controlled by Tyr kinase, since the process is induced by Genistein. Lyso-GM3 (presumably yielded by ceramidase action on GM3) strongly inhibits PKC and cell proliferation. Reprinted with permission from Hakomori S, Adv in Pharmacology 1996; 36:155-171.

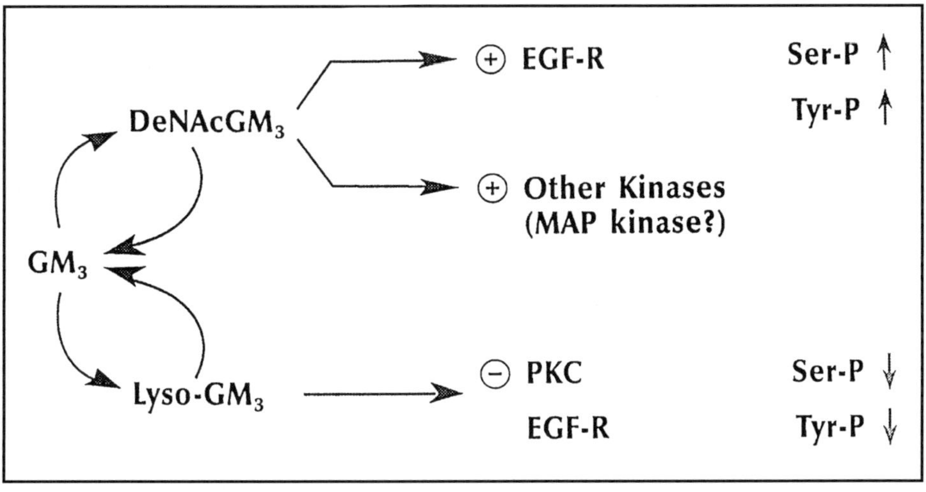

Fig. 10.7. Direct catabolites of gangliosides function as nonspecific second messengers. DeNAcGM3 and lyso-GM3 are not further degraded, but rather reconverted to GM3 by N-acetylation or N-fatty acylation. DeNAcGM3 promotes activity of EGF-RK and other kinases. Lyso-GM3 inhibits PKC and EGF-RK. Both these catabolites act as second messengers. Reprinted with permission from Hakomori S, Adv in Pharmacology 1996; 36:155-171.

PLPS AS A NOVEL TRANSMEMBRANE SIGNAL INDUCER IN NEURONAL CELLS

In order to study Sph and other cationic lipids in brain, we examined lipid fractions adsorbed on carboxymethyl Sephadex column. A small quantity of Sph and a trace of DMS were found, along with two major unknown components. No trace of psychosine (Gal-Sph) or Glc-Sph was found in the cationic lipid fraction. The two unknown compounds were identified as 3,4- and 4,6-cyclic plasmal conjugates of β-galactosyl-Sph, and were respectively termed plasmalopsychosine (PLPS) A and B.[62] The structure of PLPS-B, in comparison with psychosine and cerebroside, is shown in Figure 10.8. PLPS-B, and to a lesser extent PLPS-A, strongly activate NGF-RK (TrkA; $p140^{trk}$).[63] This leads to prolonged activation of MAPK and consequent induction of neurite outgrowth in neuronal cells. This was demonstrated in pheochromocytoma PC12 and neuroblastoma Neuro2a. Tyr kinase inhibitors K252 and staurosporine blocked the neuritogenic effect of PLPS. These findings suggest that NGF and PLPS share a common signaling pathway leading to MAPK activation.

The strong enhancing effect on Tyr kinase shown by PLPS at low concentration (1-10 μM) is unusual for GSLs. Most GSLs have inhibitory effects on receptor-associated Tyr kinase. Ganglioside alone without NGF does not induce neuritogenesis in PC12 cells. An exceptionally large quantity of GM1 (mM order) is required to enhance Tyr phosphorylation of TrkA, and this does not induce neuritogenesis.[64] In this context, the effect of PLPS to enhance Tyr phosphorylation of TrkA, leading to neuritogenesis and mimicking the effect of NGF, is very peculiar.

The major question "How does PLPS activate Trk A?" remains to be answered. PLPS does not inhibit binding of ^{125}I-labeled NGF to Trk A, which suggests that PLPS has no direct effect on Trk A, and does not bind to the same site within Trk A that NGF does. It is possible that PLPS binds to low-affinity receptor $p75^{NGFR}$ and activates Trk A pathway. It was reported recently that GM1

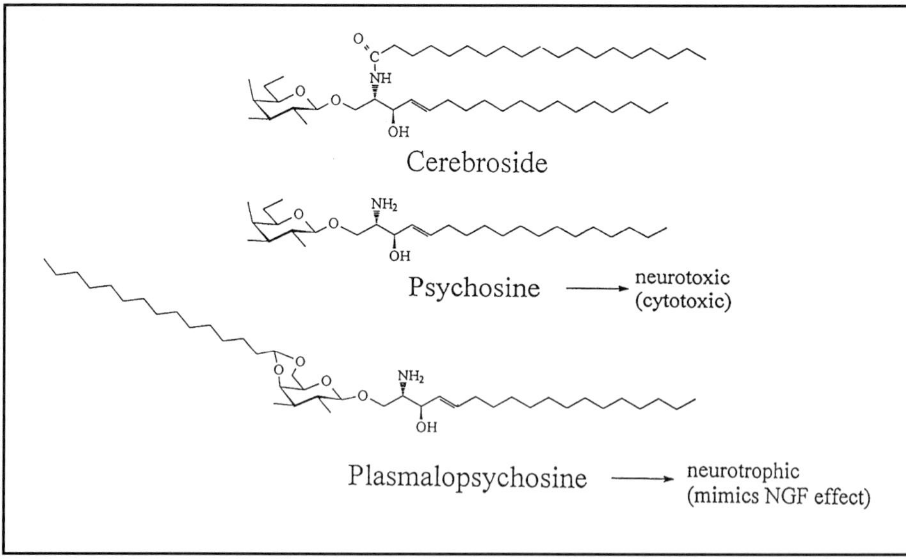

Fig. 10.8. Structure of PLPS as compared to psychosine and cerebroside. Fatty aldehyde (myristal or palmital) conjugated to a β-galactosyl residue of psychosine through 4,6 cyclic acetal is a unique feature of PLPS. Two aliphatic chains are oriented in opposite directions. PLPS is neurotrophic. In contrast, psychosine is highly neurotoxic. Reprinted with permission from Hakomori S, Adv in Pharmacology 1996; 36:155-171.

phosphorylates and binds to Trk A, but induces neurite growth only in the presence of NGF.[65] We tentatively concluded that: (1) the mechanism of Trk A phosphorylation by PLPS and GM1 differs from that by NGF; and (2) PLPS and NGF share a common signaling cascade from Trk A (see Fig. 10.9).

ROLE OF Sph-1-PHOSPHATE AS SECOND MESSENGER FOR MOTILITY CONTROL

Signaling by Sph-1-P to Inhibit Cell Motility and Ca^{2+} Mobilization

Sph-1-P has been suggested to be a PDGF receptor signaling molecule which induces Swiss 3T3 cell proliferation.[66] We reported previously that Sph-1-P as low as the nM range inhibits motility of various tumor cells,[67] although the mechanism for this phenomenon remains to be studied (see following section). We extended this observation to arterial smooth muscle cells. PDGF is known to stimulate motility and proliferation of these cells, which are important initial steps in atherosclerotic plaque formation.[68] In collaboration with E. Krebs, R. Ross, and colleagues,[69] we found that Sph-1-P has no significant effect on PDGF-dependent DNA synthesis, cell proliferation, or MAPK cascade in these cells. Thus, endogenous Sph-1-P may regulate (1) spatial and temporal changes in PI turnover leading to changes in PKA level; (2) Ca^{2+} mobilization; (3) actin filament assembly associated with chemotactic to PDGF. How the same ligand and same receptor open up different pathways which show different responses to the same SL modulator (e.g., Sph-1-P) remains to be elucidated.

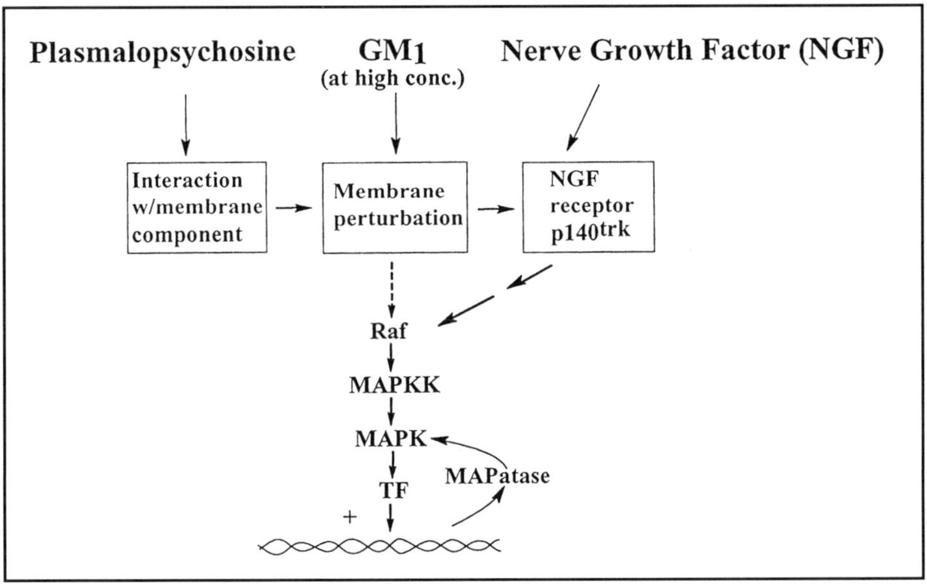

Fig. 10.9. Possible mechanism of neurotrophic effect of PLPS. Experiments described in the text clearly indicate that PLPS activates NGF receptor kinase (Trk A) and MAPK, and consequently induces neuritogenesis in PC12 cells. The effect of PLPS mimics that of NGF, and is observed only in NGF-susceptible cells (i.e., those possessing NGF receptor). PLPS and NGF share a common signaling pathway leading to activation of MAPK. We postulate that PLPS initially interacts with an unknown membrane component, leading to membrane perturbation which activates Trk A. PLPS does not bind to NGF receptor directly, because binding of ^{125}I-labeled NGF to its receptor is not affected by PLPS. Reprinted with permission from Hakomori S, Adv in Pharmacology 1996; 36:155-171.

Mechanism of Cell Motility Inhibition by Sph-1-P

Our preliminary studies indicate that Sph-1-P inhibits pseudopodium formation by blocking polymerization and reorganization of actin filaments, and significantly reduces assembly of F-actin filaments in B16/F1 cells. A pyrene-labeled actin nucleation assay revealed that Sph-1-P at 10-100 nM inhibits actin nucleation mediated by F1 cell plasma membrane. This suggests that Sph-1-P may interact with yet-unidentified components that mediate actin nucleation.[70] A likely candidate is the component Rho G-protein or its analog, an essential transducer molecule linking membrane signaling and cytoskeleton.[71] It was reported recently that Sph-1-P induces Rho-dependent neurite retraction.[72]

Sph AND DMS MODULATE MODULATOR PROTEINS (CHAPERONES) OF SIGNAL TRANSDUCTION

Sph was previously shown to activate kinases that phosphorylate various functional proteins,[73,74] including focal adhesion kinase (pp125^{FAK}) and paxillin.[75] DMS greatly enhances EGF-RK.[76] Further systematic study is necessary to correlate Sph/DMS-dependent kinase activity with cellular functions. A few examples are described below.

Sph and DMS Modulate Signal Modulator Protein 14-3-3

We studied several proteins whose phosphorylation is induced in vitro by Sph or DMS but not by Cer, Sph-1-P, GM_3, lyso-GM_3, LacCer, SPC, SM, PA, or lyso-PA. One protein (Mr 28 kDa) showing high susceptibility to Sph- and DMS-dependent kinase was purified by sequential column chromatography of cytosolic fraction of 3T3 cell extract. The protein was characterized as having an amino acid sequence identical to that of "protein 14-3-3".[50] Phosphorylation of the protein upon addition of semipurified kinase was dose-dependently induced by Sph and DMS, but not by 17 other SLs, GSLs, and phospholipids tested. 14-3-3 is known to be the modulator of a number of key protein kinases involved in regulation of intracellular signaling.[46,47]

Sph and DMS May Modulate Grp, Endoplasmin, and Calreticulin

Sph-/DMS-dependent kinase is capable of phosphorylating a protein with Mr 100 kDa, which shows high amino acid sequence homology with Grp 100 and endoplasmin. Another phosphorylated protein with Mr 58 kDa shows sequence homology with calreticulin, an endogenous lectin showing Glc-binding property. Phosphorylation of these proteins occurred only in the presence of Sph or DMS, but not Cer or other lipids (Megidish T, Takio K, Titani K, Hakomori S, unpublished data). A recent study indicates that Grp 100 is a molecular chaperon showing heparanase activity.[77] Further extensive study is necessary to elucidate Sph-/DMS-dependent induction of heparanase activity. Calreticulin plays an important role in folding two-dimensional structure of N-linked glycoprotein, i.e., it acts as a chaperon of N-linked glycoprotein. Thus, Sph and DMS may modulate chaperon proteins through Sph-/DMS-dependent kinase, as demonstrated for 14-3-3 and calreticulin.

GSL AND SL ANALOGS THAT INHIBIT CANCER PROGRESSION AND INFLAMMATORY PROCESSES: DEVELOPMENT OF "ORTHO-SIGNALING THERAPY" FOR CANCER AND INFLAMMATION

Ortho-Signaling Therapy for Cancer

Cancer invasion and progression are based on: (1) uncontrolled cell proliferation; (2) uncontrolled, active cell motility; (3) ability of cancer cells to activate platelets and endothelial cells to elicit expression of P- and E-selectin; (4) enhanced expression of collagenase and endoglycosidase at the cell surface; and (5) balance of spontaneous apoptotic death and cell proliferation.

All these processes 1-5 are controlled by transmembrane signaling, and some are susceptible to GSL or SL analogs. Sph and DMS inhibit tumor growth, possibly through inhibition of PKC[78,79] and enhanced apoptosis.[80] TMS strongly inhibits PKC[79] and P-selectin expression on platelets and endothelial cells.[81-83] Sph-1-P inhibits tumor cell motility by an unknown mechanism.[67] Dimeric and multimeric lyso-GM_3 and N-modified GM3 are also interesting candidates for ortho-signaling therapy of cancer (see section 8.1.4).

Inhibition of mouse melanoma metastasis by administration of liposomes containing both TMS and Sph-1-P

In view of the inhibitory effect of TMS on selectin expression which mediates tumor cell metastasis, and of Sph-1-P on tumor cell motility, we studied the effect of liposomes containing these two compounds on B16 melanoma cell metastasis. When such liposomes were injected intravenously or intraperitoneally, metastasis of B16 melanoma variant BL6 was greatly reduced. Liposomes containing Sph-1-P or TMS alone showed much less inhibitory effect.[84]

GSL and SL derivatives as inducers of apoptosis in human solid tumors

Gb3Cer expressed on lymphoid cells and various solid tumors induces apoptosis when verotoxin subunit B is applied.[85] This subunit by itself is not cytotoxic, but simply binds to Gb3 and induces internalization.[86,87] Expression of Le^Y GSL in various human solid tumors is correlated with degree of apoptosis, as revealed by nick-end labeling studies.[88] The apoptosis-inducing effect of C2-Cer was obvious for nonadherent hematopoietic cell lines such as HL-60, U937, and K562 in the absence of serum.[89] However, TNFα-induced HL-60 cell apoptosis was correlated with increase of Sph and concomitant decrease of Ras.[90] Under physiological, serum-containing conditions, C2-Cer addition showed little correlation with apoptosis in the majority of adherent tumor cell lines (from colonic, gastric, lung, and breast carcinomas). In contrast, human colonic and gastric carcinomas HT29, HRT18, MKN74, and Colo205 were highly susceptible to apoptosis upon addition of 5-10 μM DMS in the presence of serum.[80] Interestingly, Sph- and DMS-induced apoptosis are not observed in normal fibroblasts or endothelial cells. Tumor cells are susceptible regardless of their resistance to chemotherapeutic reagents (Shirahama T, Sweeney EA, Sakakura C, Hakomori S, Igarashi Y, unpublished data). In view of these observations, we expect that combinations of verotoxin subunit B, anti-Le^Y antibody, and possibly liposomal forms of Sph and DMS may provide effective reagents for induction of apoptosis and consequent growth suppression of a large variety of tumors in vivo.

Lyso-GSLs and their derivatives (e.g., PLPS)

Distribution and quantity of lyso-GSLs in various cell types, particularly normal vs. tumor cells, is unknown. Lyso-GSLs are known to be cytotoxic and may inhibit transmembrane signaling (e.g., through PKC inhibition).[17] PLPS is a plasmal derivative of psychosine, discovered incidentally in neuronal tissues when we performed cation exchange chromatography of brain lipid extracts.[62] This compound was found to strongly promote Tyr kinase associated with high-affinity NGF-RK (Trk A), and consequently induce neuronal differentiation.[63] We therefore expect that PLPS can serve as a differentiation inducer for common tumors derived from neuronal tissues (e.g., neuroblastoma), and thus as a useful therapeutic reagent for control of neuroblastoma and glioma.

N-modified GM_3

GM_3, GD3, and other gangliosides having neuraminic acid (i.e., de-N-acetyl sialic acid) have been implicated as promoters of cell growth. This is presumably because de-N-acetylase coupled with enhanced Tyr kinase is active in tumor cells. N-modified gangliosides, which inhibit de-N-acetylation, may therefore be useful for blocking tumor growth. In preliminary studies we found that GM_3 containing N-trifluoroacetyl neuraminic acid, when exogenously added to tumor cell culture, restored contact inhibitability and reduced cell population density.[91] This is due partially to decrease of ganglioside:sialidase activity. Whatever the mechanism, such derivatives are interesting candidates to suppress tumor cell growth.

Ortho-Signaling Therapy of Inflammatory Diseases

Acute and chronic inflammatory processes are triggered by: (1)expression of E- and P-selectin on microvascular endothelial cells; (2)recruitment to inflammatory sites of neutrophils, which overproduce O_2^- and other free radicals, and damage the tissues.[92] Since P-selectin expression on platelets and endothelial cells plays a key role in initiation of neutrophil recruitment, and is strongly inhibited by TMS,[81-83] we studied the effect of TMS in reperfusion injury of the cat cardiovascular system. TMS (60 μg/kg), administered intravenously 10 minutes before reperfusion, significantly reduced myocardial necrosis (153 vs. 314% necrosis of area at risk; $p < 0.01$) and cardiac myeloperoxidase activities, a marker of neutrophil accumulation, compared with vehicle-treated cats. Endothelium-dependent relaxation to acetylcholine in ischemic-reperfused coronary artery rings treated with TMS was also significantly preserved compared with vehicle (734 vs. 344% vasorelaxation; $p < 0.01$). Polymorphonuclear neutrophil (PMN) adherence to coronary endothelium 270 minutes after reperfusion was markedly reduced in the TMS group compared with vehicle-treated cats (375 vs. 765 PMN/mm^2; $p <0.01$). TMS also reduced upregulation of P-selectin on coronary venular endothelium by immunohistochemistry. This was consistent with in vitro findings that TMS reduces PMN adherence to thrombin-stimulated coronary endothelium and P-selectin upregulation on thrombin-stimulated cat platelets.[82]

Thus, the SL derivative TMS at physiological concentrations exerts cardioprotective actions and preserves coronary endothelial function following myocardial ischemia and reperfusion in vivo. The effects appear to be mediated by the inhibition of PMN-endothelial interaction and subsequent accumulation into the ischemic myocardium. TMS may be a useful agent in reducing myocardial reperfusion injury.

SYNOPSIS

Knowledge on functional roles of GSLs has accumulated rather slowly during the past three decades. Two major roles have been revealed:[93] (1) involvement in cell adhesion and recognition; and (2) control of signal transduction (as described in this review). Protein kinase activity associated with cell proliferation and oncogenesis is modulated by GSLs, particularly gangliosides and Sph derivatives. These protein kinases include not only growth factor receptor-associated kinases,[9-12,94] but also cytoplasmic protein kinases (e.g., PKC, src kinase, Ras kinase, Raf kinase) which control a variety of cellular phenotypes.[16,18,24,76,95]

The role of gangliosides and their degradation products as signal modulators of Tyr kinase linked receptors was initially reported in the early 1980s.[11,12,94] Hannun and Bell observed that Sph, the common backbone of SLs, inhibits PKC, whereas DAG promotes it.[16,17] This led to the currently fashionable trend of research on Cer and Sph-1-P as second messengers (see refs. 27-29 for reviews), in contrast to GSLs and gangliosides as modulators. However, function of Tyr kinase linked receptors is obviously inhibited by gangliosides, but promoted by Sph derivatives. The fact that two GM_3 ganglioside derivatives (de-N-acetyl and lyso) affect receptor function in opposite directions (positive vs. negative effect) is a provocative basis for future studies. Proliferation of basal cell layer of epidermis depends highly on GlcCer, but not on other GSLs, Cer, or Sph derivatives.[96] Likewise, arterial smooth muscle cell proliferation depends on LacCer, but not on other GSLs, Cer, or Sph.[97] The susceptibility of the cells to low concentrations (10 μM) of LacCer is ascribable to induced phosphorylation of

MAPK (p44^{MAPK}). LacCer also stimulates (seven-fold) loading of GTP on *Ras*, as catalyzed by *sos*, with concomitant stimulation of MAPKK and *Raf* within 2.5 minutes.[98] Neither Cer nor Sph derivatives affect this signaling system of arterial smooth muscle cells. Proliferative response to LacCer is also observed in a large variety of other cell types (Chatterjee S, personal communication). PDGF-dependent motility (but not proliferation) of these cells is strongly inhibited by Sph-1-P,[69] while PDGF-dependent proliferation of Swiss 3T3 cells is mediated by Sph-1-P.[66] The role of Sph-1-P as a second messenger activating AP-1 to induce DNA synthesis has been well documented in 3T3 cells.[44,99,100] However, the types of cells showing proliferative response to Sph-1-P are limited. Many cell types so far examined showed cell motility inhibition rather than cell growth promotion in response to Sph-1-P,[67] although the mechanism for this motility inhibition remains to be elucidated.

Signaling systems and their susceptibility to GSLs and SLs presumably vary depending on cell type, stage of ontogenesis, and degree of malignancy of tumor cells. It is also plausible that only slight variation exists in structure of regulatory domain (or site) of transducer molecules, and that the extensive variation observed in signaling activities in vivo may depend on quantity and quality of modulator GSLs and SLs present in various cell types.

Further studies along this line will (1)clarify the basic regulatory mechanism of signal transduction, and differences in this mechanism in benign vs. malignant tumor cells; (2)provide a means to normalize aberrant signaling in tumors or inflammatory cells through application of GSL or SL derivatives ("ortho-signaling therapy").

Acknowledgment

I thank Dr. Stephen Anderson for scientific editing and preparation of the manuscript.

References

1. Hakomori S, Murakami WT. Glycolipids of hamster fibroblasts and derived malignant-transformed cell lines. Proc Natl Acad Sci USA 1968; 59:254-261.
2. Hakomori S, Wyke JA, Vogt PK. Glycolipids of chick embryo fibroblasts infected with temperature-sensitive mutants of avian sarcoma viruses. Virology 1977; 76:485-493.
3. Tsuchiya S, Hakomori S. Cell surface glycolipids of transformed NIH 3T3 cells transfected with DNAs of human bladder and lung carcinomas. EMBO J 1983; 2:2323-2326.
4. Laine RA, Hakomori S. Incorporation of exogenous glycosphingolipids in plasma membranes of cultured hamster cells and concurrent change of growth behavior. Biochem Biophys Res Commun 1973; 54:1039-1045.
5. Lingwood CA, Hakomori S. Selective inhibition of cell growth and associated changes in glycolipid metabolism induced by monovalent antibodies to glycolipids. Exp Cell Res 1977; 108:385-391.
6. Lingwood CA, Ng A, Hakomori S. Monovalent antibodies directed to transformation-sensitive membrane components inhibit the process of viral transformation. Proc Natl Acad Sci USA 1978; 75:6049-6053.
7. Hakomori S. Cell density-dependent changes in glycolipid concentrations in fibroblasts, and loss of this response in virus-transformed cells. Proc Natl Acad Sci USA 1970; 67:1741-1747.
8. Sakiyama H, Gross SK, Robbins PW. Glycolipid synthesis in normal and virus-transformed hamster cell lines. Proc Natl Acad Sci USA 1972; 69:872-876.
9. Bremer EG, Hakomori S. GM_3 ganglioside induces hamster fibroblast growth inhibition in chemically-defined medium: Ganglioside may regulate growth factor receptor function. Biochem Biophys Res Commun 1982; 106:711-718.
10. Bremer EG, Hakomori S, Bowen-Pope DF, Raines EW, Ross R. Ganglioside-mediated modulation of cell growth, growth factor binding, and receptor phosphorylation. J Biol Chem 1984; 259:6818-6825.
11. Bremer EG, Schlessinger J, Hakomori S. Ganglioside-mediated modulation of cell growth: specific effects of GM_3 on ty-

rosine phosphorylation of the epidermal growth factor receptor. J Biol Chem 1986; 261:2434-2440.

12. Hanai N, Nores GA, Torres-Mendez C-R, Hakomori S. Modified ganglioside as a possible modulator of transmembrane signaling mechanism through growth factor receptors: a preliminary note. Biochem Biophys Res Commun 1987; 147:127-134.
13. Hanai N, Dohi T, Nores GA, Hakomori S. A novel ganglioside, de-*N*-acetyl-GM_3 ($II^3NeuNH_2LacCer$), acting as a strong promoter for epidermal growth factor receptor kinase and as a stimulator for cell growth. J Biol Chem 1988; 263: 6296-6301.
14. Hanai N, Nores GA, MacLeod C, Torres-Mendez C-R, Hakomori S. Ganglioside-mediated modulation of cell growth: specific effects of GM_3 and lyso-GM_3 in tyrosine phosphorylation of the epidermal growth factor receptor. J Biol Chem 1988; 263:10915-10921.
15. Hannun YA, Loomis CR, Merrill AH Jr, Bell RM. Sphingosine inhibition of protein kinase C activity and of phorbol dibutyrate binding in vitro and in human platelets. J Biol Chem 1986; 261:12604-12609.
16. Hannun YA, Bell RM. Lysosphingolipids inhibit protein kinase C: implications for the sphingolipidoses. Science 1987; 235: 670-674.
17. Hannun YA, Bell RM. Functions of sphingolipids and sphingolipid breakdown products in cellular regulation. Science 1989; 243:500-507.
18. Igarashi Y, Hakomori S, Toyokuni T et al. Effect of chemically well-defined sphingosine and its *N*-methyl derivatives on protein kinase C and src kinase activities. Biochemistry 1989; 28:6796-6800.
19. Igarashi Y, Hakomori S. Enzymatic synthesis of N,N-dimethyl-sphingosine: demonstration of the sphingosine:N-methyltransferase in mouse brain. Biochem Biophys Res Commun 1989; 164: 1411-1416.
20. Oishi K, Zheng B, Kuo JF. Inhibition of Na,K-ATPase and sodium pump by protein kinase C regulators sphingosine, lysophosphatidylcholine, and oleic acid. J Biol Chem 1990; 265:70-75.
21. Merrill AH Jr, Hannun YA, Bell RM. Introduction: sphingolipids and their metabolites in cell regulation. In: Bell RM, Hannun YA, Merrill AH Jr, eds. Sphingolipids: Part A: Functions and Breakdown Products (Adv Lipid Research, Vol 25). San Diego, CA: Academic Press, 1993:1-24.
22. Zhang H, Desai NN, Murphey JM, Spiegel S. Increases in phophatidic acid levels accompany sphingosine-stimulated proliferation of quiescent Swiss 3T3 cells. J Biol Chem 1990; 265:21309-21316.
23. Zhang H, Desai NN, Olivera A, Seki T, Brooker G, Spiegel S. Sphingosine-1-phosphate, a novel lipid, involved in cellular proliferation. J Cell Biol 1991; 114:155-167.
24. Ghosh TK, Bian J, Gill DL. Intracellular calcium release mediated by sphingosine derivatives generated in cells. Science 1990; 248:1653-1656.
25. Spiegel S, Olivera A, Carlson RO. The role of sphingosine in cell growth regulation and transmembrane signaling. In: Bell RM, Hannun YA, Merrill AH Jr, eds. Sphingolipids: Part A: Functions and breakdown products (Adv Lipid Research, Vol 25). San Diego, CA: Academic Press, 1993:105-129.
26. Okazaki T, Bell RM, Hannun YA. Sphingomyelin turnover induced by vitamin D3 in HL-60 cells: role in cell differentiation. J Biol Chem 1989; 264:19076-19080.
27. Hannun YA, Linardic CM. Sphingolipid breakdown products: anti-proliferative and tumor-suppressor lipids. Biochim Biophys Acta 1993; 1154:223-236.
28. Kolesnick RN, Golde DW. The sphingomyelin pathway in tumor necrosis factor and interleukin-1 signaling. Cell 1994; 77:325-328.
29. Spiegel S, Foster D, Kolesnick RN. Signal transduction through lipid second messengers. Curr Opin Cell Biol 1996; 8:159-167.
30. Igarashi Y, Kitamura K, Zhou Q, Hakomori S. A role of lyso-phosphatidylcholine in GM3-dependent inhibition of epidermal growth factor receptor autophosphorylation in A431 plasma mem-

branes. Biochem Biophys Res Commun 1990; 172:77-84.
31. Ueki K, Matsuda S, Tobe K et al. Feedback regulation of mitogen-activated protein kinase kinase kinase activity of c-Raf-1 by insulin and phorbol ester stimulation. J Biol Chem 1994; 269: 15756-15761.
32. Kolch W, Heidecker G, Kochs G et al. Protein kinase Cα activates RAF-1 by direct phosphorylation. Nature 1993; 364:249-252.
33. Wu J, Dent P, Jelinek T, Wolfman A, Weber MJ, Sturgill TW. Inhibition of the EGF-activated MAP kinase signaling pathway by adenosine 3',5'-monophosphate. Science 1993; 262:1065-1068.
34. Graves LM, Bornfeldt KE, Raines EW et al. Protein kinase A antagonizes platelet-derived growth factor-induced signaling by mitogen-activated protein kinase in human arterial smooth muscle cells. Proc Natl Acad Sci USA 1993; 90:10300-10304.
35. Caputto RI, Maccioni AHR, Caputto BL. Activation of deoxycholate solubilized adenosine triphosphatase by ganglioside and asialoganglioside preparations. Biochem Biophys Res Commun 1977; 74: 1046-1052.
36. Spiegel S, Handler JS, Fishman PH. Gangliosides modulate sodium transport in cultured toad kidney epithelia. J Biol Chem 1986; 261:15755-15760.
37. Goldenring JR, Otis LC, Yu RK, DeLorenzo RJ. Calcium/ganglioside-dependent protein kinase activity in rat brain membrane. J Neurochem 1985; 44:1229-1234.
38. Chan K-FJ. Ganglioside-modulated protein phosphorylation: partial purification and characterization of a ganglioside-inhibited protein kinase in brain. J Biol Chem 1988; 263:568-574.
39. Partington CR, Daly JW. Effect of gangliosides on adenylate cyclase activity in rat cerebral cortical membranes. Mol Pharmacol 1979; 15:484-491.
40. Yajima M, Hosoda K, Kanbayashi Y, Nakamura T, Takahashi I, Ui M. Biological properties of islets-activating protein (IAP) purified from the culture medium of *Bordetella pertussis*. J Biochem (Tokyo) 1978; 83:305-312.
41. Hoshino S, Kikkawa S, Takahashi K et al. Identification of sites for alkylation by N-ethylmaleimide and pertussis toxin-catalyzed ADP-ribosylation on GTP-binding proteins. FEBS Lett 1990; 276: 227-231.
42. Aktories K, Mohr C, Koch G. *Clostridium botulinum* C3 ADP-ribosyltransferase. Curr Top Microbiol Immunol 1992; 175:115-131.
43. Hara-Yokoyama M, Hirabayashi Y, Irie F et al. Identification of gangliosides as inhibitors of ADP-ribosyltransferases of pertussis toxin and exoenzyme C3 from *Clostridium botulinum*. J Biol Chem 1995; 270:8115-8121.
44. Goodemote KA, Mattie ME, Berger A, Spiegel S. Involvement of a pertussis toxin-sensitive G protein in the mitogenic signaling pathways of sphingosine 1-phosphate. J Biol Chem 1995; 270: 10272-10277.
45. Aitken A, Howell S, Jones D, Madrazo J, Patel Y. 14-3-3 α and δ are the phosphorylated forms of Raf-activating 14-3-3 β and ζ: In vivo stoichiometric phosphorylation in brain at a Ser-Pro-Glu-Lys motif. J Biol Chem 1995; 270:5706-5709.
46. Morrison D. 14-3-3: modulators of signaling proteins? Science 1994; 266:56-57.
47. Aitken A. 14-3-3 proteins on the MAP. Trends Biochem Sci (TIBS) 1995; 20: 95-97.
48. Freed E, Symons M, MacDonald SG, McCormick F, Ruggieri R. Binding of 14-3-3 proteins to the protein kinase Raf and effects on its activation. Science 1994; 265:1713-1716.
49. Fu H, Xia K, Pallas DC et al. Interaction of the protein kinase Raf-1 with 14-3-3 proteins. Science 1994; 266:126-129.
50. Megidish T, White T, Takio K, Titani K, Igarashi Y, Hakomori S. The signal modulator protein 14-3-3 is a target of sphingosine-or *N,N*-dimethylsphingosine-dependent kinase in 3T3(A31) cells. Biochem Biophys Res Commun 1995; 216:739-747.
51. Higashi H, Omori A, Yamagata T. Calmodulin, a ganglioside-binding protein: Binding of gangliosides to calmodulin in the presence of calcium. J

Biol Chem 1992; 267:9831-9838.
52. Higashi H, Yamagata T. Mechanism for ganglioside-mediated modulation of a calmodulin-dependent enzyme: modulation of calmodulin-dependent cyclic nucleotide phosphodiesterase activity through binding of gangliosides to calmodulin and the enzyme. J Biol Chem 1992; 267:9839-9843.
53. Dyer CA, Benjamins JA. Glycolipids and transmembrane signaling: antibodies to galactocerebroside cause an influx of calcium in oligodendrocytes. J Cell Biol 1990; 111:625-633.
54. Yatomi Y, Igarashi Y, Hakomori S. Effects of exogenous gangliosides on intracellular Ca^{2+} mobilization and functional responses in human platelets. Glycobiology 1996; 6:347-353.
55. Weis FMB, Davis RJ. Regulation of epidermal growth factor receptor signal transduction: Role of gangliosides. J Biol Chem 1990; 265:12059-12066.
56. Tsuruoka T, Tsuji T, Nojiri H, Holmes EH, Hakomori S. Selection of a mutant cell line based on differential expression of glycosphingolipid, utilizing anti-lactosylceramide antibody and complement. J Biol Chem 1993; 268:2211-2216.
57. Nojiri H, Stroud MR, Hakomori S. A specific type of ganglioside as a modulator of insulin-dependent cell growth and insulin receptor tyrosine kinase activity: possible association of ganglioside-induced inhibition of insulin receptor function and monocytic differentiation induction in HL60 cells. J Biol Chem 1991; 266:4531-4537.
58. Ullrich A, Schlessinger J. Signal transduction by receptors with tyrosine kinase activity. Cell 1990; 61:203-212.
59. Zhou Q, Hakomori S, Kitamura K, Igarashi Y. G_{M3} directly inhibits tyrosine phosphorylation and de-*N*-acetyl-G_{M3} directly enhances serine phosphorylation of epidermal growth factor receptor, independently of receptor-receptor interaction. J Biol Chem 1994; 269:1959-1965.
60. Sjoberg ER, Chammas R, Ozawa H et al. Expression of de-*N*-acetyl-gangliosides in human melanoma cells is induced by genistein or nocodazole. J Biol Chem 1995; 270:2921-2930.
61. Barnes D, Sato G. Methods for growth of cultured cells in serum-free medium. Anal Biochem 1980; 102:255-270.
62. Nudelman ED, Levery SB, Igarashi Y, Hakomori S. Plasmalopsychosine, a novel plasmal (fatty aldehyde) conjugate of psychosine with cyclic acetal linkage: isolation and characterization from human brain white matter. J Biol Chem 1992; 267:11007-11016.
63. Sakakura C, Igarashi Y, Anand JK, Sadozai KK, Hakomori S. Plasmalopsychosine of human brain mimics the effect of nerve growth factor by activating its receptor kinase and mitogen-activated protein kinase in PC12 cells: Induction of neurite outgrowth and prevention of apoptosis. J Biol Chem 1996; 271:946-952.
64. Ferrari G, Anderson BL, Stephens RM, Kaplan DR, Greene LA. Prevention of apoptotic neuronal death by G_{M1} ganglioside: involvement of Trk neurotrophin receptors. J Biol Chem 1995; 270:3074-3080.
65. Mutoh T, Tokuda A, Miyada T, Hamaguchi M, Fujiki N. Ganglioside GM1 binds to the Trk protein and regulates receptor function. Proc Natl Acad Sci USA 1995; 92:5087-5091.
66. Olivera A, Spiegel S. Sphingosine-1-phosphate as second messenger in cell proliferation induced by PDGF and FCS mitogens. Nature 1993; 365:557-560.
67. Sadahira Y, Ruan F, Hakomori S, Igarashi Y. Sphingosine 1-phosphate, a specific endogenous signaling molecule controlling cell motility and tumor cell invasiveness. Proc Natl Acad Sci USA 1992; 89:9686-9690.
68. Ross R. The pathogenesis of atherosclerosis: A perspective for the 1990s. Nature 1993; 362:801-809.
69. Bornfeldt KE, Graves LM, Raines EW et al. Sphingosine-1-phosphate inhibits PDGF-induced chemotaxis of human arterial smooth muscle cells: spatial and temporal modulation of PDGF chemotactic signal transduction. J Cell Biol 1995; 130:193-206.
70. Yamamura S, Sadahira Y, Ruan F, Hakomori S, Igarashi Y. Sphingosine-1-phosphate inhibits actin nucleation and

pseudopodium formation to control cell motility of mouse melanoma cells. FEBS Lett 1996; 382:193-197.

71. Machesky LM, Hall A. Rho: a connection between membrane receptor signalling and the cytoskeleton. Trends Cell Biol 1996; 6:304-310.
72. Postma FR, Jalink K, Hengeveld T, Moolenaar WH. Sphingosine-1-phosphate rapidly induces Rho-dependent neurite retraction: action through a specific cell surface receptor. EMBO J 1996; 15:2388-2395.
73. Faucher MF, Girones N, Hannun YA, Bell RM, Davis RJ. Regulation of the epidermal growth factor receptor phosphorylation state by sphingosine in A431 human epidermoid carcinoma cells. J Biol Chem 1988; 263:5319-5327.
74. Pushkareva MY, Khan WA, Alessenko AV, Sahyoun N, Hannun YA. Sphingosine activation of protein kinases in Jurkat T cells: in vitro phosphorylation of endogenous protein substrates and specificity of action. J Biol Chem 1992; 267:15246-15251.
75. Seufferlein T, Rozengurt E. Sphingosine induces $p125^{FAK}$ and paxillin tyrosine phosphorylation, actin stress fiber formation, and focal contact assembly in Swiss 3T3 cells. J Biol Chem 1994; 269:27610-27617.
76. Igarashi Y, Kitamura K, Toyokuni T et al. A specific enhancing effect of *N,N*-dimethylsphingosine on epidermal growth factor receptor autophosphorylation: demonstration of its endogenous occurrence (and the virtual absence of unsubstituted sphingosine) in human epidermoid carcinoma A431 cells. J Biol Chem 1990; 265:5385-5389.
77. DeVouge MW, Yamazaki A, Bennett SA et al. Immunoselection of GRP94/endoplasmin from a KNRK cell-specific lambda gt11 library using antibodies directed against a putative heparanase amino-terminal peptide. Int J Cancer 1994; 56: 286-294.
78. Endo K, Igarashi Y, Nisar M, Zhou Q, Hakomori S. Cell membrane signaling as target in cancer therapy: Inhibitory effect of *N,N*-dimethyl and *N,N,N*-trimethyl sphingosine derivatives on in vitro and in vivo growth of human tumor cells in nude mice. Cancer Res 1991; 51:1613-1618.
79. Okoshi H, Hakomori S, Nisar M et al. Cell membrane signaling as target in cancer therapy II: inhibitory effect of *N,N,N*-trimethylsphingosine on metastatic potential of murine B16 melanoma cell line through blocking of tumor cell-dependent platelet aggregation. Cancer Res 1991; 51:6019-6024.
80. Sweeney EA, Sakakura C, Shirahama T et al. Sphingosine and its methylated derivative N,N-dimethylsphingosine (DMS) induce apoptosis in a variety of human cancer cell lines. Int J Cancer 1996; 66:358-366.
81. Handa K, Igarashi Y, Nisar M, Hakomori S. Downregulation of GMP-140 (CD62 or PADGEM) expression on platelets by *N,N*-dimethyl and *N,N,N*-trimethyl derivatives of sphingosine. Biochemistry 1991; 30:11682-11686.
82. Murohara T, Buerke M, Margiotta J et al. Myocardial and endothelial protection by TMS in ischemia-reperfusion injury. Am J Physiol (Heart Circ Physiol) 1995; 269:H504-H514.
83. Scalia R, Murohara T, Delyani JA, Nossuli TO, Lefer AM. Myocardial protection by *N,N,N*-trimethylsphingosine in ischemia reperfusion injury is mediated by inhibition of P-selectin. J Leukocyte Biol 1996; 59:317-324.
84. Park YS, Ruan F, Hakomori S, Igarashi Y. Cooperative inhibitory effect of N,N,N-trimethylsphingosine and sphingosine-1-phosphate, co-incorporated in liposomes, on B16 melanoma cell metastasis: Cell membrane signaling as a target in cancer therapy IV. Int J Oncol 1995; 7:487-494.
85. Mangeney M, Rousselet G, Taga S, Tursz T, Wiels J. The fate of human CD77+ germinal center B lymphocytes after rescue from apoptosis. Mol Immunol 1995; 32:333-339.
86. Maloney MD, Lingwood CA. CD19 has a potential CD77 (globotriaosyl ceramide)-binding site with sequence similarity to verotoxin B-subunits: implications of molecular mimicry for B cell adhesion and enterohemorrhagic *Escherichia coli* pathogenesis. J Exp Med 1994;

180:191-201.

87. Farkas-Himsley H, Hill R, Rosen B, Arab S, Lingwood CA. The bacterial colicin active against tumor cells in vitro and in vivo is verotoxin 1. Proc Natl Acad Sci USA 1995; 92:6996-7000.
88. Hiraishi K, Suzuki K, Hakomori S, Adachi M. Le^y antigen expression is correlated with apoptosis (programmed cell death). Glycobiology 1993; 3:381-390.
89. Obeid LM, Linardic CM, Karolak LA, Hannun YA. Programmed cell death induced by ceramide. Science 1993; 259: 1769-1771.
90. Ohta H, Sweeney EA, Masamune A, Yatomi Y, Hakomori S, Igarashi Y. Induction of apoptosis by sphingosine in human leukemic HL-60 cells: a possible endogenous modulator of apoptotic DNA fragmentation occurring during phorbol ester-induced differentiation. Cancer Res 1995; 55:691-697.
91. Hakomori S, Young WW Jr, Patt LM, Yoshino T, Halfpap L, Lingwood CA. Structure and function of gangliosides. In: Svennerholm L, Dreyfus H, Urban PF, eds. Cell Biological and Immunological Significance of Ganglioside Changes Associated with Transformation. New York, NY: Plenum Publ. Corp., 1980:247-261.
92. Harlan JM, Liu DY, eds. Adhesion: its Role in Inflammatory Disease. New York, NY: W.H. Freeman & Co., 1992.
93. Hakomori S. Bifunctional role of glycosphingolipids: Modulators for transmembrane signaling and mediators for cellular interactions. J Biol Chem 1990; 265: 18713-18716.
94. Bremer EG, Hakomori S. Cell growth regulation through glycosphingolipids: GM_3 ganglioside may regulate fibroblast growth factor reception. In: Galeotti T, Cittadini A, Neri G, Papa S, eds. Membranes in Tumor Growth. Amsterdam: Elsevier Biomedical Press, 1982:419-429.
95. Kreutter D, Kim JYH, Goldenring JR et al. Regulation of protein kinase C activity by gangliosides. J Biol Chem 1987; 262:1633-1637.
96. Marsh NL, Elias PM, Holleran WM. Glucosylceramides stimulate murine epidermal hyperproliferation. J Clin Invest 1995; 95:2903-2909.
97. Chatterjee SK. Lactosylceramide stimulates aortic smooth muscle cell proliferation. Biochem Biophys Res Commun 1991; 181:554-561.
98. Bhunia AK, Han H, Snowden A, Chatterjee S. Lactosylceramide stimulates Ras-GTP loading, kinases (MEK, Raf), p44 mitogen-activated protein kinase, and c-*fos* expression in human aortic smooth muscle cells. J Biol Chem 1996; 271:10660-10666.
99. Su Y, Rosenthal D, Smulson M, Spiegel S. Sphingosine 1-phosphate, a novel signaling molecule, stimulates DNA binding activity of AP-1 in quiescent Swiss 3T3 fibroblasts. J Biol Chem 1994; 269: 16512-16517.
100. Wu J, Spiegel S, Sturgill TW. Sphingosine 1-phosphate rapidly activates the mitogen-activated protein kinase pathway by a G protein-dependent mechanism. J Biol Chem 1995; 270:11484-11488.
101. Hannun YA. The sphingomyelin cycle and the second messenger function of ceramide. J Biol Chem 1994; 269:3125-3128.
102. Sun H, Tonks NK. The coordinated action of protein tyrosine phosphatases and kinases in cell signaling. Trends Biochem Sci (TIBS) 1994; 19:480-485.

CHAPTER 11

Ceramide Synthase and Ceramidases in the Regulation of Sphingoid Base Metabolism

Mariana Nikolova-Karakashian, Teresa R. Vales, Elaine Wang, David S. Menaldino, Christopher Alexander, Jane Goh, Dennis C. Liotta and Alfred H. Merrill, Jr.

Ceramides and sphingoid bases (such as sphingosine and sphinganine) are not only intermediates of complex sphingolipid biosynthesis and turnover, but also highly bioactive compounds that have been proposed to mediate cellular responses to growth factors, cytokines, and other agents (including γ-irradiation and some anti-cancer drugs). In studying the involvement of these compounds in growth, differentiation, diverse cell functions, and cell death, sphingoid bases and/or ceramides are often added to cells exogenously, or the amounts of the endogenous compounds are analyzed. A complicating factor in such analyses is that sphingoid bases and ceramides can be interconverted by acylation and deacylation reactions, respectively, catalyzed by ceramide synthase and ceramidase. This review describes the characteristics of ceramide synthesis and turnover, properties of the enzymes responsible for these reactions, and how inhibitors are useful in studying the function of these compounds. Furthermore, one class of ceramide synthase inhibitor (the fumonisins) are responsible for severe diseases of plants and animals; therefore information is presented about how fumonisins disrupt sphingolipid metabolism, alter cell behavior, and cause disease.

CERAMIDE BIOSYNTHESIS AND TURNOVER

Ceramide is synthesized by two routes: introduction of a 4,5-trans-double bond into the sphinganine backbone of dihydroceramide (N-acyl-sphinganine) (reaction b in Fig. 11.1) and reacylation of sphingosine that is recycled from sphingolipid turnover (reaction a in Fig. 11.1, with sphingosine as the substrate).[1] It is presumed that complex sphingolipid synthesis proceeds in the order: dihydroceramide→ceramide→sphingomyelin, glucosylceramide and other complex sphingolipids as shown in Figure 11.1, based on the intermediates that are observed when cells in culture are incubated with [^{14}C]serine.[2,3] This scheme has been substantiated by the direct demonstration of conversion of dihydroceramides to ceramides.[4]

Sphingolipid-Mediated Signal Transduction, edited by Yusuf A. Hannun.

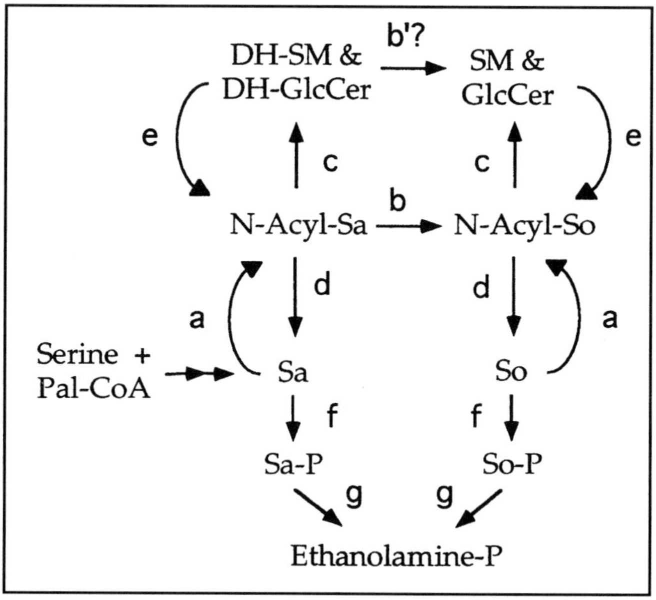

Fig. 11.1. Pathways of ceramide biosynthesis and turnover. The initial reaction of de novo ceramide synthesis is the condensation of serine and palmitoyl-CoA (Pal-CoA) to form 3-ketosphinganine (not shown), which is reduced to sphinganine (Sa). Sa is acylated (reaction a) to N-acyl-sphinganine (N-Acyl-Sa), which can be converted (reaction c) to complex dihydrosphingomyelin (DH-SM) or dihydroglucosylceramide (DH-GlcCer) with the sphinganine backbone, or oxidized (reaction b) to ceramide (N-acyl-sphingosine, N-Acyl-So), which is further metabolized to SM and GlcCer. It is not known if there is oxidation of dihydrocomplex sphingolipids to produce the ceramide backbone (reaction b′). Complex sphingolipid turnover (reaction e) produces (dihydro)ceramides that can be reutilized or hydrolyzed to Sa or So by ceramidases (reaction d). The sphingoid bases can also be reutilized, or phosphorylated (reaction f) by sphingosine kinase to the 1-phosphates (Sa-P and So-P), which can be cleaved to fatty aldehydes and ethanolamine phosphate (reaction g).

Nonetheless, it is difficult to rule out the alternative possibility wherein headgroups can be added to dihydroceramides and the double bond is introduced later (as shown by reaction b' in Fig. 11.1). In studies of J774 cells,[3] it was evident that [^{14}C]serine was initially incorporated into [^{14}C]dihydroceramide and complex sphingolipids containing a [^{14}C]sphinganine backbone, but complex sphingolipids later contained mostly [^{14}C]sphingosine, which is usually the predominant sphingoid base. A similar metabolic profile has been seen in more recent studies (JW Kok, personal communication) where several cell lines were incubated with a fluorescent analog (N-[6-(7-nitrobenz-2-oxa-1,3-diazol-4-yl)amino]hexanoyl-D-*erythro*-sphinganine, or NBD-dihydroceramide), which was first metabolized to glucosyl-NBD-dihydroceramide followed by glucosyl-NBD-ceramide. In this study, however, it was possible to distinguish between pathways b and b' because the cells could be incubated with glucosyl-NBD-dihydroceramide under conditions where either turnover (reaction e) or resynthesis (reaction c) were inhibited, and in both cases the formation of glucosyl-NBD-ceramide was blocked. Therefore, these observations suggest that cells readily incorporate dihydroceramides into complex sphingolipids, with introduction of the 4,5-trans-double bond of ceramide later in the pathway, perhaps during turnover of the initially synthesized sphingolipid and recycling of the dihydroceramide backbone.

The reasons for this circuitous pathway of ceramide synthesis are not known; however, ceramides are highly bioactive compounds that can induce cell death (via apoptosis) in some cell types, whereas dihydroceramides appear to be less potent cell death signals.[5] Therefore, this might provide a way for cells to avoid accidentally triggering apoptosis when de novo sphingolipid synthesis is increased (for example, during cell growth). If this is the case, then an important "decision" must be made at the point where dihydroceramide-containing sphingolipids are turned over as to whether the lipid backbone is oxidized to ceramide or degraded. Very little is known about these fundamental aspects of sphingolipid metabolism.

Properties of Ceramide Synthase

Ceramide synthase catalyzes the acylation of long-chain sphingoid bases by transfer of a fatty acid from the coenzyme A thioester (reaction a in Fig. 11.1). On the basis of the preceding discussion, it would seem to be more appropriate for this name to apply to the enzyme(s) that introduce the 4,5-trans-double bond into dihydroceramides (reaction b in Fig. 11.1); however, the pathway of de novo sphingolipid biosynthesis was first thought to be sphinganine→sphingosine→ceramide, which explains this nomenclature. The synthesis of ceramide by a noncoenzyme A-dependent reaction has also been seen in vitro[6,7] however, the role of this reaction in vivo is not clear.

Studies using mouse brain microsomes have found that ceramide synthase activity depends on the nature of the fatty acyl-CoA in the order: stearoyl→lignoceroyl→palmitoyl→oleoyl-CoA, with a ratio of activities of 60:12:3:1.[8] The properties of the activities that utilize stearoyl-CoA and lignoceroyl-CoA differ considerably, which has been interpreted as evidence for more than one enzyme.[9] Some stereoselectivity toward the naturally occurring *erythro* forms of sphingosine and sphinganine has been reported;[10] however, other similar studies[11] have failed to show stereoselectivity toward the long-chain base.

We have conducted studies of the in vitro activity of ceramide synthase with stearoyl-CoA using rat liver and brain microsomes (T Vales, Masters thesis from Emory University, 1995) and will summarize these results. As the first observation, liver microsomes were found to contain 12-30 pmol of sphingosine/mg protein and about half this amount of sphinganine, whereas, brain microsomes contained similar amounts of sphingosine but 427 ± 74 pmol of sphinganine/mg protein. Furthermore, the brain preparations have some intrinsic ceramidase activity, so that incubation of brain microsomes in 25 mM potassium phosphate buffer (pH 7.4) for 10 minutes increased the endogenous sphingosine content to 77 ± 44 pmol/mg protein. The presence of endogenous sphingoid bases should be kept in mind, but probably had little effect on the kinetics since only 40 μg of microsomal protein was used in each assay (see legend to Fig. 11.2).

Studies of the acylation of chemically synthesized[12] D-*erythro*-sphinganine and -sphingosine revealed that both sphingoid bases yielded the same maximal activity (Fig. 11.2). On the basis of Vmax/Km, sphinganine (with a Vmax/Km of 248) is a better substrate than sphingosine (with a Vmax/Km of 90). D-*erythro*-cis-sphingosine was also acylated with a Vmax of 97 pmol/min/mg of microsomal protein and an apparent Km of 1.7 μM, which further suggests that ceramide synthase has little selectivity for the conformation at positions 4 and 5 of the sphingoid base. Studies of the 4 stereoisomers (Fig. 11.2) established that all are acylated (Fig. 11.2) with the following activities relative to D-*erythro* sphingosine: D-*erythro* sphinganine (99%); D-*erythro* sphingosine (100%); D-*threo* sphingosine (34%); L-*erythro* sphingosine (13%); and L-*threo* sphingosine (14%). DL-*threo*-sphinganine is often used as an inhibitor of sphingosine kinase;[13,14] therefore, it should be borne in mind that this results in synthesis of N-acyl-DL-*threo*-sphinganine (and presumably some inhibition of the acylation of endogenous sphingoid bases).

The subcellular localization and topology of ceramide synthesis has been studied, and the active site(s) of ceramide synthase localized to the cytosolic aspect of the endoplasmic reticulum.[15] Nonetheless,

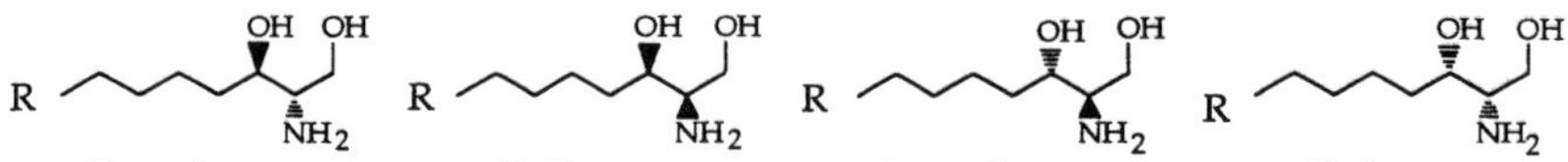

D-*erythro*	D-*threo*	L-*erythro*	L-*threo*
	Sphingosine		
	V_{max} (pmol/min per mg microsomal protein)		
126 ± 46	43 ± 2	16	18 ± 2
	Apparent K_m (μM)		
1.4 ± 0.6	1.3 ± 0.1	6.8	1.8 + 0.3
	Sphinganine		
	V_{max} (pmol/min per mg microsomal protein)		
124 ± 35	36	7.1 at 5 μM	28
	Apparent K_m (μM)		
0.5 ± 0.2	1.5 ± 0.3	ND	2.1

Fig. 11.2. Ceramide synthase activity with different sphinganine and sphingosine stereoisomers. The assays were conducted using 40 μM stearoyl-CoA, 10 μM egg phosphatidylcholine to deliver the ceramide and 40 μg of microsomal protein, with resolution of the products by high-performance liquid chromatography or thin-layer chromatography. The Vmax and apparent Km were analyzed by linear regression analyses of double reciprocal plots of the data; several different experiments were conducted and the results were used when the correlation coefficients for the linear regression analyses were > 0.96 (when several plots met this criterion, the results are given as the mean ± SD).

relatively little is known about the intracellular trafficking of ceramides made de novo (in contrast to fluorescent sphingolipids, which have been studied fairly extensively).[16]

Properties of Ceramidases

Ceramide is metabolized by several routes: synthesis of more complex sphingolipids, phosphorylation to ceramide 1-phosphate, acylation at the primary hydroxyl, and cleavage by ceramidases.[1] Only the latter reactions will be discussed here.

Ceramidases catalyze the hydrolysis of ceramides to the free sphingoid base and fatty acid, and are widely distributed in animal tissues such as brain, kidney, spleen and liver.[17] In rat liver, acidic,[18] alkaline[18] and neutral (plasma membrane-bound)[19] activities have been described. Except for the acidic enzyme, these activities are not distinguished by sharp pH maxima. The acidic ceramidase has been purified, cloned, and sequenced.[20]

Ceramidases have become implicated in sphingolipid-mediated signaling as a necessary intermediate in the hyrolysis of complex sphingolipids to ceramide, followed by cleavage of the ceramide to sphingosine, which can be phosphorylated to sphingosine 1-phosphate—a proposed trigger for release of intracellular calcium.[21] This pathway has been studied most extensively for the stimulation of cell growth by platelet derived growth factor with Swiss 3T3 cells[22] and vascular smooth muscle cells.[23] In the latter case, sphingolipids were prelabeled with [^{3}H-serine] and platelet-derived growth factor stimulated ceramide degradation and sphingosine accumulation. A decrease in complex sphingolipid labeling and increase in sphingosine has also been observed in human fibroblasts treated with the pro-inflammatory peptide bradykinin.[24] In contrast, cytokines (such as tumor necrosis factor-α and interleukin 1β) have been shown to induce turnover of sphingomyelin (and apparently a distinct pool of sphingomyelin)[25] to ceramide, and it has been suggested[26] that the activation of ceramidase is a deciding factor in whether ceramide or downstream metabolites are produced in response to agonists.

Agonist-inducible ceramidase activities (at neutral and alkaline pH) have been assayed in vitro using a 70,000 x g pellet from lysates of messangial cells treated with platelet derived growth factor, and maximal activities were seen 30-60 minutes after growth factor administration. Addition of inhibitors of phosphotyrosine dephosphorylation (such as vanadate ion) was critical for the detection of activity. Genistein (an inhibitor of some tyrosine protein kinases) abolished the activation; therefore, it is likely that tyrosine phosphorylation/dephosphorylation (either of ceramidase or an upstream factor) is involved in the regulation of ceramidase(s).

We have analyzed the turnover of sphingomyelin induced by interleukin 1β in rat liver hepatocytes[27] and noted that the change in ceramide mass did not correspond with the interleukin 1β dose-response for sphingomyelin hydrolysis (i.e., ceramide mass was increased at higher concentrations of interleukin 1β than were needed to stimulate sphingomyelin hydrolysis); furthermore, interleukin 1β caused a 2-fold increase in sphingosine.[27] Therefore, it appeared likely that ceramidase activation was occurring at low doses of this cytokine, which we have subsequently found to be the case (Nikolova-Karakashian et al, submitted for publication). Interleukin 1β treatment of rat liver hepatocytes increases ceramidase activity measured in vitro at acidic, neutral, and alkaline pH, with the greatest increase (almost 10-fold) in the activity at neutral pH. Consistent with the dose response of ceramide accumulation seen in the initial study,[27] above 2.5 ng/ml of interleukin 1β, ceramidase activities are not increased, and may decrease.

These observations indicate that sphingolipid signaling consists of not only the turnover of complex sphingolipids (namely, sphingomyelin) to ceramide; but also, activation of additional enzymes (ceramidase and sphingosine kinase) to produce several bioactive metabolites—and that the "switch" that determines the latter pathway is dependent not only on the type of stimulus (e.g., growth factors versus cytokines)[26] but also on the levels of the stimulus (i.e., low versus high levels of interleukin 1β in our recent studies). It is also likely that the balance of these enzymatic activities varies according to the cell-type. In this regard, it warrants mention that increases in sphingosine have been seen in a number of cell types,[28] and that one often sees greater changes in primary cells (such as neutrophils) than in transformed hemopoietic cell lines (i.e., HL-60 cells, which exhibit elevations in ceramide with little or no sphingosine accumulation).[29] One factor that could contribute to this difference is that HL-60 cells are much more active in acylation of sphingosine than are neutrophils.[30] Therefore, the availability of ceramide versus sphingosine (and downstream metabolites) for signaling is likely to be affected by the activities of both ceramidase(s) and ceramide synthase.

CERAMIDE VERSUS SPHINGOSINE IN CELL SIGNALING

It is very difficult to distinguish between sphingosine and ceramide as the bioactive species responsible for a biological response because these compounds are so closely interrelated by metabolism. Few studies have measured the levels of both ceramides and sphingosine (and in the case of the latter, the analyses are complicated by the further metabolism of sphingosine to sphingosine 1-phosphate and, possibly, N,N-dimethylsphingosine). The addition of exogenous analogs (such as short-chain ceramides) is often used to confirm that a given species accounts for the biological response; however, this can also be misleading because the analogs can be metabolized unless a nonmetabolizable analog is used, or an inhibitor of metabolism is included. What is needed are tools that can block ceramide hydrolysis to sphingosine and vice versa; fortunately, both have recently been developed.

Inhibitors of Ceramidase

D-*e*-MAPP (1S,2R-D-*erythro*-2-N-myristoylamino-1-phenyl-1-propanol) (Fig. 11.3) is an inhibitor of the alkaline ceramidase (IC_{50} ~5 μM), but has little effect on the acidic enzyme (less than 10% inhibition at 50 μM).[31] The *threo*-stereoisomer is also an inhibitor, but is a ceramidase substrate that is cleaved as readily as N-palmitoylsphingosine. D-*e*-MAPP does not inhibit sphingomyelin synthase or GlcCer synthase, nor activate sphingomyelinase, cerebrosidase, or the ceramide-activated protein phosphatase. Addition of D-*e*-MAPP to HL-60 cells elevated endogenous ceramide levels and suppressed growth; therefore, this compound is a very promising tool for studies of sphingolipid signaling.

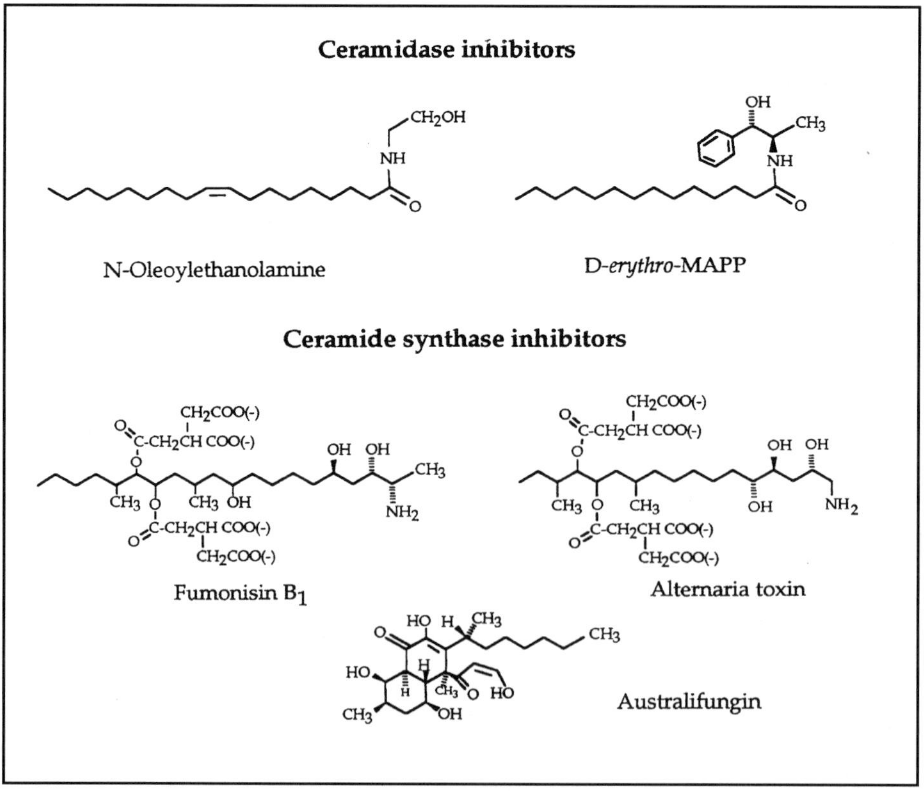

Fig. 11.3. Inhibitors of ceramide synthase and of ceramidase.

N-Oleoylethanolamine (Fig. 11.3) is also a ceramidase inhibitor,[32] but only at fairly high concentrations (the IC_{50} is 500-700 μM).[31,32] Although little is known about the effects of N-oleoylethanolamine on other ceramide metabolizing enzymes, it has been shown to inhibit growth factor-induced mitogenesis,[26] presumably because it blocked this upstream step in the formation of sphingosine 1-phosphate.

Inhibitors of Ceramide Synthase

Nature has been generous in providing potent inhibitors of ceramide synthase, three of which are shown in Figure 11.3. Fumonisin B_1 (FB_1) is produced by some strains of *Fusarium moniliforme*,[33] Alternaria toxins are produced by *Alternaria alternata*,[34] and australifungins are made by *Sporomiella australis*.[35] Of these, FB_1 is the most potent and specific and will, therefore, be described in some detail here.

FUMONISINS AS INHIBITORS OF CERAMIDE SYNTHASE IN VITRO AND IN VIVO

Studies with Ceramide Synthase In Vitro

FB_1 inhibits ceramide synthase in vitro with apparent Ki in the range of 0.05-0.1 μM, depending on the concentrations of sphinganine (or sphingosine) and fatty acyl-coenzyme A (several of which have been examined as the cosubstrate; activity was inhibited for each fatty acyl-CoA tested).[36-39] Although it is difficult to interpret the kinetics of enzymes and substrates that are membrane associated, the inhibition of ceramide synthase by FB_1 is competitive versus the long-chain (sphingoid) base and mixed with the fatty-acyl-CoA,[37] which indicates that fumonisins inhibit ceramide synthase by interacting with the binding sites for both substrates, as shown schematically in Figure 11.4. Consistent with this model, removal of the tricarballylic acid sidechains reduces the potency of FB_1 by approximately 10-fold.[38]

As an additional way to evaluate the importance of the tricarballylic acids, we have synthesized (using the approach described in ref. 12) the 2-amino-,3,5-dihydroxyoctadecane with the same *threo* stereochemistry at each stereocenter as fumonisin B_1 (shown in Fig. 11.3),[40-43] and find that this analog is acylated well (Vmax, 18 pmol/min/mg microsomal protein; apparent Km, 5 μM; Vmax/Km, 4) comparably to the *threo*-sphinganines (Fig. 11.2). The Ki of 2-amino-,3,5-dihydroxyoctadecane as an inhibitor of the acylation of sphingosine was similar to its Km as a substrate; hence, this analog is one to two orders of magnitude less potent than fumonisin B_1 as an inhibitor of ceramide synthase. For comparison, the same 1-deoxy-5 hydroxy analogs of sphinganine were prepared with the D-*erythro* stereochemistry at positions 2 and 3, and they had kinetic properties comparable to sphingosine (Vmax ~100 pmol/min/mg protein and Km of 1-2 μM); therefore, the removal of the 1-hydroxyl group and the presence of a hydroxyl at position 5 have little effect on the utilization of sphingoid bases by ceramide synthase.

Studies with Cells in Culture

Ceramide synthase inhibition has been characterized with numerous eukaryotic cells in culture (e.g., hepatocytes, neurons, renal cells, macrophages, plants, yeast inter alia).[39] FB_1 blocks the incorporation of radiolabelled serine into the sphingoid base backbone of (dihydro)ceramides and complex sphingolipids, causes sphinganine (and sphinganine 1-phosphate) to be elevated, and prevents the conversion of sphinganine to sphingosine via addition of the 4,5-trans-double bond, which agrees with the conclusion that acylation must occur before introduction of the double bond (as shown in Fig. 11.1). The increased formation of sphinganine 1-phosphate leads to production of substantial amounts of ethanolamine 1-phosphate (plus fatty aldehyde) via sphingoid bases, such that one third of the ethanolamine in phosphatidylethanolamine has been calculated to come from this source.[3]

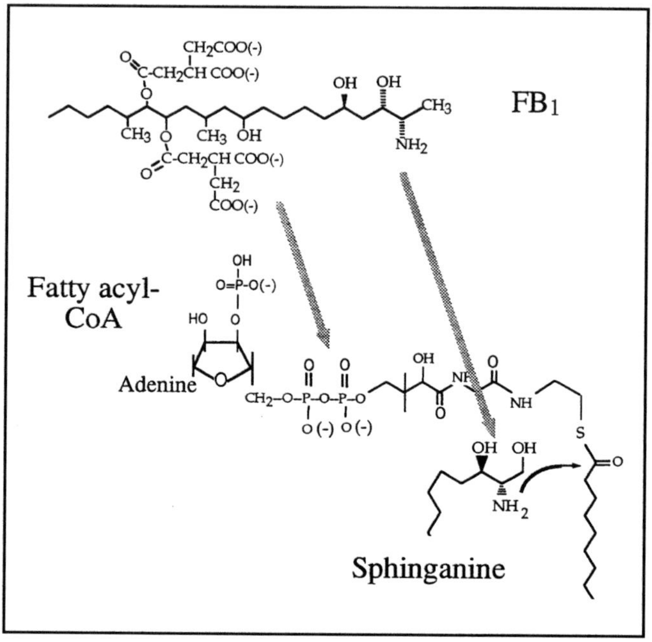

Fig. 11.4. A model for the structural features of fumonisin B_1 that may interact with the substrate binding sites for sphinganine and fatty acyl-CoAs of ceramide synthase.

FB_1 has been used in numerous studies of the role of sphingoid bases versus ceramides in cell regulation[39] because it can be added to cells to block the acylation of exogenous sphingoid bases. It can also help distinguish whether an exogenously added (short-chain) ceramide analog is acting directly or via hydrolysis to the sphingoid base backbone that is reacylated with a long-chain fatty acid. Since little is known about how FB_1 is taken up by cells (it is a highly charged compound and presumably requires transport), it is usually added to cells for up to an hour before introduction of the sphingolipid of interest.

FB_1 can also be used in studies of the functions of endogenous sphingolipids,[39] although the concentration dependence and time course of inhibition varies among different cell types. In general, FB_1 causes a more rapid elevation in sphinganine than depletion of cellular complex sphingolipids, since the latter is governed by the rate of membrane synthesis versus turnover. In cerebellar neurons, for example, no significant depletion of total sphingolipid was seen over several days,[37] whereas, in growing LLC-PK1 cells, the amount was reduced by half.[44] Furthermore, because the levels of sphinganine and sphinganine 1-phosphate are elevated, whereas the amounts of complex sphingolipids are reduced, the role of these separate events may need to be evaluated. The latter can be achieved using inhibitors of the first step of this pathway responsible for the observed cell changes.[45,46] FB_1 has also proven useful in determining whether exogenously added sphingoid bases must be metabolized to ceramides to elicit their cellular effects (such as induction of retinoblastoma protein dephosphorylation), since this conversion can be blocked with FB_1.[47] The possibility that FB_1 may have effects other than inhibition of sphingolipid metabolism must always be considered, and a recent report has described the inhibition of protein phosphatases,[48] but at much higher concentrations than are required for inhibition of ceramide synthase.

In Vivo Effects of Fumonisins

One of the exceptional features of inhibition of ceramide synthase by fumonisin B_1 is that it can be readily achieved in vivo. This is not surprising since these compounds were discovered because they cause diseases of plants and animals that are exposed to *Fusarium moniliforme* (Sheldon) and related fungi, which are common contaminants of corn (*Zea maize*), sorghum and related grains throughout the world. Fumonisins were initially discovered by WFO Marasas and co-workers as part of their investigations into the cause of the high incidence of esophageal cancer in the Transkei region of South Africa.[49] Early in these investigations, they concluded that a likely factor in this disease was preparation of beer from corn contaminated with *F. moniliforme* (Sheldon), and proceeded to isolate fumonisin B_1,[33] which met the criteria of being both toxic and carcinogenic in animal studies. Fumonisins are now known to be important in at least two other diseases, equine leukoencephalomalacia and porcine pulmonary edema, and have been shown to be nephrotoxic, hepatotoxic, embryotoxic, immunostimulatory and immunosuppressive, which makes it likely that they could be involved in additional, as yet undiscovered, diseases (for reviews see refs. 50 and 51).

Animals treated with fumonisins exhibit elevation of sphinganine (and sometimes sphingosine) in tissues, blood and urine, and this can be used as a biomarker for exposure.[52] There is also a reduction in the amount of complex sphingolipids in serum,[53] as would be expected since the liver secretes sphingolipids with very-low density lipoproteins.[54] In recent studies, we have identified N-acetyl-sphinganine and N-acetyl-sphingosine as components of both serum and tissues of animals that have consumed fumonisins (E Wang, unpublished observations).

POSSIBLE MECHANISMS OF ACTION OF FUMONISINS

Such diverse effects of fumonisins are not surprising given the many cellular functions that would be expected to result from disruption of sphingolipid metabolism (Fig. 11.5).[39] Because of the large number of systems that are affected, it is difficult to identify the direct links between fumonisins and changes in sphingolipid amounts and times with cellular responses and disease; however, notable changes in cell behavior that are likely to be important are:

- alteration of the behavior of cell-surface proteins, particularly GPI-anchored proteins,[55-58] and changes in cell morphology[59] and cell-cell interactions;[60]
- inhibition of protein kinases[61] and phosphoprotein phosphatases;[48]
- inhibition of cell growth and induction of cell death,[44,62-64] and in some systems, stimulation of growth;[45]
- changes in the metabolism of other lipids.[3,65]

At the level of signal transduction, the most extensive studies have been conducted recently by Clinton Jones and colleagues using CV-1 cells (African green monkey kidney cells).[61] FB_1 causes repression of protein kinase C and transcription dependent on AP-1 transcription factor, as well as (JR Ciacci-Zanella et al, personal communication) dephosphorylation of retinoblastoma protein and cell cycle arrest. They have also found lower activities of cyclin dependent kinase 2 (CDK2) and cyclin dependent kinase 4 (CDK4), lower cyclin E protein levels, and induction of two cyclin dependent kinase inhibitors (Kip 1 and Kip 2). It is already known that sphingoid bases are potent inhibitors of protein kinase C[66] and induce dephosphorylation of retinoblastoma protein.[47,67] Therefore, it is likely that the cell cycle arrest induced by fumonisins is downstream from the elevation of cellular sphingoid bases.

The mechanism of the cytotoxicity of fumonisins is not known, but FB_1 has been recently shown to induce apoptosis in cell culture.[68,69] Ceramides are thought to be mediators of apoptosis;[70] therefore, one might expect the inhibition of ceramide synthesis to have an opposite effect. However, recent studies have shown that ceramide-induced apoptosis is inhibited by Bcl-2[71] and that sphingosine suppresses Bcl-2,[72] which raises the possibility that fumonisin B_1 makes cells more susceptible to apoptosis by elevation of sphingoid bases that suppress Bcl-2.

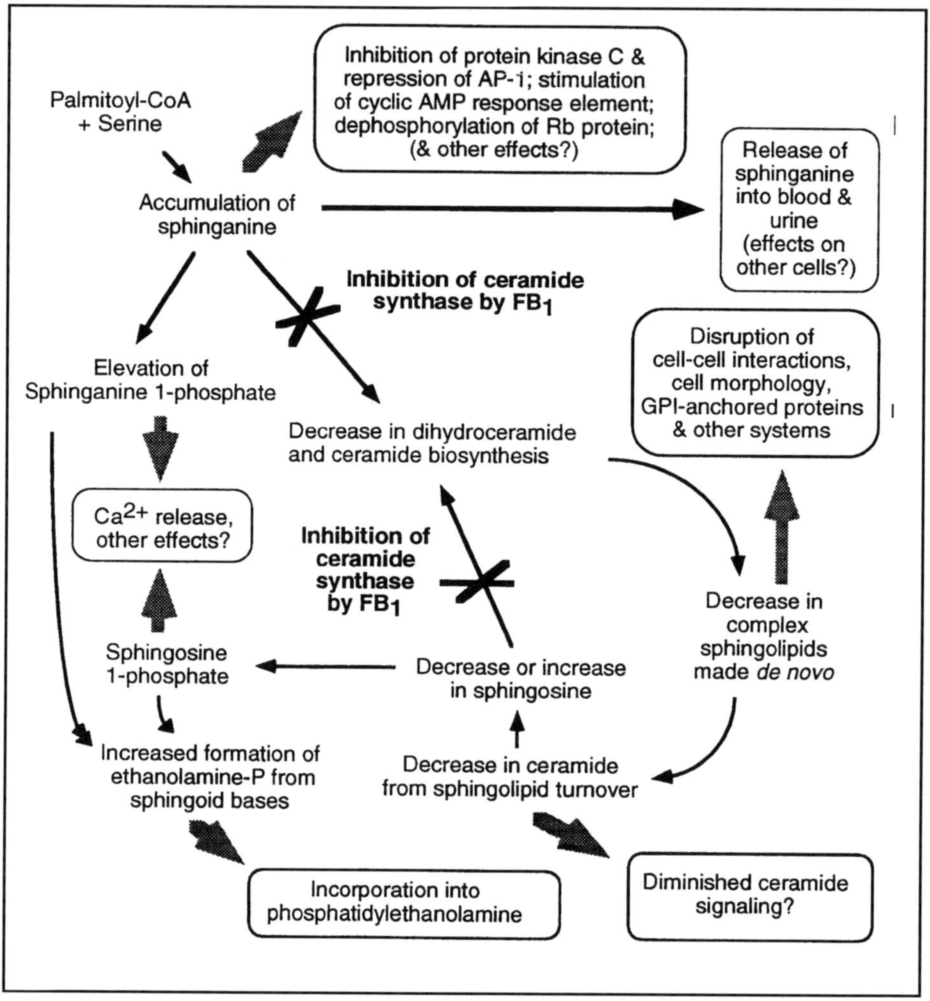

Fig. 11.5. Effects of fumonisin B_1 on sphingolipid metabolism, and cellular systems that would be predicted to be altered by the increases in sphinganine and sphingoid base 1-phosphates, and the reduction in formation of ceramides and complex sphingolipids.

The stimulation of DNA synthesis by fumonisin B_1, which has been seen in growth-arrested Swiss 3T3 cells,[45] is not well understood, but might involve the same signaling events that account for growth stimulation by sphingosine in this cell type.[73,74]

FUTURE PERSPECTIVES

Cells appear to have developed sophisticated mechanisms to minimize the risks of their utilization of some classes of compounds as metabolic intermediates and as signals for important regulatory processes. This makes it somewhat difficult to obtain definitive information by perturbing these systems with exogenous compounds, although major advances have certainly arisen from both the targeted (and serendipitous) discovery of inhibitors of ceramide synthase and ceramidase(s). More agents are clearly needed to provide even greater potency and selectivity in the inhibition of the desired enzymes.

Pathogenic organisms have utilized the development of such compounds to tip the subtle balance of sphingolipid synthesis and turnover to their own advantage. Considering the widespread occurrence of sphingolipids, and their potent biological activities, it would not be surprising if many more compounds of this type exist, and that their presence in food (and, perhaps, some traditional medicines) would help explain some of the relationships between diet and health.

Acknowledgments

The work by the co-authors that has been described in this review was accomplished with the vital help of many additional collaborators, who can be identified through the references. Much of this research was conducted with support from NIH grant GM46368, USDA grant 91-37204-6684, and a Focused Giving Award from the Johnson & Johnson Foundation.

References

1. Merrill AH Jr, Sweeley CC. Sphingolipid metabolism and cell signalling. In: Vance DE, Vance J, eds. Biochemistry of Lipids, Lipoproteins, and Membranes. Amsterdam: Elsevier Science Pub, 1996: chapter 12.
2. Merrill AH Jr, Wang E. Biosynthesis of long-chain (sphingoid) bases from serine by LM-cells. Evidence for introduction of the 4-*trans*-double bond after de novo biosynthesis of N-acylsphinganine(s). J Biol Chem 1986; 261:3764-3769.
3. Smith ER, Merrill AH Jr. Differential roles of de novo sphingolipid biosynthesis and turnover in the "burst" of free sphingosine and sphinganine, and their 1-phosphates and N-acyl-derivatives, that occurs upon changing the medium of cells in culture. J Biol Chem 1995; 270: 18749-18758.
4. Rother J, van Echten G, Schwarzmann G et al. Biosynthesis of sphingolipids: dihydroceramide and not sphinganine is desaturated by cultured cells. Biochem Biophys Res Commun 1992; 189:14-20.
5. Bielawska A, Linardic CM, Hannun YA. Ceramide-mediated biology: determination of structural and stereospecific requirements through the use of N-acyl-phenylaminoalcohol analogs. J Biol Chem 1992; 267:18493-18497.
6. Gatt SJ. Enzymic synthesis and hydrolysis of ceramide. J Biol Chem 1963; 238:PC3131-PC3133.
7. Stoffel W, Kruger E, Melzner I. Studies on the biosynthesis of ceramide. Does the reverse ceramidase reaction yield ceramide? Hoppe-Seyler's Z Physiol Chem 1980; 361: 773-779.
8. Morell P, Radin N. Specificity of ceramide biosynthesis from long chain bases and various fatty acyl CoAs by brain microsomes. J Biol Chem 1970; 245: 342-350.
9. Ullman M, Radin N. Enzymatic formation of hydroxy-ceramide and comparison with enzymes forming nonhydroxy-ceramides. Arch Biochem Biophys 1972; 152:767-777.
10. Sagisaka T, Nakano M, Fujino Y. Reactivity of sphingosine base in enzymatic synthesis of ceramide. Agric Biol Chem 1972; 36:1983-1987.
11. Sribney M. Enzymatic synthesis of ceramide. Biochim Biophys Acta 1966; 125: 542-547.
12. Nimkar S, Menaldino D, Merrill AH Jr et al. A stereoselective synthesis of sphingosine, a protein kinase C inhibitor. Tetrahedron Lett 1988; 29: 3037-3040.
13. Buehrer BM, Bell RM. Sphingosine kinase: properties and cellular functions. Adv Lipid Res 1993; 26:59-67.
14. Olivera A, Zhang H, Carlson RO et al. Stereospecificity of sphingosine-induced intracellular calcium mobilization and cellular proliferation. J Biol Chem 1994; 289:17924-17930.
15. Hirschberg K, Rodger J, Futerman A. The long-chain sphingoid base is acylated at the cytosolic surface of the endoplasmic reticulum in rat liver. Biochem J 1993; 290:751-757.
16. Rosenwald AG, Pagano RE. Intracellular transport of ceramide and its metabolites at the Golgi complex: insights from short-chain analogs. Adv Lipid Res 1993; 26:101-118.
17. Haasler D, Bell R. Ceramidases: enzymology and metabolic roles. Adv Lipid Res 1993; 25:49-59.
18. Spence MW, Beed S, Kook HW. Acid

and alkaline ceramidases of rat tissues. Biochem Cell Biol 1986; 64:400-404.
19. Slife CW, Wang E, Hunter R et al. Free sphingosine formation from endogenous substrates by a liver plasma membrane system with a divalent cation dependence and a neutral pH optimum. J Biol Chem 1989; 264: 10371-10377.
20. Bernardo K, Hurwitz R, Zenk T et al. Purification, characterization, and biosynthesis of human acid ceramidase. J Biol Chem 1995; 270:11098-11102.
21. Ghosh TK, Bian J, Gill DL. Intracellular calcium release mediated by sphingosine derivatives generated in cells. Science 1990; 248:1653-1656.
22. Olivera A, Spiegel S. Sphingosine-1-phosphate as second messenger in cell proliferation induced by PDGF and FCS mitogens. Nature 1993; 365:557-560.
23. Jacobs LS, Kester M. Sphingolipids as mediators of effects of platelet-derived growth factor in vascular smooth muscle cells. Am J Physiol 1993; 265:C740-747.
24. Meacci E, Vasta V, Fernando M et al. Bradykinin increases ceramide and sphingosine content in human fibroblasts: possible involvement of glycosphingolipids. Biochem Biophys Res Comm 1996; 221:1-7.
25. Linardic CM, Hannun YA. Identification of a distinct pool of sphingomyelin involved in the sphingomyelin cycle. J Biol Chem 1994; 269:23530-23537.
26. Coroneos E, Martinez M, McKenna S et al. Differential regulation of sphingomyelinase and ceramidase activities by growth factors and cytokines—Implications for cellular proliferation and differentiation. J Biol Chem 1995; 270:23305-23309.
27. Chen J, Nikolova-Karakashian M, Merrill AH Jr et al. Regulation of cytochrome P450 2CII (CYP2C11) gene expression by interleukin-1, sphingomyelin hydrolysis, and ceramides in rat hepatocytes. J Biol Chem 1995; 270:25233-25238.
28. Merrill AH Jr, Liotta DC, Riley RE Bioactive properties of sphingosine and structurally related compounds. In: Bell RM, ed. Handbook of Lipid Research. New York: Plenum Press, 1996; 8:205-237.
29. Okazaki T, Bielawska A, Bell RM et al. Role of ceramide as a lipid mediator of 1α-25-dihydroxyvitamin D_3-induced HL-60 cell differentiation. J Biol Chem 1990; 265:15823-15831.
30. Wilson E, Wang E, Mullins RE et al. Modulation of the free sphingosine levels in human neutrophils by phorbol esters and other factors. J Biol Chem 1988; 263:9304-9309.
31. Bielawska A, Greenberg MS, Perry D et al. (1S,2R)-D-erythro-2-(N-myristoylamino)-1-phenyl-1-propanol as an inhibitor of ceramidase. J Biol Chem 1996; 271:12646-12654.
32. Sugita M, Williams M, Dulaney JT, Moser HW. Ceramidase and ceramide synthesis in human kidney and cerebellum. Description of a new alkaline ceramidase. Biochim Biophys Acta 1975; 398:125-131.
33. Bezuidenhout CS, Gelderblom WCA, Gorstallman CP et al. Structure elucidation of the fumonisins, mycotoxins from *Fusarium moniliforme.* J Chem Soc Commun 1988;743-745.
34. Bottini AT, Bowen JR, Gilchrist DG. Phytotoxins II. Characterization of a phytotoxic fraction from *Alternaria alternata f. sp. lycopersici,* Tetrahedron Lett 1981; 22:2723-2726.
35. Mandala SM, Thornton RA, Frommer BR et al. The discovery of australifungin, a novel inhibitor of sphinganine N-acyltransferase from *Sporomiella australis.* J Antibiotics 1995; 48:349-356.
36. Wang E, Norred WP, Bacon CW et al. Inhibition of sphingolipid biosynthesis by fumonisins: implications for diseases associated with *Fusarium moniliforme.* J Biol Chem 1991; 266:14486-14490.
37. Merrill AH Jr, vanEchten G, Wang E et al. Fumonisin B_1 inhibits sphingosine (sphinganine) N-actyltransferase and de novo sphingolipid biosynthesis in cultured neurons in situ. J Biol Chem 1993; 268:27299-27306.
38. Merrill AH Jr, Wang E, Gilchrist DG et al. Fumonisins and other inhibitors of de novo sphingolipid biosynthesis. Adv Lipid Res 1993; 26:215-234.
39. Merrill AH Jr, Liotta DC, Riley RT. Fumonisins: fungal toxins that shed light

on sphingolipid function. Trends Cell Biol 1996; 6:218-223.

40. ApSimon JW, Blackwell BA, Edwards OE et al. Relative configuration of the C-1 to C-5 fragment of fumonisin B_1. Tetrahedron Lett 1994; 35:7703-7706.
41. Harmange J-C, Boyle CD, Kishi Y. Relative and absolute stereochemistry of the fumonisin B_2 backbone. Tetrahedron Lett 1994; 35:6819-6822.
42. Hoye TR, Jimenez JI, Shier WT. Relative and absolute configuration of the fumonisin B_1 backbone. J Am Chem Soc 1994; 116:9409-9410.
43. Poch GK, Powell RG, Plattner RD et al. Relative stereochemistry of fumonisin B_1 at C-2 and C-3. Tetrahedron Lett 1994; 35:7707-7710.
44. Yoo H, Norred WP, Wang E et al. Fumonisin inhibition of de novo sphingolipid biosynthesis and cytotoxicity are correlated in LLC-PK_1 cells. Toxicol Appl Pharmacol 1992; 114: 9-15.
45. Schroeder JJ, Crane HM, Xia J et al. Disruption of sphingolipid metabolism and stimulation of DNA synthesis by fumonsin B_1: a molecular mechanism for carcinogenesis associated with *Fusarium moniliforme*. J Biol Chem 1994; 269: 3475-3481.
46. Miyake Y, Kozutsumi Y, Nakamura S et al. Serine palmitoyltransferase is the primary target of a sphingosine-like immunosuppressant, ISP-1/myriocin. Biochem Biophys Res Commun 1995; 211: 396-403.
47. Pushkareva M, Chao R, Bielawska A et al. Stereoselectivity of induction of the retinoblastoma gene product (pRb) dephosphorylation by D-*erythro*-sphingosine supports a role for pRb in growth suppression by sphingosine. Biochemistry 1995; 34:1885-1892.
48. Fukuda H, Shima H, Vesonder RF et al. Inhibition of protein serine threonine phosphatases by fumonisin B1, a mycotoxin. Biochem Biophys Res Comm 1996; 220:1160-165.
49. Marasas WFO. Fumonisins: history, world-wide occurrence and impact. Adv Exp Med Biol 1996; 392:1-17.
50. Riley RT, Voss KA, Yoo H-S et al. Mechanism of fumonisin toxicity and carcinogenicity. J Food Protect 1994; 57:638-645.
51. Riley RT, Wang E, Schroeder JJ et al. Evidence for disruption of sphingolipid metabolism as a contributing factor in the toxicity and carcinogenicity of fumonisins. Natural Toxins 1996; 4:3-15.
52. Riley RT, Wang E, Merrill AH Jr. Liquid chromatography of sphinganine and sphingosine: use of sphinganine to sphingosine ratio as a biomarker for consumption of fumonisins. J AOAC International 1993; 77: 533-540.
53. Wang E, Ross PF, Wilson TM et al. Increases in serum sphingosine and sphinganine and decreases in complex sphingolipids in ponies given feed containing fumonisins, mycotoxins produced by *Fusarium moniliforme*. J Nutr 1992; 122:1706-1716.
54. Merrill AH Jr, Lingrell S, Wang E et al. Sphingolipid biosynthesis de novo by rat hepatocytes in culture: ceramide and sphingomyelin are associated with, but not required for, very low density lipoprotein secretion. J Biol Chem 1995; 270: 13834-13841.
55. Horvath A, Sutterlin C, Manning-Krieg U et al. Ceramide synthesis enhances transport of GPI-anchored proteins to the Golgi apparatus in yeast. EMBO J 1994; 13:3687-3695.
56. Hanada K, Izawa K, Nishijima M et al. Sphingolipid deficiency induces hypersensitivity of CD14, a glycosyl phosphatidyl-inositol-anchored protein, to phosphatidylinositol-specific phospholipase C. J Biol Chem 1993; 268:13820-13823.
57. Futerman AH. Inhibition of sphingolipid synthesis: effect of glycosphingolipid GPI-anchored protein microdomins. Trends Cell Biol 1995; 5:377-380;
58. Mays RW, Siemers KA, Fritz BA et al. Hierarchy of mechanisms involved in generating Na/K-ATPase polarity in MDCK epithelial cells. J Cell Biol 1995; 130:1105-1115.
59. Furuya S, Ono K, Hirabayashi Y. Sphingolipid biosynthesis is necessary for dendrite growth and survival of cerebellar Purkinje cells in culture. J Neurochem 1995; 65:1551-1561.
60. Ramasamy S, Wang E, Hennig B et al. Fumonisin B_1 alters sphingolipid me-

tabolism and disrupts the barrier function of endothelial cells in culture. Toxicol & Applied Pharm 1995; 133:343-348.
61. Huang C, Dickman M, Henderson G et al. Repression of protein kinase C and stimulation of cyclic AMP response elements by fumonisin, a fungal encoded toxin which is a carcinogen. Cancer Res 1995; 55:1655-1659.
62. Norred WP, Plattner RD, Vesonder RF, Bacon CW, Voss KA. Effects of selected secondary metabolites of *Fusarium moniliforme* on unscheduled synthesis of DNA by rat primary hepatocytes. Food Chem Toxicol 1992; 30:233-237.
63. Cawood ME, Gelderblom WC, Alberts JF et al. Interaction of 14C-labelled fumonisin B mycotoxins with primary rat hepatocyte cultures. Food Chem Toxicol 1994; 32:627-632.
64. Gelderblom WC, Snyman SD, Van der Westhuizen L et al. Mitoinhibitory effect of fumonisin B_1 on rat hepatocytes in primary culture. Carcinogenesis 1995; 16:625-631.
65. Wu W-I, McDonough M, Nickels JT Jr et al. Regulation of lipid biosynthesis in *Saccharomyces cerevisiae* by fumonisin B_1. J Biol Chem 1995; 270:13171-13178.
66. Hannun YA, Loomis CR, Merrill AH Jr et al. Sphingosine inhibition of protein kinase C activity and of phorbol bibutyrate binding in vitro and in human platelets. J Biol Chem 1986; 261:12604-12609.
67. Dbaibo GS, Wolff RA, Obeid LM et al. Activation of a retinoblastoma-protein-dependent pathway by sphingosine. Biochem J 1995; 310:453-459.
68. Tolleson WH, Melchior WB Jr, Morris SM et al. Apoptotic and anti-proliferative effects of fumonisin B_1 in human keratinocytes, fibroblasts, esophageal epithelial cells and hepatoma cells. Carcinogenesis 1996; 17:239-249.
69. Wang HC, Jones C, Ciacci-Zanella J et al. Fumonisins and *Alternaria alternata* lycopersici toxins: sphinganine analog mycotoxins induce apoptosis in monkey kidney cells. Proc Natl Acad Sci USA 1996; 93:3461-3465.
70. Hannun YA, Obeid LM. Ceramide: an intracellular signal for apoptosis. Trends Biochem Sci 1995; 20:73-77.
71. Zhang JD, Alter N, Reed JC et al. Bcl-2 interrupts the ceramide-mediated pathway of cell death. Proc Natl Acad Sci USA 1996; 93:5325-5328.
72. Sakakura C, Sweeney EA, Shirahama T et al. Suppression of Bcl-2 gene expression by sphingosine in the apoptosis of human leukemic HL-60 cells during phorbol ester-induced terminal differentiation. FEBS Lett 1996; 379:177-180.
73. Spiegel S. Sphingosine and sphingosine 1-phosphate in cellular proliferation: relationship with protein kinase C and phosphatidic acid. J Lipid Mediators 1993; 8:169-175.
74. Spiegel S, Merrill AH Jr. Sphingolipid metabolism and growth regulation: a state-of-the-art review. FASEB J 1996; 10:1388-1397.

CHAPTER 12

Ceramidase and Signal Transduction

Charles P. McKay

The deacylation of ceramide to form sphingosine in cells is the result of the activity of a group of enzymes referred to as the ceramidases (EC 3.5.1.23). There is a lysosomal form with an acidic pH optimum, a form with an alkaline pH optimum that may be associated with the plasma membrane and a form with a neutral pH optimum with a more limited tissue distribution. These enzymes are important in the degradative pathway of ceramide and are the key enzymes in the formation of sphingosine. The activity of these enzymes therefore are major determinates of the relative concentrations of these two biologically active lipids in the cell. Since the publication of a recent review of ceramidase,[1] studies have demonstrated that modulation of ceramidase results in changes in cell growth and apoptosis. Despite the potential importance of ceramidases, there is actually little known about their enzymology, although one acid form and one alkaline form have recently been purified. In addition there is nothing known of the molecular biology of ceramidase. It is the intent of this chapter to briefly review the known enzymology of ceramidases, describe briefly the effect of deficiency of the acid form (Farber's disease) and then show how inhibitors of ceramidase have begun to demonstrate its biologic importance in the control of cell growth.

ENZYMOLOGY

The first report of ceramidase activity was from a rat brain homogenate that hydrolyzed the amide linkage of ceramide, yielding ceramide and a free fatty acid.[2] The brain was homogenized with sucrose and EDTA and the ceramidase activity partitioned with the particulate fraction after 15,000 x g spin. In an in vitro assay at pH 5.0, in the presence of Tween 20, Triton X-100 and sodium cholate, a soluble enzyme preparation released either fatty acid or sphingosine from radiolabeled precursors synthesized by the author. Activity was dependent on the presence of cholate in the assay. The optimum pH was found to be 5.0 and there was no increase in activity with the addition of calcium, magnesium or manganese ions. Other compounds that had no effect included digitonin, inositides, cerebroside phosphate, and phosphorylcholine. There was product inhibition noted with the addition of

Sphingolipid-Mediated Signal Transduction, edited by Yusuf A. Hannun.

fatty acid or sphingosine. Preliminary studies of substrate specificity suggested the preference for ceramides with unsaturated over those with saturated fatty acids (with no activity for C-2 or C-24 fatty acids) as well as a preference for ceramides with sphingosine over those with dihydrosphingosine. The enzyme preparation was found to also catalyze the reverse reaction. This reverse reaction was not activated by the presence of magnesium, ATP, or coenzyme A. Thus these studies suggested the presence of an acid pH, probably lysosomal ceramidase that required no cofactors but could also act as a ceramide synthase as well.

After the initial description and purification of ceramidase,[1] a subsequent report by the same group further purified acid ceramidase from rat brain 200-fold by taking a choline extract of the particulate fraction of the homogenate, precipitating it with ammonium sulfate and applying the 30-60% fraction to a Sephadex G-150 column.[3] Though the enzyme was presumed to be lysosomal ceramidase with an acid pH optimum, the assay for ceramidase in these studies was performed at pH 7.4 because of solubility concerns with sodium cholate in the reaction buffer. Trypsin and chymotrypsin treatment of enzyme preparation did not significantly impair enzyme activity, nor did heating the enzyme for 5 minutes at 55°.[3] Preincubation with *p*-hydroxymercuribenzoate resulted in 50% inhibition at 1 mM and complete inhibition at 3-5 mM.[3] The investigators were unable to separate the ceramidase activity from the ceramide synthase activity using any of these procedures. Estimate of enzyme size by gel-filtration through Bio-Gel P-150 was 150,000 before proteolysis and 100,000 after proteolysis.

The best purification to date of ceramidase is of human acid ceramidase concentrated from human urine.[4] Concentrated human urine subjected to ammonium sulfate precipitation, octyl-Sepharose, concanavalin-A Sepharose, blue-Sepharose and DEAE-Sepharose columns resulted in 4,454-fold purification and a single band on SDS-PAGE analysis. The purification required the presence of Triton X-100 throughout the scheme to allow its purification. This protein had an apparent molecular mass of ~50 kDa that could be reduced by β-mercaptoethanol to 2 polypeptides of ~13 and ~40 kDa, respectively. Glycosidase treatment reduced the size of the nonreduced protein to two peptides of ~40 and 46 kDa. Antibodies raised against the purified protein immunoprecipitated an ~47 kDa peptide secreted by human fibroblasts but not an ~50 kDa heterodimeric precursor that was thought to be the mature intracellular enzyme. Kinetic studies performed on this purified acid ceramidase will be discussed below.

The purification of two membrane-bound ceramidases, CDase I and CDase II, from guinea pig skin has been described with pH optima in the 7-10 and 8-9 range, respectively.[5] This extraction required 0.5% sodium cholate, but in contrast to studies of acid ceramidase,[4] Triton X-100 resulted in loss of activity. These ceramidases were purified to 1133-fold and 404-fold respectively, using combinations of steps including DEAE-Sepharose, HCA-hydroxyapatite and TSK-3000SW column chromatography and isoelectric focusing.[5] The apparent molecular weights for CDase I and II were 150,000 and 62,000 kDa, respectively. Whereas the CDase I was shown to migrate as a single band on SDS-PAGE, the CDase II migrated as 5 bands, indicating only partial purification. Substrate specificity and other enzymology of these ceramidases will be discussed below.

Since the original description of ceramidase with an optima pH in the acid range, ceramidases have been largely characterized by their pH optimum in in vitro assays. The majority of the reports have been of acid ceramidase with optimum pH of 4.0-5.0 that is presumably of lysosomal origin.[2-4,6-9] It is this form of ceramidase that is presumably deficient in Farber's disease or lipogranulomatosis.[8,10-12] In contrast, there are reports of ceramidase in a number of tissues that have an optimum pH for ceramidase activity in in vitro assays in the highly alkaline pH of 8.0-9.0.[5,8,9,13-16] Finally, there are reports of ceramidase with pH optimum in the in vitro assays in the neutral range from liver plasma membrane,[15] rat liver microsomes,[17] and the rat intestinal brush border membrane.[18] In some reports, the pH range has been rather broad,[9,15] whereas in others a fairly sharp peak of activity with pH changes have been documented.[8,10,16] With only human acid ceramidase and a guinea pig alkaline ceramidase being purified, it is far from clear how many forms of ceramidase exist and what their tissue distribution is. The acid form of ceramidase is thought to play a housekeeper role as a lysosomal enzyme and is assumed to be ubiquitous, and the detection of an alkaline activity in cells from Farbers disease patients demonstrate that the alkaline activity is clearly distinct from the acid enzyme.[8,10]

There is conflicting evidence regarding cofactor requirement for in vitro assays of ceramidase activity. Detergents such as sodium cholate or Triton X-100 are required in the assays, although this is thought to be due to the membrane bound nature of the enzyme rather than any specific effect.[2,6] In contrast to the original suggestion of no effect of divalent cations on ceramidase activity, other studies have suggested activation by calcium and magnesium.[15,16] A small group of potential cofactors including cephalins, inositides, cerebroside sulfate and phosphorylcholine were found to have no effect.[2] A group of 4 small glycoproteins designated sphingolipid activator proteins or saposins have been identified that are required for efficient hydrolysis of sphingolipids in lysosomes. Recently, saposin D has been shown to activate acid ceramidase in vivo,[19] as well as in vitro in some studies[20] but this is controversial.[4] Saposin D seems to have no effect on alkaline ceramidase and saposins A, B and C do not affect ceramidase activity.[6,20] The significance of a patient with deficiency of a form of saposin A that displayed features of both Gaucher diseases and Farber's disease is not clear.[21]

There is evidence of product inhibition of ceramidase by sphingosine and fatty acids.[2,5,13,22] In further studies of potential inhibitors of in vitro ceramidase activity, it was found that the simple sphingolipids, ceramide galactoside and ceramide glucoside have a Ki of 1.3 and 2.5 x 10^{-3} M, respectively for acid ceramidase from human renal tissue.[13] In addition, *N*-oleoylethanolamine was found to be a more potent inhibitor of acid ceramidase with a Ki of 7.0 x 10^{-4} M. More recent studies confirmed that *N*-oleoylethanolamine as an inhibitor of acid ceramidase in vitro with an identical Ki. It is also an inhibitor in vivo when added at a concentration of 0.5 mM.[23] It is interesting that an *N*-acylethanolamine acts as an inhibitor in the light of the data that suggest that *N*-acylethanolamine is a product of the "reversed ceramidase reaction" in vitro when ethanolamine is present in the reaction buffer.[24] Studies with the purified alkaline ceramidases demonstrated mild inhibition (< 20%) with phosphatidylinositol and phosphatidylethanolamine, moderate inhibition (25-50%) with phosphatidylglycerol, phosphatidylserine, phosphatidic acid, and a variety of lysophospholipids; finally, the greatest degree of inhibition (> 50%) was seen with phosphatidylcholine and sphingomyelin.[5] In all of these cases inhibition was most marked at concentrations of 0.5-2.0 mM.

The ceramide analog, (1S,2R)-D-*erythro*-2-(*N*-myristoylamino)-1-phenyl-1-propanol (D-*e*-MAPP) has been shown in a recent study to be a potent inhibitor of alkaline ceramidase extracted from HL-60 cells.[16] This synthetic ceramide analog based on the N-acylated phenylaminopropanols was compared to its enantiomer, L-*e*-MAPP in vitro as well as in vivo studies which will be discussed below. Whereas L-*e*-MAPP has the same absolute configuration of D-*erythro*-ceramide, the natural substrate for ceramidase, D-*e*-MAPP is its mirror image.[16] Consistent with the hypothesis that the stereochemistry of these compounds plays an important role was the finding that L-*e*-MAPP but not D-*e*-MAPP was a substrate for ceramidase. Furthermore, D-*e*-MAPP was found to be an effective inhibitor of alkaline but not acid ceramidase in vitro, and displayed much more powerful inhibition of alkaline ceramidase than N-oleoylethanolamine with an apparent Ki of 2-3 x 10^{-6} M. The in vivo results as discussed below were consistent with these in vitro studies.[16]

The original description of acid ceramidase demonstrated a preference for ceramides containing an unsaturated fatty acid as well as the preference for a double bond in the long-chain base and little ceramidase activity when the fatty acid is very short (C-2 or C-6).[2] These findings were confirmed by a subsequent study by the same group.[22] Momoi et al,[8] in a more detailed study of acid ceramidase, also demonstrated that, as the number of double bonds increased from 0-3 in a C-18 fatty acid moiety of ceramide, ceramidase activity also increased. They also confirmed that sphingosine is preferred over dihydrosphingosine as the choice for the long-chain base. In addition, these authors demonstrated decreasing order of activity as the fatty acid chain of ceramide went from C-12, C-14, C-16, C-18. The preference of acid ceramidase for *N*-laurylsphingosine has been suggested to be so characteristic as to be useful as a diagnostic test for Farber disease.[26] In studies with purified acid ceramidase, C-12 was again demonstrated as the preferred substrate with decreasing activity as the fatty acid chain grew longer.[4] Whereas C-8 ceramide was found to be a moderately good substrate, ceramides containing C-2 and C-6 fatty acids again were shown to be poor substrates. These studies are fairly consistent in the demonstration of the preference by acid ceramidase for ceramides containing sphingosine over dihydrosphingosine, unsaturated fatty acids and finally preference for intermediate length fatty acids. Whether this was an artifact of the in vitro assay is not clear. When the fatty acid composition of ceramide from kidney and cerebellum from a patient with Farber's disease was compared with non-Farber's disease patients, no clues as to the in vivo substrate specificity of acid ceramidase were revealed.[25]

The substrate specificity of alkaline ceramidase is less clear. In one study it was found to be similar to acid ceramidase with the saturated fatty acid containing ceramides being poor substrates and, in the C-18 series fatty acids, the more double bonds from 0-3 resulting in greater ceramidase activity.[8] The studies with the purified alkaline ceramidase demonstrated activity in decreasing order of linoleoylsphingosine, oleoylsphingosine, and palmitoylsphingosine.[5] Finally, the neutral microsomal ceramidase displayed quite different activity in regard to fatty acid chain length with C-12 ceramide being the worst substrate and increases as chain lengths decreased to C-2 or increased to C-16.[17]

Due to its strong acidic pH optimum, acid ceramidase has always been assumed to be of lysosomal origin.[2,22] Further support for a lysosomal location for acid ceramidase comes from enzyme enrichment results from the fraction associated with acid ceramidase activity.[15] In addition, lysosomal origin for the degradation of ceramide was supported in studies with fibroblasts loaded with either fluorescent or radiolabeled ceramide.[7,27] Finally, Farber's disease is associated with abnormalities in acid ceramidase,[10-12] lysosomal inclusions by electron microscopy[28] and abnormalities in the turnover of LDL-associated radioactive sphingomyelin which is normally taken up and metabolized by lysosomes.[29,30] Unlike the fair certainty of the origin of acid ceramidase, the origin of neutral and alkaline ceramidases are not so clear. Whereas a neutral ceramidase from rat liver was suggested originally by the cell fractionation method to be of microsomal origin,[17] a microsomal fraction from rat liver characterized by enzyme enrichment displayed little ceramidase activity and a plasma membrane fraction displayed neutral as well as alkaline ceramidase activity.[15] The alkaline activity purified from guinea pig skin was isolated from a crude lysosomal fraction.[5]

It is assumed that most if not all normal tissues display acid ceramidase activity which performs a vital function in the metabolism of sphingolipids in cells and tissues. Acid ceramidase has been specifically identified in rat brain,[1,2] rat liver,[15] rat kidney,[9] human skin fibroblasts,[4,7,8] human spleen,[6] human kidney,[13] human brain,[10,13] human leukocytes[26] and porcine epidermidis.[14] In contrast, there is an apparent selective distribution of alkaline ceramidase in such tissues as rat cerebellum but not rat kidney,[13] HL-60 cells[16] but not Molt-4 cells (unpublished observations), but also in porcine epidermis,[14] guinea pig skin,[5] human fibroblasts,[8] rat liver[15] and human brain.[13] Neutral ceramidase activity is described only in rat liver[15,17] and porcine intestine.[18] The extent of the suggested tissue and species specificity of ceramidases has to be considered unclear at this point due to the crude methods of detection and overall lack of purification and characterization of the ceramidases.

There are a number of published enzyme kinetics for ceramidase, but as discussed by Yavin and Gatt[3] these values have to be considered in most cases to be "apparent" rather than true equilibrium constants. This is because of the nature of the enzyme assays employed, in which the actual amount of substrate available to the enzyme is not known, in a reaction mixture containing sodium cholate, Triton X-100, unspecified cellular lipids and proteins. In addition, there may be operation of the reverse reaction of ceramide synthesis in the reaction mixture.[2,3,13] Nevertheless, there are a number of published enzyme kinetics for both the acid and alkaline forms of the enzyme (Table 12.1). The range of Kms reported are generally slightly above 10^{-4} M with substrate specificity demonstrated in human fibroblasts[8] with laurylsphingosine being the preferred substrate over oleoylsphingosine. The range of Vmax is broader with a range from 6-170 nmol/mg/hr.

The standard assay for ceramidase activity is largely based on the original description by Gatt,[2] in which the ceramide substrate is labeled either on the fatty acid or sphingosine moiety and allowed to react in a buffered solution containing Triton X-100, sodium cholate and the enzyme extract.[31] The labeled product is then isolated and counted. Other methods employed have included the formation of sphingosine as measured by HPLC.[15] Variations of these methods have been used in the diagnosis of Farber's disease, in particular its prenatal diagnosis.[12,32-34]

CLINICAL AND BIOLOGICAL SIGNIFICANCE

The predominant clinical condition associated with ceramidase is lipogranulomatosis or Farber's disease, which is the congenital abnormality of lysosomal acid ceramidase.[10,11] In most cases this is due to a primary impairment that is not absolute[29] whereas it has also been described in the absence of normal saposin D.[21] Our lack of knowledge of the molecular biology of acid

Table 12.1.

Tissue/ Source	pH Dependence	Km	Vmax	Ref.
Rat brain	acid	3×10^{-4} M	170 nmol/mg/hr	22
Human spleen	acid	2.2×10^{-4} M	57 nmol/mg/hr	6
Pig skin	acid,alkaline	1.1×10^{-4} M		14
Human fibroblasts	acid,alkaline	12:0 2.4×10^{-4}M 18:1 12×10^{-4}M	19.2 nmol/mg/hr 6.23 nmol/mg/hr	8
Rat brain, kidney	acid-alkaline (broad range)	pH5 1.5×10^{-4}M pH8 2.5×10^{-4}M		9
Human urine	acid	1.49×10^{-4} M	136 nmol/mg/hr	4

ceramidase prevents us from further describing the exact defects in each of these conditions. This rare condition that was first described in 1952 is characterized by hoarseness, painful swollen joints, disseminated subcutaneous nodules, progressive cachexia and eventual death.[35] The first description of ceramidase in a disease other than Farber's disease comes from a recent report that ceramidase activity is increased in aged skin.[36] Ceramide comprises greater than 50% of the lipids in the stratum corneum and contributes to the water retention and barrier functions of the stratum corneum. Consistent with studies of decreased ceramide in dry skin seen with aging, it was found that the activity of alkaline ceramidase was significantly increased in the aged skin.[36] The mechanism for control of ceramidase in skin is not well understood but seems to differ from that of control of phospholipid metabolism.[37]

The importance of the sphingomyelin cycle as outlined in this volume is now being recognized with ceramidase as well. Until just a few years ago, ceramidase was largely thought of as being only a degradative pathway for complex sphingolipids. The demonstration by Coroneos et al[23] that a membrane bound alkaline ceramidase in rat glomerular mesangial cells were activated by platelet derived growth factor which then led to evidence of mitogenesis places ceramidase squarely in the role of regulating cell growth and mitogenesis. The mechanism by which growth factors regulate alkaline ceramidase was secondary to tyrosine kinase phosphorylation, as evidenced by potentiation by vanadate, a tyrosine phosphatase inhibitor, and inhibition by genistin, a tyrosine kinase inhibitor. The effect on activation of ceramidase could be mimicked by other growth factors such as FGF, insulin-like growth factor and EGF but not the cytokines TNF or IL-1. In contrast, the cytokines activated sphingomyelinase to a greater degree than PDGF. *N*-oleoylethanolamine, an inhibitor of ceramidase suppressed the ability of PDGF to activate mitogenesis, further supporting the role of ceramidase in this process. The dose of *N*-oleoylethanolamine required was 0.5 mM, consistent with it being a relatively poor inhibitor of alkaline ceramidase.[16] The conclusion that ceramidase plays an important role in growth factor, in particular PDGF, induction of mitogenesis through the formation of sphingosine and its metabolites such as sphingosine-1-phosphate is consistent with previous studies demonstrating sphingosine-1-phosphate formation and mitogenesis in response to PDGF[38] and that the differential effects of PDGF on sphingosine and ceramide during activation of mitogenesis suggested a role for ceramidase.[39]

Studies in HL-60 cells with (1S,2R)-D-*erythro*-2-(*N*-myristoylamino)-1-phenyl-1-propanol or D-*e*-MAPP further demonstrated the importance of alkaline ceramidase in the control of the cell cycle and cell growth.[16] Increases in cellular ceramide by inhibition of alkaline ceramidase by 5 μM D-*e*-MAPP (1% of the dose of *N*-oleoylethanolamine used in the Coroneos study[23]) led to cell cycle arrest and growth suppression. The dose used in vivo was in the range of the in vitro Ki as discussed above. The effect was due not to direct activation of ceramide activated protein phosphatase or inhibition of glucosylceramide synthase, but rather due to inhibition of alkaline ceramidase. The specificity of this effect was further demonstrated by the lack of effect of the enantiomer of D-*e*-MAPP, L-*e*-MAPP, which unlike D-*e*-MAPP is a substrate for alkaline ceramidase and was metabolized in vivo as well as in vitro. One subtle feature of these observations was the change induced by D-*e*-MAPP in the distribution of the ceramide species as recognized by the diacylglycerol kinase assay. Whereas in Farber's disease there is a relative increase in the upper spot,[33] D-*e*-MAPP resulted in a greater change in the lower spot.[16] This not only supports the effect of D-*e*-MAPP on inhibition of alkaline and not acid ceramidase, but suggests that certain ceramide species may play a particular role in the control of cell growth and these species may be regulated at least in part by alkaline ceramidase.

A role for ceramidase has been demonstrated in other studies. The effect of glucocorticoids like betamethasone on surfactant synthesis through activation of CTP:choline cytidyltransferase was shown to involve inhibition of acid sphingomyelinase and alkaline ceramidase.[40] Sphingosine inhibits CTP:choline cytidyltransferase and betamethasone decreased cellular sphingosine in association with 33% inhibition of acid sphingomyelinase and 21% inhibition of alkaline ceramidase. Other investigators have found that dexamethasone inhibited increased neutral sphingomyelinase activity in 3T3-L1 cells but did not affect ceramidase activity at acid, neutral or alkaline pH.[41,42] Therefore the role of glucocorticoids in the control of ceramidase is not yet clear. A more recent study demonstrated that dexamethasone increased ceramide in cells but without evidence of sphingomyelin breakdown suggesting possible ceramide synthesis.[43] Further support of the fact that this was not due to inhibition of alkaline ceramidase was the fact that both spots on the ceramidase assay were equally increased in contrast to the findings with D-*e*-MAPP.[16] Induction of apoptosis by staurosporine involves increases in ceramide, but this effect was thought to be secondary to activation of sphingomyelinase.[44] Though ceramidase was not primarily implicated, the effect was enhanced by 50-100 μ *N*-oleoylethanolamine, presumably through inhibition of acid ceramidase. Finally, it has been reported that the phorbol ester PMA induced apoptosis in HL-60 cells by increasing sphingosine formation, presumably by ceramidase activation.[45] Neither ceramide concentrations nor in vitro ceramidase activity were measured but the conversion of exogenously administered radiolabeled ceramide to sphingosine was enhanced by PMA.

We are just beginning to understand the important role the sphingomyelin cycle plays in the control of cell growth and other cellular functions. Studies over the past few years provide exciting evidence that ceramidase plays an important role in the control of those processes. Our lack of knowledge of the molecular biology of these enzymes has prevented us from defining the role of the ceramidases in biology. With knowledge of the protein structure of the ceramidases, development of antibodies, and identification of the cDNAs we will be able to truly answer the questions of species specificity, organ specificity, and subcellular distribution and function. Increased characterization of these enzymes should allow us to develop molecular probes that act as sensitive and specific ceramidase inhibitors or activators that will modulate not only total ceramide content of cells but specific ceramide species that may have specific cellular effects. Eventually we will have antisense probes and will be able to study these changes in activity on a molecular biology basis.

The central role for ceramidase in the modulation of the ceramide mediated effects on cell biology hold the promise for the potential treatment of a variety of disorders including cancer through its ability to modulate cell growth and apoptosis. In concert with the ability to modulate ceramide levels, control of ceramidase activity in cells will allow us to regulate a variety of compounds of the sphingomyelin cycle including sphingosine, sphingosine-1-phosphate and probably yet to be recognized biologically active compounds.

References

1. Hassler DF, Bell RM. Ceramidases: enzymology and metabolic roles. Adv Lipid Res 1993; 26:49-57.
2. Gatt S. Enzymic hydrolysis and synthesis of ceramides. J Biol Chem 1963; 238:PC3131-PC3133.
3. Yavin E, Gatt S. Enzymatic hydrolysis of sphingolipids. VIII. Further purification and properties of rat brain ceramidase. Biochemistry 1969; 8:1692-1698.
4. Bernardo K, Hurwitz R, Zenk T et al. Purification, characterization and biosynthesis of human acid ceramidase. J Biol Chem 1995; 270:11098-11102.

5. Yada Y, Higuchi K, Imokawa G. Purification and biochemical characterization of membrane-bound epidermal ceramidases from guinea pig skin. J Biol Chem 1995; 270:12677-12684.
6. Al BJM, Tiffany CW, de Mesquita DSG et al. Properties of acid ceramidase from human spleen. Biochim Biophys Acta 1989; 1004:245-251.
7. Chen WW, Moser AB, Moser HW: role of lysosomal acid ceramidase in the metabolism of ceramide in human skin fibroblasts. Arch Biochem Biophys 1981; 208:444-455.
8. Momoi T, Ben-Yoseph Y, Nadler HL. Substrate-specificities of acid and alkaline ceramidases in fibroblasts from patients with Farber disease and controls. Biochem J 1982; 205:419-425.
9. Spence MW, Beed S, Cook HW. Acid and alkaline ceramidases of rat tissues. Biochem Cell Biol 1986; 64:400-404.
10. Sugita M, Dulaney JT, Moser HW. Ceramidase deficiency in Farber's disease (lipogranulomatosis). Science 1972; 178: 1100-1102
11. Dulaney JT, Milunsky A, Sidbury JB et al. Diagnosis of lipogranulomatosis (Farber disease) by use of cultured fibroblasts. J Pediatr 1976; 89:59-61.
12. Fensom AH, Benson PF, Neville BRG et al. Prenatal diagnosis of Farber's disease. Lancet 1979; 2:990-992
13. Sugita M, Williams M, Dulaney JT et al. Ceramidase and ceramide synthesis in human kidney and cerebellum. Biochim Biophys Acta 1975; 398:125-131.
14. Wertz PW, Downing DT. Ceramidase activity in porcine epidermis. FEBS Lettr 1990; 268:110-112.
15. Slife CW, Wang E, Hunter R et al. Free sphingosine formation from endogenous substrates by a liver plasma membrane system with a divalent cation dependence and a neutral pH optimum. J Biol Chem 1989; 264:10371-10377.
16. Bielawska A, Greenberg MS, Perry D et al. (1S,2R)-D-erythro-2-(N-myristoylamino)-1-phenyl-1-propanol as an inhibitor of ceramidase. J Biol Chem 1996; 271:12646-12654.
17. Stoffel W, Melzner I. Studies in vitro on the biosynthesis of ceramide and sphingomyelin. A reevaluation of proposed pathways. Hoppe-Seyler's Z Physiol Chem 1980; 361:755-771.
18. Nilsson A. The presence of sphingomyelin- and ceramide-cleaving enzymes in the small intestinal tract. Biochim Biophys Acta 1969; 176:339-347.
19. Klein A, Hensler M, Klein C et al. Sphingolipid activator protein D (sap-D) stimulates the lysosomal degradation of ceramide in vivo. Biochem Biophys Res Comm 1994; 200:1440-1448.
20. Azuma N, O'Brien JS, Moser HW et al. Stimulation of acid ceramidase activity by saposin D. Arch Biochem Biophys 1994; 311:354-357.
21. Harzer K, Paton BC, Poulos A et al. Sphingolipid activator protein deficiency in a 16-week-old atypical Gaucher disase patient and his fetal sibling: biochemical signs of combined sphingolipidoses. Eur J Pediatr 1989; 149:31-39.
22. Gatt S. Enzymatic hydrolysis of sphingolipids. I. Hydrolysis and synthesis of ceramides by an enzyme from rat brain. J Biol Chem 1966; 241:3724-3730.
23. Coroneos E, Martinez M, McKenna S et al. Differential regulation of sphingomyelinase and ceramidase activities by growth factors and cytokines. J Biol Chem 1995; 270:23305-23309.
24. Stoffel W, Kruger E, Melzner I. Studies on the biosynthesis of ceramide. Does the reversed ceramidase reaction yield ceramides? Hoppe-Seyler's Z Physiol Chem 1980; 361:773-779.
25. Sugita M, Connoly P, Dulaney JT et al. Fatty acid composition of kidney and cerebellum from a patient with Farber's disease. Lipids 1973; 8:401-406.
26. Ben-Yoseph Y, Gagne R, Parvathy MR et al. Leukocyte and plasma N-laurylsphingosine deacylase (ceramidase) in Farber disease. Clin Genet 1989; 36:38-42.
27. Chen WW, Decker GL. Abnormalities of lysosomes in human diploid fibroblasts from patients with Farber's disease. Biochim Biophys Acta 1982; 718:185-192.
28. Zappatini-Tommasi L, Dumontel C, Guibaud P et al. Farber disease: an ultrastructural study. Virchows Arch Pathol Anat 1992; 420:281-290.
29. Levade T, Tempesta MC, Salvayre R. The

in situ degradation of ceramide, a potential lipid mediator, is not completely impaired in Farber disease. FEBS Lett 1993; 329:306-312.

30. Levade T, Tempesta MC, Moser HW et al. Sulfatide and sphingomyelin loading of living cells as tools for the study of ceramide turnover by lysosomal ceramidase-implications for the diagnosis of Farber disease. Biochem Molec Med 1995; 54:117-125.
31. Gatt S, Yavin E. Ceramidase from rat brain. Methods in Enzymol 1969; 14: 139-144.
32. Mitsuo K, Kobayashi T, Shinnoh N et al. A high-performance lipid chromatographic assay for acid-ceramidase activity in cultured fibroblasts from patients with Farber's disease and from controls. Clin Chim Acta 1988; 173:281-288.
33. Levade T, Enders H, Schliephacke M et al. A family with combined Farber and Sandhoff, isolated Sandhoff and isolated fetal Farber disease; postnatal exclusion and prenatal diagnosis of Farber disease using lipid loading tests on intact cultured cells Eur J Pediatr 1995; 154: 643-648.
34. Ahkhunov, Gargaun SS, Krasnopolskaya XD. First-trimester enzyme exclusion of Farber disease using a micromethod with [^{3}H]ceramide. J Inher Metab Dis 1995; 18:616-619.
35. Toppet M, Vamos-Hurwitz, Jonniaux G et al. Farber's disease as a ceramidosis; clinical, radiological and biochemical aspects. Acta Paed Scand 1978; 67:113-119.
36. Jin K, Higaki Y, Takagi Y et al. Analysis of beta-glucocerebrosidase and ceramidase activities in atopic and aged dry skin. Acat Derm Venereol 1994; 74: 337-340.
37. Kondoh H, Kanoh H, Ono T. Deacylation of ceramide, triacylglycerol and phospholipids in guinea pig epidermal cells. Biochim Biophys Acta 1983; 753: 97-106.
38. Olivera A, Spiegel S. Sphingosine-1-phosphate as second messenger in cell proliferation induced by PDGF and FCS mitogens. Nature 1993; 365:557-560.
39. Jacobs LS, Kester M. Sphingolipids as mediators of effects of platelet-derived growth factor in vascular smooth muscle cells. Am J Physiol 1993; 265:C740-C747.
40. Mallampalli RK, Mathur SN, Warnock LJ et al. Betamethasone modulation of sphingomyelin hydrolysis upregulates CTP:cholinephosphate cytidyltransferase activity in adult rat lung. Biochem J 1996; 318:333-341.
41. Ramachandron CK, Murray DK, Nelson DH. Dexamethasone increases neutral sphingomyelinase activity and sphingosine levels in 3T3-L1 fibroblasts. Biochem Biophys Res Comm 1990; 167: 607-613.
42. Ramachandran CR, Murray DK, Nelson DH. Enzymatic hydrolysis of sphingomyelin in 3T3-L1 cells: modulation by dexamethasone. Biochemical Arch 1992; 8:369-377.
43. Quintans J, Kilkus J, McShan et al. Ceramide mediates the apoptotic response of WEHI 231 cells to anti-immunoglobulin, corticosteroids and irradiation. Biochem Biophys Res Comm 1994; 202:710-714.
44. Wiesner DA, Dawson G. Staurosporine induces programmed cell death in embryonic neurons and activation of the ceramide pathway. J Neurochem 1996; 66:1418-1425.
45. Ohta H, Sweeney EA, Masamune A et al. Induction of apoptosis by sphingosine in human leukemic HL-60 cells: a possible endogenous modulator of apoptotic DNA fragmentation occurring during phorbol ester-induced differentiation. Cancer Res 1995; 55:691-697.

INDEX

J

K

L

M

N

O

P

R

S